PRINCIPLES OF THE QUANTUM CONTROL OF MOLECULAR PROCESSES

PRINCIPLES OF THE QUANTUM CONTROL OF MOLECULAR PROCESSES

MOSHE SHAPIRO
PAUL BRUMER

WILEY-INTERSCIENCE

A JOHN WILEY & SONS PUBLICATION

Library of Congress Cataloging-in-Publication Data:

Brumer, Paul.
    Principles of the quantum control of molecular processes/Paul Brumer, Moshe Shapiro.
        p. cm.
    Includes bibliographical references and index.
    ISBN 0-471-24184-9 (cloth)
    1. Quantum optics. 2. Coherence (Optics). 3. Molecular dynamics. I. Shapiro, Moshe.
II. Title.

QC446.2 .B78 2002
535'.2--dc21

Printed in the United States of America

10 9 8 7 6 5 4 3 2 1

IN MEMORY OF OUR PARENTS
AND
TO OUR WIVES
RACHELLE AND ABBEY, נשי חיל

CONTENTS

PREFACE

Despite its maturity, quantum mechanics remains one the most intriguing of subjects. Since its emergence over 75 years ago, each generation has discovered, investigated, and utilized different attributes of quantum phenomena. In this book we introduce results from research over the past 15 years that demonstrate that quantum attributes of light and matter afford the possibility of unprecedented control over the dynamics of atomic and molecular systems. This subject is the result of extensive investigations in chemistry and physics since 1985 and has seen enormous growth and interest over the past years. This growth reflects a confluence of developments—the maturation of quantum mechanics as a tool for chemistry and physics, the development of new laser devices that afford extraordinary facility in manipulating light, and the recognition that coherent laser light can be used to imprint information on atoms and molecules in a manner such that their subsequent dynamics leads to desirable goals. As such, an appreciation of coherent control requires input from optical physics, physical chemistry, atomic and molecular physics, and quantum mechanics. This book aims to provide this background in a systematic manner, allowing the reader to gain expertise in the area.

We have written this monograph with the mature chemistry or physics graduate student in mind; the development is systematic, starting with the fundamental principles of light–matter interactions and concluding with a wide variety of specific topics. We endeavor to include a sufficient number of steps throughout the book to allow self-study or use in class. To retain the focus on the role of quantum interference in control, we tend to utilize examples from our own research, while including samples from that of others. This focus is partially made possible by the recent appearance of a comprehensive survey of the field by Rice and Zhao

(*Optical Control of Molecular Dynamics*, Wiley, New York, 2000). It is our expectation that the two books will complement one another.

This book is organized, after a discussion of light and light–matter interactions in Chapter 1, in order of increasing incident electromagnetic field strength. Chapters 2 to 8 primarily deal with molecular dynamics and control where the field strengths are such that perturbation theory is applicable. Emphasis is placed on the principle of coherent control, that is, control via quantum interference between simultaneous indistinguishable pathways to the same final state. From the viewpoint of chemistry, the vast majority of control work has thus far been done on photodissociation processes. As a consequence, we provide a thorough introduction to the dynamics of photodissociation in Chapter 2 and discuss its control in Chapters 2, 4, and 6. The extension of quantum control to bimolecular collision processes is provided in Chapter 7 and to the control of chirality (and asymmetric synthesis) in Chapter 8.

Applications of control using moderate fields are discussed in Chapters 9 to 11. These fields allow for new physical phenomena in both bound state and continuum problems, including adiabatic population transfer in both regimes, electromagnetically induced transparency in bound systems, as well as additional unimolecular and bimolecular control scenarios.

Strong fields introduce yet another set of phenomena allowing for the controlled manipulation of matter. Examples of light-induced potentials and the controlled focusing, alignment, and deposition of molecules are discussed in Chapter 12, after the introduction of the quantized electromagnetic field.

All of the quantum control scenarios involve a host of laser and system parameters. To obtain maximal control in any scenario necessitates a means of tuning the system and laser parameters to optimally achieve the desired objective. This topic, optimal control, is introduced and discussed in Chapters 4 and 13. The role of quantum interference effects in optimal control are discussed as well, providing a uniform picture of control via optimal pulse shaping and coherent control.

By definition, quantum control relies upon the unique quantum properties of light and matter, principally the wavelike nature of both. As such, maintenance of the phase information contained in both the matter and light is central to the success of the control scenarios. Chapter 5 deals with *decoherence*, that is, the loss of phase information due to the influence of the external environment in reducing the system coherence. Methods of countering decoherence are also discussed.

This book has benefited greatly from the research support that we have received over the past years. First, we acknowledge the ongoing support by the U.S. Office of Naval Research through the research program of Dr. Peter J. Reynolds. We are also grateful to NSERC Canada, Photonics Research Ontario, the Israel Science Foundation, and the Minerva Foundation, Germany. Equally importantly, we thank the many students and colleagues who have taken part in the development of coherent control and have contributed so much to the field. We wish to acknowledge Ignacio Franco, Einat Frishman, David Gerbasi, Michal Oren, and Alexander Pegarkov for comments on various parts of the manuscript, Ms. Susan Arbuckle for unstinting assistance with copyediting and indexing, Daniel Gruner for expert assistance on puzzling TeX issues, and Amnon Shapiro for preparing many of the Postscript figure files. On a personal note, P.B. thanks Meir and Malka Cohen–Nehemia for training in the Mitzvah Technique that allowed him to counter the debilitating effects of back pain and body misuse.

None of this work would be possible without our wives, Rachelle and Abbey, who have provided the environment and support so necessary to allow productive science to be done. We are more than grateful to them both, as indicated in the dedication.

Finally, we welcome, at our email addresses below, any suggested corrections or additions to this book.

תושלב"ע

Moshe Shapiro (Moshe.Shapiro@weizmann.ac.il)

Paul Brumer (pbrumer@tikva.chem.utoronto.ca)

CHAPTER 1

PRELIMINARIES OF THE INTERACTION OF LIGHT WITH MATTER

In this chapter we introduce some of the fundamental concepts needed to understand how light interacts with matter. We start by examining a system of classical charged particles that interacts with a pulse of electromagnetic radiation. We then quantize the particle variables and develop the semiclassical theory of light interacting with quantized particles. The details of the derivations are not required for subsequent chapters. However, the resultant equations [Eqs. (1.50) to (1.52)] form the basis for the theoretical development presented in Chapter 2, which deals with both the interaction of weak lasers with molecules and with photodissociation processes.

1.1   CLASSICAL ELECTRODYNAMICS OF A PULSE OF LIGHT

1.1.1   Classical Hamiltonian

Consider a system of charged particles interacting with a pulse of light. The dynamics of the particles and of electric field $\mathbf{E}(\mathbf{r}, t)$ and magnetic field $\mathbf{B}(\mathbf{r}, t)$ are determined by combining Maxwell's equations for the fields [1]

$$
\begin{aligned}
\nabla \cdot \mathbf{E}(\mathbf{r}, t) &= \frac{1}{\epsilon_0} \rho(\mathbf{r}, t), \\
\nabla \cdot \mathbf{B}(\mathbf{r}, t) &= 0, \\
\nabla \times \mathbf{E}(\mathbf{r}, t) &= -\frac{1}{c} \frac{\partial}{\partial t} \mathbf{B}(\mathbf{r}, t), \\
\nabla \times \mathbf{B}(\mathbf{r}, t) &= \frac{1}{c} \frac{\partial}{\partial t} \mathbf{E}(\mathbf{r}, t) + \frac{1}{\epsilon_0 c} \mathbf{j}(\mathbf{r}, t),
\end{aligned}
\tag{1.1}
$$

with Lorentz's equation for a charged particle moving in an electric and magnetic field:

$$m_i \frac{d\mathbf{v}_i}{dt} = \mathbf{F}(\mathbf{r}_i, t) = q_i \left[\mathbf{E}(\mathbf{r}_i, t) + \frac{\mathbf{v}_i}{c} \times \mathbf{B}(\mathbf{r}_i, t) \right]. \tag{1.2}$$

Here m_i, q_i, $\mathbf{r}_i$, and $\mathbf{v}_i$ are, respectively, the mass, the charge, the position, and the velocity of the ith particle. The quantities ρ and $\mathbf{j}$ are the charged-particle density and the current of the charged particles, defined by

$$\rho(\mathbf{r}, t) = \sum_i q_i \delta[\mathbf{r} - \mathbf{r}_i(t)],$$
$$\mathbf{j}(\mathbf{r}, t) = \sum_i q_i \mathbf{v}_i \delta[\mathbf{r} - \mathbf{r}_i(t)]. \tag{1.3}$$

The symbol ϵ_0 denotes the permittivity of free space and is a constant, as is c, the speed of light. In atomic units (a.u.) $\hbar = 1$, $m_e = 1$, $q_e = -1$, where m_e is the mass, q_e is the charge of the electron, and $\epsilon_0 = 1/(4\pi)$.

It is advantageous to reexpress the electrodynamics in terms of the vector and scalar potentials $\mathbf{A}(\mathbf{r}, t)$ and $\Phi(\mathbf{r}, t)$, which are related to the electric and magnetic fields by the following relations:

$$\mathbf{E} = -\frac{1}{c} \frac{\partial \mathbf{A}}{\partial t} - \nabla \Phi,$$
$$\mathbf{B} = \nabla \times \mathbf{A}. \tag{1.4}$$

These equations do not completely specify the vector and scalar potentials. Rather, it is possible to define different forms of the vector and scalar potentials, the so-called *gauges*, that give the same electric and magnetic fields. Specifically, it follows from Eq. (1.4) that given $\mathbf{A}$ and Φ we can construct other potentials $\mathbf{A}'$ and Φ' as

$$\mathbf{A}' = \mathbf{A} + \nabla \chi,$$
$$\Phi' = \Phi - \frac{1}{c} \frac{\partial \chi}{\partial t}, \tag{1.5}$$

where χ is a scalar field, which result in $\mathbf{E}$ and $\mathbf{B}$ fields that satisfy Maxwell's equations.

The choice that is often made, called the *Coulomb gauge*, is defined by choosing χ such that $\nabla^2 \chi = -\nabla \cdot \mathbf{A}$, which means, given Eq. (1.5), that

$$\nabla \cdot \mathbf{A}' = 0. \tag{1.6}$$

Requiring $\nabla^2 \chi = -\nabla \cdot \mathbf{A}$ still does not completely determine $\mathbf{A}$ and Φ because additional gauge transformations, defined by different choices of the scalar fields χ' that satisfy $\nabla^2 \chi' = 0$, also satisfy Eq. (1.6). We shall make use of this flexibility later.

The total energy of a particle-plus-field system (i.e., the Hamiltonian H) is the sum of the *kinetic energy* of the particles and the energy of the field. That is,

$$H = \sum_i \frac{1}{2} m_i \left(\frac{d\mathbf{r}_i}{dt} \right)^2 + \frac{\epsilon_0}{2} \int d^3 r [\mathbf{E}^2(\mathbf{r}, t) + \mathbf{B}^2(\mathbf{r}, t)]. \tag{1.7}$$

Notice that, remarkably, this form does not contain an explicit contribution from the potential energy of the particles. Rather, the potential energy will arise [Eq. (1.40)] naturally by accounting for the way in which the particle density and the particle current affect the electric and magnetic field via Maxwell equations, and the way in which the electric and magnetic fields modify the particle position and velocity through the Lorentz equation. Hence, in this sense, the kinetic energy is a more fundamental quantity than is the potential energy of interaction.

To obtain this potential energy contribution, we start with the fact that the electric field can be written as a sum of longitudinal (L) and transverse (R) components:

$$\mathbf{E} = \mathbf{E}_L + \mathbf{E}_R, \tag{1.8}$$

defined by the relations

$$\nabla \times \mathbf{E}_L = 0, \qquad \nabla \cdot \mathbf{E}_R = 0. \tag{1.9}$$

It is clear from Eq. (1.9) that the first of Maxwell's equation pertains only to $\mathbf{E}_L$. It also follows from the second of Maxwell's equation that $\mathbf{B}$ is a purely transverse vector field.

Substituting Eq. (1.8) into Eq. (1.7) gives

$$H = \sum_i \frac{1}{2} m_i \left(\frac{d\mathbf{r}_i}{dt} \right)^2 + H_L + H_R, \tag{1.10}$$

where we define

$$H_L \equiv \frac{\epsilon_0}{2} \int d^3 r \, \mathbf{E}_L^2(\mathbf{r}, t), \tag{1.11}$$

and

$$H_R \equiv \frac{\epsilon_0}{2} \int d^3 r [\mathbf{E}_R^2(\mathbf{r}, t) + \mathbf{B}^2(\mathbf{r}, t)]. \tag{1.12}$$

Using Eqs. (1.4) and (1.6), according to which $\mathbf{E}_L = -\nabla \Phi$, we can write H_L, with the aid of the first of Maxwell's equations [Eq. (1.1)], as

$$H_L = \frac{\epsilon_0}{2} \int d^3 r (\nabla \Phi)^2 = -\frac{\epsilon_0}{2} \int d^3 r \, \Phi \nabla^2 \Phi = \frac{\epsilon_0}{2} \int d^3 r \, \Phi \nabla \cdot \mathbf{E}_L = \frac{1}{2} \int d^3 r \, \Phi \rho. \tag{1.13}$$

Using the form of the charge density ρ [Eq. (1.3)], we obtain that

$$H_L = \frac{1}{2} \sum_i q_i \Phi(\mathbf{r}_i),$$

(1.14)

with $\Phi(\mathbf{r})$ being the potential induced by all the charged particles,

$$\Phi(\mathbf{r}) = \frac{1}{4\pi\epsilon_0} \sum_j \frac{q_j}{|\mathbf{r} - \mathbf{r}_j|}.$$

(1.15)

We see that the H_L gives rise, and is identical to, the electrostatic potential energy of the particles.

Equations (1.14) and (1.15) contain divergent terms that are independent of the particles' positions. These terms occur whenever $\mathbf{r} = \mathbf{r}_i$ and represent the electrostatic interaction of each particle with itself. However, because these terms are independent of the particles' positions, subtracting them from the electrostatic energy is equivalent to a simple redefinition of the zero-point energy of the particles. Since all forces in nature derive from *changes* in energy, such a redefinition of the zero point of energy is of no dynamical consequence.

In subtracting the divergent terms, we find that the H_L contribution to H is replaced by the *Coulomb potential* V_C, defined as

$$V_C = \sum_i q_i \Phi_i(\mathbf{r}_i),$$

(1.16)

where

$$\Phi_i(\mathbf{r}) = \frac{1}{4\pi\epsilon_0} \sum_{j<i} \frac{q_j}{|\mathbf{r} - \mathbf{r}_j|}.$$

(1.17)

In contrast to $\Phi(\mathbf{r})$ [Eq. (1.15)], each $\Phi_i(\mathbf{r})$, $i = 1, \ldots, N$ term is the potential due to particles whose indices j are less than that of the ith particle.

The Hamiltonian of the particle + radiation system now assumes the form

$$H = \sum_i \frac{1}{2} m_i \left(\frac{d\mathbf{r}_i}{dt}\right)^2 + V_C + H_R.$$

(1.18)

1.1.2   Free Light Field

In the absence of particles, $\mathbf{j}(\mathbf{r}, t) = 0$ and $\Phi(\mathbf{r}) = 0$, and the fourth Maxwell equation [Eq. (1.1)] takes on the form

$$-\nabla^2 \mathbf{A} + \frac{1}{c^2} \frac{\partial^2 \mathbf{A}}{\partial t^2} = 0.$$

(1.19)

Here we have made use of Eq. (1.4) and the identity $\nabla \times \nabla \times \mathbf{A} = \nabla(\nabla \cdot \mathbf{A}) - \nabla^2\mathbf{A}$, which simplifies in the Coulomb gauge to $\nabla \times \nabla \times \mathbf{A} = \nabla^2\mathbf{A}$.

As particular solutions of Eq. (1.19) we can choose plane waves (also called "field modes")

$$\mathbf{A_k}(\mathbf{r}, t) = \mathbf{A_k}(t)\exp(i\mathbf{k} \cdot \mathbf{r}), \tag{1.20}$$

where $\mathbf{k}$ is an arbitrary vector (the "wave vector") that determines the direction of propagation of the plane wave. Upon substitution of Eq. (1.20) in Eq. (1.19), each expansion coefficient $\mathbf{A_k}(t)$ is seen to be a solution of the differential equation for a harmonic oscillator

$$k^2\mathbf{A_k}(t) + \frac{1}{c^2}\frac{d^2\mathbf{A_k}(t)}{dt^2} = 0, \tag{1.21}$$

with solution

$$\mathbf{A_k}(t) = \mathbf{A_k}\exp(\mp i\omega_k t). \tag{1.22}$$

That is, each field mode is represented by a classical harmonic oscillator of mode frequency $\omega_k \equiv ck$.

Each field mode must also satisfy the Coulomb gauge condition, $\nabla \cdot \mathbf{A_k}(\mathbf{r}, t) = 0$, which, when substituted into Eq. (1.20), implies that

$$\mathbf{k} \cdot \mathbf{A_k} = 0. \tag{1.23}$$

We can ensure the validity of Eq. (1.23) by writing $\mathbf{A_k}$ as a product of a unit vector $\hat{\varepsilon}_\mathbf{k}$, called the *polarization*, and a complex scalar amplitude $A_\mathbf{k}$

$$\mathbf{A_k} \equiv \hat{\varepsilon}_\mathbf{k} A_\mathbf{k}, \tag{1.24}$$

and require that

$$\mathbf{k} \cdot \hat{\varepsilon}_\mathbf{k} = 0. \tag{1.25}$$

Since, according to Eq. (1.25), the polarization vector is perpendicular to the direction of propagation, each field mode can have only two independent polarization directions.

Because Eq. (1.19) is linear, its general solution can be expressed as a sum over all the field modes and their complex conjugates when the radiation field is in a cavity of a finite volume. That is, since $\mathbf{A}$ must be real, it satisfies

$$\mathbf{A}(\mathbf{r}, t) = \sum_\mathbf{k} \hat{\varepsilon}_\mathbf{k}\{A_\mathbf{k}\exp(-i\omega_k t + i\mathbf{k} \cdot \mathbf{r}) + A_\mathbf{k}^*\exp(i\omega_k t - i\mathbf{k} \cdot \mathbf{r})\}. \tag{1.26}$$

If the field is in an infinite volume, then the sum in Eq. (1.26) is replaced by an integral. It follows from Eq. (1.4) that the electric field (which in the absence of particles has only the transverse component) and the magnetic field are given as

$$\mathbf{E}_R(\mathbf{r}, t) = i \sum_{\mathbf{k}} k \hat{\varepsilon}_{\mathbf{k}} \{A_{\mathbf{k}} \exp(-i\omega_k t + i\mathbf{k} \cdot \mathbf{r}) - A_{\mathbf{k}}^* \exp(i\omega_k t - i\mathbf{k} \cdot \mathbf{r})\}, \tag{1.27}$$

$$\mathbf{B}(\mathbf{r}, t) = i \sum_{\mathbf{k}} \mathbf{k} \times \hat{\varepsilon}_{\mathbf{k}} \{A_{\mathbf{k}} \exp(-i\omega_k t + i\mathbf{k} \cdot \mathbf{r}) - A_{\mathbf{k}}^* \exp(i\omega_k t - i\mathbf{k} \cdot \mathbf{r})\}. \tag{1.28}$$

Equations (1.27) and (1.28) represent general time-dependent pulses of light.

We see [with the definition of the magnetic field adopted in Eq. (1.1)] that the electric and magnetic fields are two mutually perpendicular vector fields with the same amplitude. Hence, the contribution of each field to the radiation energy is, according to Eq. (1.12), the same. Using this fact and Eq. (1.27) we can write Eq. (1.12) for an infinite cavity as

$$H_R = -\epsilon_0 \int d^3 r \, d^3 k \, d^3 k' \, kk' \hat{\varepsilon}_{\mathbf{k}} \cdot \hat{\varepsilon}_{\mathbf{k}'} \{A_{\mathbf{k}} \exp(-i\omega_k t + i\mathbf{k} \cdot \mathbf{r}) - A_{\mathbf{k}}^* \exp(i\omega_k t - i\mathbf{k} \cdot \mathbf{r})\}$$

$$\times \{A_{\mathbf{k}'} \exp(-i\omega_{k'} t + i\mathbf{k}' \cdot \mathbf{r}) - A_{\mathbf{k}'}^* \exp(i\omega_{k'} t - i\mathbf{k}' \cdot \mathbf{r})\}. \tag{1.29}$$

Using the expression

$$\int d^3 r \, \exp[i(\mathbf{k} - \mathbf{k}') \cdot \mathbf{r}] = (2\pi)^3 \delta(\mathbf{k} - \mathbf{k}'), \tag{1.30}$$

we obtain

$$H_R = (2\pi)^3 \epsilon_0 \int d^3 k \, k^2 \{A_{\mathbf{k}} A_{-\mathbf{k}} \exp(-2i\omega_k t) + A_{\mathbf{k}}^* A_{-\mathbf{k}} \exp(2i\omega_k t) + 2|A_{\mathbf{k}}|^2\}. \tag{1.31}$$

Consider now the integral of H_R over time. Using the time analog of Eq. (1.30), we see that the first two terms in the curly bracket average to zero when we integrate Eq. (1.31) over a cycle of time. The *cycle-averaged* radiation energy is therefore given by

$$\bar{H}_R = 2(2\pi)^3 \epsilon_0 \int d^3 k \, k^2 |A_{\mathbf{k}}|^2. \tag{1.32}$$

For the finite case of a cavity, of volume V, an analogous derivation, coupled with

$$\int_V d^3 r \, \exp[i(\mathbf{k} - \mathbf{k}') \cdot \mathbf{r}] = V \delta_{\mathbf{k},\mathbf{k}'}, \tag{1.33}$$

for a finite cavity of volume V, gives

$$\bar{H}_R = 2\epsilon_0 V \sum_{\mathbf{k}} k^2 |A_{\mathbf{k}}|^2. \tag{1.34}$$

When considering a coherent *pulse* of light, it is necessary to superimpose a collection of plane waves, as in Eqs. (1.27) and (1.28). In doing so it is reasonable to make the simplifying assumption that all the modes of the pulse propagate in the same direction (chosen as the z axis) and that all the pulse modes have the same polarization direction $\hat{\varepsilon}$. We can therefore eliminate the integration over the $\hat{\mathbf{k}}$ directions and write Eq. (1.27) (in an infinite volume) as

$$\mathbf{E}(z, t) = \hat{\varepsilon} \int_0^\infty d\omega \left\{ \epsilon(\omega) \exp\left[i\omega\left(\frac{z}{c} - t\right)\right] + \epsilon^*(\omega) \exp\left[-i\omega\left(\frac{z}{c} - t\right)\right] \right\}$$

$$\equiv \mathbf{E}_+(\tau) + \mathbf{E}_-(\tau) \equiv \mathbf{E}(\tau) = \hat{\varepsilon} \int_{-\infty}^\infty d\omega\, \epsilon(\omega) \exp(-i\omega\tau). \tag{1.35}$$

Here $\epsilon(\omega) \equiv ikA_k/c$ and τ is the so-called retarded time,

$$\tau \equiv t - z/c. \tag{1.36}$$

Here and below, we denote $\mathbf{E}_R$, the transverse radiative electric field, by $\mathbf{E}$ since the longitudinal component of $\mathbf{E}$ will be associated exclusively with the material charge-density ρ according to the first Maxwell equation [Eq. (1.1)].

Each mode amplitude in Eq. (1.35) is a complex number,

$$\epsilon(\omega) = |\epsilon(\omega)| \exp[i\phi(\omega)], \tag{1.37}$$

where $\phi(\omega)$ are frequency-dependent phases. The fact that $\mathbf{E}(z, t)$ is real ensures that $\epsilon(-\omega) = \epsilon^*(\omega)$ and hence that

$$\phi(-\omega) = -\phi(\omega), \qquad |\epsilon(-\omega)| = |\epsilon(\omega)|. \tag{1.38}$$

As explained below, the phase $\phi(\omega)$ plays a central role in coherent control theory. However, individual phase values depend on an (arbitrary) definition of the origin of time and space. Therefore only the relative phases, which are the only phase factors that can actually be measured, are of any consequence physically.

Equation (1.35) describes a pulse of *coherent* light, where $\mathbf{E}(z, t)$ is represented by an analytic function. In cases of partially coherent light either the phase or amplitude acquires a random component, and an analytic expression for $\mathbf{E}(z, t)$ no longer exists. Appropriate descriptions of partially incoherent light interacting with molecules are discussed in Section 5.3.

1.2   DYNAMICS OF QUANTIZED PARTICLES AND CLASSICAL LIGHT FIELDS

Consider now the transition from classical mechanics to the quantum mechanics of the particles in the presence of a classical field. (The case of quantized particles in the presence of a quantized field is discussed in Chapter 12.)

To quantize the dynamics of the particles first requires that we express the velocities of the particles in terms of canonical momenta. In the presence of electromagnetic fields, the canonical momenta are not merely $m_i(d\mathbf{r}_i/dt)$. Rather, in order to incorporate Lorentz's velocity-dependent forces into Hamilton's formulation of classical mechanics, the canonical momenta are given by [2]

$$\mathbf{p}_i = m_i \frac{d\mathbf{r}_i}{dt} + \frac{q_i}{c} \mathbf{A}(\mathbf{r}_i, t). \tag{1.39}$$

It follows from Eqs. (1.18) and (1.39) that

$$H = \sum_i \frac{1}{2m_i} \left[\mathbf{p}_i - \frac{q_i}{c} \mathbf{A}(\mathbf{r}_i, t) \right]^2 + V_C + H_R. \tag{1.40}$$

Having expressed the Hamiltonian in terms of the canonical momenta, we can readily quantize the particles' dynamics. To do so we replace each particle's canonical momentum by the momentum operator in the coordinate representation,

$$\mathbf{p}_j \rightarrow -i\hbar \nabla_j. \tag{1.41}$$

The quantized Hamiltonian then assumes the form

$$H = \sum_j \frac{1}{2m_j} \left[-i\hbar \nabla_j - \frac{q_j}{c} \mathbf{A}(\mathbf{r}_j, t) \right]^2 + V_C + H_R = H_M + H'(t) + H_R, \tag{1.42}$$

with H_M, the material Hamiltonian, given by

$$H_M = \sum_j \frac{-\hbar^2}{2m_j} \nabla_j^2 + V_C, \tag{1.43}$$

and where $H'(t)$, the interaction Hamiltonian, is

$$H'(t) = \sum_j \frac{iq_j\hbar}{m_jc} \nabla_j \cdot \mathbf{A}(\mathbf{r}_j, t) + \frac{q_j^2}{2m_jc^2} A^2(\mathbf{r}_j, t) = \sum_j \frac{iq_j\hbar}{m_jc} \mathbf{A}(\mathbf{r}_j, t) \cdot \nabla_j + \frac{q_j^2}{2m_jc^2} A^2(\mathbf{r}_j, t). \tag{1.44}$$

Here we have used the fact that, in the Coulomb gauge, $\nabla_j \cdot \mathbf{A}(\mathbf{r}_j, t)\langle \mathbf{R}|\psi\rangle = \mathbf{A}(\mathbf{r}_j, t) \cdot \nabla_j\langle \mathbf{R}|\psi\rangle$, where $\mathbf{R} \equiv \mathbf{r}_1, \ldots, \mathbf{r}_N$, with N being the total number of particles. Equation (1.44) is often referred to as being in the *velocity gauge*.

Given the Hamiltonian of Eq. (1.42), the dynamics of the particles in the presence of the field are obtained by solving for the wave function $\Psi(\mathbf{R}, t)$ via the time-dependent Schrödinger equation:

$$i\hbar \frac{\partial \Psi(\mathbf{R}, t)}{\partial t} = \left\{ \sum_j \frac{1}{2m_j}\left[-i\hbar\nabla_j - \frac{q_j}{c}\mathbf{A}(\mathbf{r}_j, t) \right]^2 + V_C \right\} \Psi(\mathbf{R}, t). \tag{1.45}$$

Here H_R does not contribute since it is a function of the field variables only.

Equation (1.45) may be further simplified by noting that the variation of $\mathbf{A}$ over a typical displacement $\mathbf{r}_j$ of a particle is small. For example, for visible light, a typical wavelength of the field is $5000\,\text{Å}$, whereas the particle displacements within a molecule vary over 1 to $10\,\text{Å}$. It is therefore reasonable to replace all of the $\mathbf{r}_j$ displacements in $\mathbf{A}$ by the position of the center of mass of the molecule. For a plane wave, only the z projection of the center-of-mass position is relevant, and we can approximate $\mathbf{A}$ as

$$\mathbf{A}(\mathbf{r}_j, t) \approx \mathbf{A}(z, t). \tag{1.46}$$

For reasons that will become evident below, this is called the *dipole approximation*.

Given this approximation, we can transform the Hamiltonian of Eq. (1.44) from the velocity gauge to the so-called *length gauge* in which the matter–radiation interaction term contains only the dot product of the dipole moment and the electric field. In order to do so we choose χ [Eq. (1.5)] as

$$\chi = -\sum_i \mathbf{r}_i \cdot \mathbf{A}(z, t). \tag{1.47}$$

Clearly, due to the neglect of the $\mathbf{r}_i$ dependence in $\mathbf{A}$ in Eq. (1.46), this gauge transformation leaves $\mathbf{A}(z, t)$ within the Coulomb gauge [Eq. (1.6)] since $\nabla^2\chi = 0$.

Using the definition V_C [Eq. (1.16)] and χ [Eq. (1.47)] in the Schrödinger equation [Eq. (1.45)], and noting that $\nabla\chi = -\mathbf{A}$, we obtain that

$$i\hbar \frac{\partial \Psi(\mathbf{R}, t)}{\partial t} = \sum_j \left[\frac{-\hbar^2}{2m_j}\nabla_j^2 + q_j\Phi_j(\mathbf{r}_j) + \frac{q_j}{c}\mathbf{r}_j \cdot \frac{\partial \mathbf{A}(z, t)}{\partial t} \right] \Psi(\mathbf{R}, t). \tag{1.48}$$

Using Eq. (1.4) we can write the last term in the square brackets as $-q_j\mathbf{r}_j \cdot \mathbf{E}(z, t)$, where we have used the fact that gauge transformations do not change the electric field, which was calculated from the untransformed vector potential.

We obtain that

$$i\hbar \frac{\partial \Psi(\mathbf{R}, t)}{\partial t} = \sum_j \left[\frac{-\hbar^2}{2m_j}\nabla_j^2 + q_j\Phi_j(\mathbf{r}_j) - q_j\mathbf{r}_j \cdot \mathbf{E}(z, t) \right] \Psi(\mathbf{R}, t), \tag{1.49}$$

In this form, both the vector potential and the gradient operator no longer appear. Instead, a scalar potential, proportional to the scalar product of the transverse field and the displacement of each particle from the origin, has been added to the Coulomb potential.

Equation (1.49) can be written in a more concise form as

$$i\hbar \frac{\partial \Psi(t)}{\partial t} = H(t)\Psi(t) = [H_M + H_{MR}(t)]\Psi(t), \qquad (1.50)$$

where $H(t) = H_M + H_{MR}$ is the total Hamiltonian, H_M is the material Hamiltonian, given in Eq. (1.43), and H_{MR} is the matter–radiation interaction in the dipole approximation, given by

$$H_{MR} = -\mathbf{d} \cdot \mathbf{E}(z, t), \qquad (1.51)$$

where $\mathbf{d}$ is the molecular dipole moment,

$$\mathbf{d} \equiv \sum_j q_j \mathbf{r}_j. \qquad (1.52)$$

It is possible to go beyond the dipole approximation in the length gauge and treat the interactions between higher multipoles with the field derivatives, which is relevant when the variation of the field with $\mathbf{r}_j$ cannot be neglected [3]. However, we do not pursue these extensions here because, in all the applications discussed below, the dipole approximation will be found to suffice. Equations (1.50), (1.51), and (1.52) are the central expressions used below to describe molecule–light interactions. Extensions of this approach to include quantization of the electromagnetic field are described in Chapter 12.

CHAPTER 2

WEAK-FIELD PHOTODISSOCIATION

As will be shown throughout this book, quantum control of molecular dynamics has been applied to a wide variety of processes. Within the framework of chemical applications, control over reactive scattering has dominated. In particular, the two primary chemical processes focused upon are photodissociation, in which a molecule is irradiated and dissociates into various products, and bimolecular reactions, in which two molecules collide to produce new products. In this chapter we formulate the quantum theory of photodissociation, that is, the light-induced breaking of a chemical bond. In doing so we provide an introduction to concepts essential for the remainder of this book. The quantum theory of bimolecular collisions is also briefly discussed.

A number of issues preliminary to questions of control and process selectivity are also discussed. In particular we ask: What determines the final outcome of a photodissociation process? Although in quantum mechanics the fate of a system can only be known in a probabilistic sense, the linear time dependence of the Schrödinger equation does guarantee that the probability of future events is completely determined by the probability of past events. (That is, quantum mechanics is a deterministic theory of distributions of various observables). Hence by identifying attributes of the quantum state at earlier times we learn what is required to alter, that is, control, system dynamics in the future.

In addition to basic concepts in photodissociation, we address a number of more subtle issues such as the precise definition of the concept of a "lifetime." We show that this attribute is not a pure property of the system. Rather it is intimately related to the way the system was prepared (see Appendix 2A).

Throughout this chapter we utilize perturbation theory, assuming that the light field is "weak"; "strong" light fields are addressed in Chapter 10. The approach that we advocate applies to both pulsed and continuous-wave (cw) excitation sources.

This allows us to compare and contrast these photodissociation schemes and to identify aspects of photodissociation that are consistent with, or contrary to, our classical intuition.

2.1 PHOTOEXCITATION OF A MOLECULE WITH A PULSE OF LIGHT

Consider a molecule interacting with a pulse of coherent light, where the light is described by a purely classical field of Eq. (1.35) and the molecule is treated quantum mechanically. The dynamics of the radiation-free molecule is fully described by the (discrete or continuous) set of energy eigenvalues and eigenfunctions, denoted, respectively, as E_n and $|E_n\rangle$, of the material Hamiltonian H_M [Eq. (1.43)],

$$H_M|E_n\rangle = E_n|E_n\rangle. \tag{2.1}$$

(Here the eigenfunctions are denoted $|E_n\rangle$, with the understanding that the notation will be extended to include additional quantum numbers when energy degeneracies exist.)

Given E_n and $|E_n\rangle$, the full time-dependent Schrödinger equation [Eq. (1.50)] can be solved by expanding $|\Psi(t)\rangle$ in terms of $|E_n\rangle$, that is,

$$|\Psi(t)\rangle = \sum_n b_n(t)|E_n\rangle \exp\left(\frac{-iE_nt}{\hbar}\right), \tag{2.2}$$

with unknown coefficients $b_n(t)$. To obtain these coefficients we use the orthonormality of the $|E_n\rangle$ basis functions, and substitute Eq. (2.2) in Eq. (1.50), giving a set of ordinary differential equations for $b_n(t)$:

$$\frac{db_m(t)}{dt} = \left(\frac{1}{i\hbar}\right)\sum_n b_n(t)\exp(i\omega_{m,n}t)\langle E_m|H_{MR}(t)|E_n\rangle. \tag{2.3}$$

Here the transition frequency, $\omega_{m,n}$, is given by

$$\omega_{m,n} \equiv (E_m - E_n)/\hbar. \tag{2.4}$$

Consider first the case where the molecule is initially ($t = -\infty$) in a single state $|E_1\rangle$, that is, where

$$b_1(t = -\infty) = 1, \quad \text{and} \quad b_k(t = -\infty) = 0 \quad \text{for} \quad k \neq 1, \tag{2.5}$$

and where the perturbation is weak. The latter condition implies that

$$\int_{-\infty}^{\infty} dt|\langle E_i|H_{MR}(t)|E_j\rangle \exp(i\omega_{i,j}t)|/\hbar \ll 1. \tag{2.6}$$

Under these circumstances we obtain, in first-order perturbation theory, that the expansion coefficients in Eq. (2.2) are given by

$$b_m(t) = -\frac{d_{m,1}}{i\hbar} \int_{-\infty}^{t} dt' \exp[i\omega_{m,1}t']\varepsilon(z, t')$$

$$= -\frac{d_{m,1}}{i\hbar} \int_{-\infty}^{\infty} d\omega \, \bar{\epsilon}(\omega) \int_{-\infty}^{t} dt' \exp[i(\omega_{m,1} - \omega)t'],$$

(2.7)

where

$$d_{m,1} \equiv \langle E_m | \hat{\varepsilon} \cdot \mathbf{d} | E_1 \rangle.$$

(2.8)

Here $\hat{\varepsilon} \cdot \mathbf{d}$ is the projection of the transition dipole operator along the electric field direction. In Eq. (2.7) we have introduced $\varepsilon(z, t) = \int_{-\infty}^{\infty} d\omega \, \bar{\epsilon}(\omega) \exp(-i\omega t)$ as the length of the $\mathbf{E}(z, t) = \varepsilon(z, t)\hat{\varepsilon}$ vector with

$$\bar{\epsilon}(\omega) \equiv \epsilon(\omega) \exp(i\omega z/c) = |\epsilon(\omega)| \exp[i(\phi(\omega) + \omega z/c)].$$

(2.9)

Hence,

$$H_{MR}(t) = -\mathbf{E} \cdot \mathbf{d} = -\varepsilon(z, t)\hat{\varepsilon} \cdot \mathbf{d}.$$

(2.10)

Equation (2.7) provides the expansion coefficients at any time t. If our interest is in observing or controlling the *final* product states (as it is in photodissociation), then we only require the wave function $\Psi(t)$ as $t \to +\infty$. In this limit we can insert the equality

$$\int_{-\infty}^{\infty} dt' \exp[i(\omega_{m,1} - \omega)t'] = 2\pi\delta(\omega_{m,1} - \omega)$$

(2.11)

into Eq. (2.7) to obtain

$$b_m(+\infty) = \frac{2\pi i}{\hbar} \bar{\epsilon}(\omega_{m,1}) d_{m,1} = \frac{2\pi i}{\hbar} |\epsilon(\omega_{m,1})| d_{m,1} \exp\left[i\left(\phi(\omega_{m,1}) + \left(\frac{\omega_{m,1}z}{c} \right) \right) \right].$$

(2.12)

Equation (2.12) clearly shows that in preparing the state $|E_m\rangle$, the light field has imparted both a magnitude as well as phase to $\Psi(t)$. Similar information is contained in the finite time result as well, but in a somewhat more complex fashion (see Section 2.2).

We note, for use later, that if $E_m > E_1$ (corresponding to light absorption), then $\omega_{m,1} > 0$, and the phase acquired by b_m from the laser is positive. Alternatively, when $E_1 > E_m$ (corresponding to stimulated emission), then $\omega_{m,1} < 0$, and by Eq. (1.38) $\phi(\omega_{m,1}) \equiv -\phi(|\omega_{m,1}|)$, and the acquired phase is negative. Hence we have the rule that light absorption imparts the laser phase evaluated at the frequency of transition to b_m, whereas stimulated emission imparts the negative of the laser phase evaluated at the frequency of emission.

Equation (2.12) also defines the *resonance* (energy conservation) condition: A material energy state $|E_m\rangle$ only absorbs or emits light, at infinite time, for which $\omega = \omega_{m,1}$ or $\omega = -\omega_{m,1}$. Equation (2.11) suggests that it takes an infinite time to establish this resonance condition. However, in the case of pulsed light, no transitions can take place after the pulse is over, no matter how short. Hence the resonance condition must actually be established in the finite time by which the pulse is over. These issues, as well as others related to dynamics during a pulse, are discussed in the next section.

2.2 STATE PREPARATION DURING THE PULSE

To explore the behavior of the system while the pulse is on, we express the integrals over t', in Eq. (2.7), as follows:

$$
A(t) \equiv \lim_{T \to \infty} \int_{-T}^{t} dt' \exp[i(\omega_{m,1} - \omega)t'] = \frac{\exp[i(\omega_{m,1} - \omega)t]}{i(\omega_{m,1} - \omega)}
$$

$$
- \lim_{T \to \infty} \frac{\exp[-i(\omega_{m,1} - \omega)T]}{i(\omega_{m,1} - \omega)}.
$$

(2.13)

Positive ω gives the so-called *rotating-wave* (rw) contribution whereas negative ω gives the *counterrotating wave* (crw) contribution.

We can show, using contour integration, that when we insert Eq. (2.13) in Eq. (2.7) and integrate over ω, the contribution from the second term in Eq. (2.13) vanishes. To do so we modify the path along the real axis to include an infinitesimally small semicircle in the upper-half complex ω plane with $\omega = \omega_{m,1}$ as its center. The contour is then closed by adding a large semicircle in the upper-half plane. The contour, shown in Figure 2.1, thus excludes the $\omega = \omega_{m,1}$ pole. Because of this, and provided we can deform the contour to exclude any existent complex poles of $\bar{\epsilon}(\omega)$ in the upper half ω plane, the integral over the closed contour is zero. It remains to be shown that the integral over the large semicircle is also zero. If this is the case, the contribution from the real-line segment [i.e., the part appearing in Eq. (2.7)] is also zero.

To show this, write $\omega - \omega_{m,1} = Re^{i\theta}$; hence, $e^{iT(\omega - \omega_{m,1})} = e^{iRT\cos\theta - RT\sin\theta}$. Since in the upper half of the complex plane, $\theta > 0$, $e^{iT(\omega - \omega_{m,1})}$ vanishes on the large semicircle as $T \to \infty$, a result that holds for all R (since by definition $R > 0$). We can therefore deform the large upper-circle portion of the contour to exclude all the poles of $\bar{\epsilon}(\omega)$. Thus, the contribution from the second term in Eq. (2.13), to the integral in Eq. (2.7), is zero, irrespective of the form of $\bar{\epsilon}(\omega)$.

The first term in Eq. (2.13) *does* contribute to $b_m(t)$ since $\exp[i(\omega_{m,1} - \omega)t]$ is nonzero on the large semicircle in the upper plane. Substituting the first term of Eq.

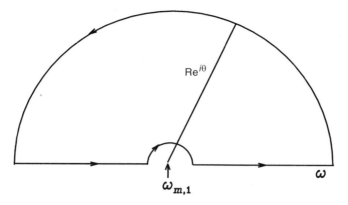

Figure 2.1   Contour to evaluate ω integral.

(2.13) into Eq. (2.7), we obtain

$$b_m(t) = (i/\hbar)\mathrm{d}_{m,1}[\bar{\epsilon}(\omega_{m,1})c_{m,1}^+(\tau) + \bar{\epsilon}(-\omega_{m,1})c_{m,1}^-(\tau)], \qquad (2.14)$$

where $c_{m,1}^{\pm}(t)$ are radiative preparation coefficients, defined as

$$c_{m,1}^{\pm}(t) \equiv \frac{1}{\bar{\epsilon}(\pm\omega_{m,1})} \int_0^{\infty} d\omega \; \epsilon(\omega) \frac{\exp[i(\omega_{m,1} \mp \omega)t]}{i(\omega_{m,1} \mp \omega)}. \qquad (2.15)$$

The subscript 1 defines the initial state, but this subscript is suppressed for the remainder of this chapter. The coefficient $c_m^+(t)$ results from the rw term and $c_m^-(t)$ results from the crw term. We show below that $c_m^-(t)/c_m^+(t) \to 0$, as $t \to \infty$, that is, only the rw term contributes asymptotically, and that at finite times this ratio is smallest for $\omega \approx \omega_{m,1}$.

To gain insight into the character of the preparation coefficients, we study pulses whose frequency profiles are Gaussian functions,

$$\epsilon(\omega) = \pi^{-1/2}\epsilon_0\delta_t \exp\{-[\delta_t(\omega - \omega_0)]^2\}. \qquad (2.16)$$

Their time dependence is also described by Gaussian functions:

$$\varepsilon(t) = \epsilon_0 \exp(-\Gamma^2 t^2) \exp(-i\omega_0 t), \qquad (2.17)$$

where $\Gamma \equiv \frac{1}{2}\delta_t$. It follows from Eqs. (2.15) and (2.16) and that the preparation coefficients for such pulses are given as

$$c_m^+(t) = \mathrm{sgn}(t)2\pi\{\theta(t) - (\tfrac{1}{2}\exp[\beta_+^2]W[\mathrm{sgn}(t)\beta_+]\}, \qquad (2.18)$$

and

$$c_m^-(t) = \text{sgn}(t)\pi \, \exp[\beta_-^2]W[\text{sgn}(t)\beta_-], \tag{2.19}$$

where

$$\beta_{\pm} \equiv \delta_t(\omega_{m,1} \mp \omega_0) + i\frac{t}{2\delta_t}, \tag{2.20}$$

and $W[z]$ is the complex error function:

$$W[z] \equiv \exp(-z^2)[1 - \text{erf}(-iz)], \tag{2.21}$$

[see Ref. [4], Eqs. (7.1.3) and (7.1.8)].

Given that $\text{erf}(z) \to 1$ as $z \to \infty$, $|\arg z| < \pi/4$, it follows from Eqs. (2.18) and (2.21) that

$$c_m^{\pm}(t) \to 0, \quad \text{for } t \ll \frac{-1}{\Gamma},$$

and

$$c_m^+(t) \to 2\pi, \quad c_m^-(t) \to 0, \quad \text{for } t \gg \frac{|(\omega_{m,1} - \omega_0)|}{3.7\Gamma^2}, \tag{2.22}$$

that is,

$$b_m(t) = (2\pi i/\hbar)d_{m,1}\bar{\epsilon}(\omega_{m,1}). \tag{2.23}$$

Thus, the crw coefficients do not contribute after the pulse is over; they are pure transients. By contrast, the rw coefficients $c_m^+(t)$ can be nonzero after the pulse is over, with magnitude depending, as shown below, on the detuning $\omega_{m,1} - \omega_0$. Equation (2.22) thus gives a criterion, for the Gaussian pulse, as to the time required to establish the resonance condition [Eq. (2.12)]. It follows from Eq. (2.22) that the relevant parameter is the pulse duration $1/\Gamma$. This quantity can, in principle, be *shorter* than a single optical cycle, but the resonance condition still holds.

To see the character of the pulse preparation, we display the quantity

$$c_m'(t) \equiv c_m^+(t)|\epsilon(\omega_{m,1})/\epsilon(\omega_0)|, \tag{2.24}$$

using Eqs. (2.18) to (2.20). The results for $|c_m'(t)|$, Re $c_m'(t)$ and Im $c_m'(t)$ for a Gaussian pulse whose intensity bandwidth $[2(\ln 2)^{1/2}/\delta_t]$ is $120 \, \text{cm}^{-1}$, for different detunings $\Delta_0 \equiv \omega_{m,1} - \omega_0 \equiv E/\hbar$, are presented in Figures 2.2a to 2.2c. It is evident from Figure 2.2a that although the amplitude for populating a state with transition frequency $\omega_{m,1}$ at the end of the pulse is proportional to $|\epsilon(\omega_{m,1})|$, the time-dependent path leading to this value varies with $\omega_{m,1}$. For example, for $\omega_{m,1}$ near the line

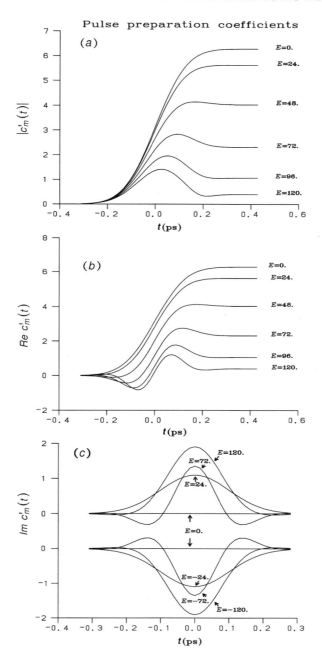

Figure 2.2 Time evolution of $c'_m(t)$ coefficients at different detunings from the center of pulse for a Gaussian pulse with full-width at half maximum (FWHM) of 120 cm^{-1}. (a) $|c'_m(t)|$, (b) Re[$c'_m(t)$], (c) Im[$c'_m(t)$]. [E is to be replaced by $\hbar(\omega_{m,1} - \omega_0)$]. Note that Re denotes the real part, and Im denotes the imaginary part, of the argument that follows.

center the $c'_m(t)$ coefficients rise monotonically to their asymptotic values whereas at off-center energies ($\omega_{m,1} - \omega_0 \neq 0$), this is not the case. In essence, what happens is that (by Fourier's theorem) the pulse appears to have a much broader frequency profile at short times than it does at long times. This causes all the off-center $c'_m(t)$ coefficients to rise uniformly at short times. As time progresses and as the true nature of the $\epsilon(\omega_{m,1})$ profile becomes apparent, the system responds by depleting the off-center $c'_m(t)$ coefficients until they become proportional to $\epsilon(\omega_{m,1})$. In particular, in the case of extreme detuning where $\epsilon(\omega_{m,1}) \approx 0$, the coefficients $c'_m(t)$ must vanish at the end of the pulse but may be nonzero during the pulse. This means that such levels get populated and completely depopulated during the pulse. States of this kind are usually termed *virtual* states, although according to this description they are simply highly detuned (with respect to the pulse center) real states that become transiently populated and depopulated during the pulse.

Understanding how the material phase develops during and after the excitation process will prove to be important for control purposes. In accord with Eq. (2.12) the phase of $b_m(t)$ [see Eq. (2.14)] at the end of the pulse is that of $d_{m,1}\bar{\epsilon}(\omega_{m,1})$, which means [see also Eqs. (2.18) and (2.20)] that $c'_m(t)$ is real at the end of the pulse. As Figure 2.2b shows, $c'_m(t)$ is real at all times for zero detuning ($\omega_{m,1} - \omega_0 = 0$) and complex during the pulse for finite detunings. In fact, the phase of $c'_m(t)$ changes linearly with time at the early stages of the pulse, with a slope given by $\omega_{m,1}$. This time dependence counteracts the natural time evolution of the wave packet of excited states, given by Eq. (2.2). Thus, during the buildup phase of the wave packet, that is, during the early part for which the preparation phase changes linearly, the wave packet hardly moves: It merely grows in size while changing its shape due to the changes in $|c_m(t)|$. As the pulse wanes, the time dependence of the preparation phases becomes less and less pronounced, until they become constant. When this occurs, the wave packet of excited states is "freed" to move naturally since nothing counteracts the factors of $\exp(-iE_n t/\hbar)$ in Eq. (2.2).

It is also of interest to look at $(d/dt)|c'_m(t)|^2$, which is proportional to the *rate* of populating the mth level. We see from Figure 2.3 that this rate, which depends on the detuning, is far from being constant. This result contradicts the celebrated Fermi "golden rule" formula, introduced by most textbooks (cf. [5]), according to which $(d/dt)|c'_m(t)|^2$ *averaged over the pulse modes* is constant. Such averaging is permissible if the action of each pulse mode on the rate is additive. Quite clearly, the effect of the different pulse modes is not additive since the probability for observing each state involves first calculating $b_m(t)$, which, according to Eqs. (2.14) and (2.15), is given as an integral over all the pulse modes, and then squaring the result. By contrast, in the derivation of the golden rule, one first calculates the rate $(d/dt)|b_m(t)|^2$ for each mode and *then* integrates over the pulse modes. This procedure is permissible only when the $\phi(\omega)$ is in some sense a random variable, which corresponds to a pulse that is "incoherent."

Figures 2.2 and 2.3 only display the rw coefficients. However, it can be shown from Eq. (2.19) that the behavior of crw coefficients resemble that of the highly detuned rw coefficients, save for the fact that the crw coefficients rigorously vanish in the long time limit. Thus, the $c_m^-(t)$ coefficients make a noticeable contribution

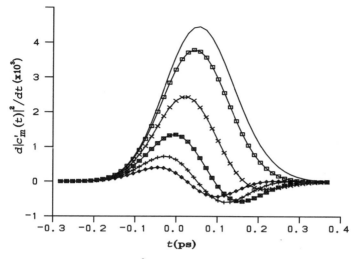

Figure 2.3 Excitation rate $(d|c'_m(t)|^2/dt)$ for 120-cm^{-1}-wide pulse at different detunings from pulse center. Line corresponds to $\Delta_0 = 0$, boxes to $\Delta_0 = 24$ cm^{-1}, the x's to $\Delta_0 = 48$ cm^{-1}, filled boxes to $\Delta_0 = 72$ cm^{-1}, pluses to $\Delta_0 = 96$ cm^{-1}, and diamonds to $\Delta_0 = 120$ cm^{-1}.

only at short times. This justifies the usual practice of neglecting the crw terms whenever detuning with respect to some material levels is small.

The results of Eqs. (2.2), (2.14), and (2.22) are readily summarized as follows: During the pulse the material wave function is

$$|\Psi(t)\rangle = |E_1\rangle \exp\left(\frac{-iE_1 t}{\hbar}\right) + \left(\frac{i}{\hbar}\right) \sum_m [c_m^+(\tau)\bar{\epsilon}(\omega_{m,1})$$

$$+ c_m^-(\tau)\bar{\epsilon}(-\omega_{m,1})]\mathrm{d}_{m,1}|E_m\rangle \exp\left(\frac{-iE_m t}{\hbar}\right). \qquad (2.25)$$

At the end of the pulse the wave function is given by

$$|\Psi(t \gg 1/\Gamma)\rangle = |E_1\rangle \exp\left(\frac{-iE_1 t}{\hbar}\right) + \left(\frac{2\pi i}{\hbar}\right) \sum_m \bar{\epsilon}(\omega_{m,1})\mathrm{d}_{m,1}|E_m\rangle \exp\left(\frac{-iE_m t}{\hbar}\right).$$

$$(2.26)$$

In the next section we extend these results to the case of excitation involving a continuous spectrum.

2.3 PHOTODISSOCIATION

2.3.1 General Formalism

Photodissociation results when the energy eigenstates reached by photon absorption are in the continuum. When the spectrum is continuous, we have to use the scattering wave functions as the matter states. These are defined as eigenstates $|E, \mathbf{m}\rangle$ of the material Hamiltonian with continuous eigenvalues E, that is,

$$[E - H_M]|E, \mathbf{m}\rangle = 0, \tag{2.27}$$

where $\mathbf{m}$ designates a collection of additional quantum numbers that may be necessary to completely specify the state. In particular, if we regard the state $|E, \mathbf{m}\rangle$ as representing a collisional or a dissociation process, then $\mathbf{m}$ includes the chemical identity as well as all the internal (electronic, vibrational, rotational, etc.) quantum numbers of the molecules that participate in the collision, before (or after) the event [6].

The portion of the wave packet excited to a continuous segment of the spectrum is given by

$$|\Psi'(t)\rangle = \frac{2\pi i}{\hbar} \sum_{\mathbf{n}} \int dE \, \bar{\varepsilon}(\omega_{E,1}) \langle E, \mathbf{n}|\hat{\varepsilon} \cdot \mathbf{d}|E_1\rangle |E, \mathbf{n}\rangle \exp\left(\frac{-iEt}{\hbar}\right). \tag{2.28}$$

Because we now have an integral over E in the expansion, the normalization of the constituent states $|E, \mathbf{n}\rangle$ is different than that of $|E_m\rangle$. It is given by

$$\langle E', \mathbf{m}|E, \mathbf{n}\rangle = \delta(E - E')\delta_{\mathbf{m},\mathbf{n}}. \tag{2.29}$$

Likewise, the dimension of the scattering states is different; due to Eq. (2.29), $\psi_{E,\mathbf{n}}(R) \equiv \langle R|E, \mathbf{n}\rangle$ has dimensions of $[\text{length}]^{-1/2}[\text{energy}]^{-1/2}$.

We wish now to investigate the long-time properties of Eq. (2.28). To do so we need to relate the eigenstates of H_M to the eigenstates that describe the freely moving fragments at the end of the process. Take as an example a triatomic molecule ABC, which breaks apart at the end of the process to yield, say, the A + BC channel. (The extension of this treatment to the breakup into different arrangements is discussed at the end of this chapter. Here the term *arrangement* is used to denote the way in which the particles are bound to one another.) Factoring out the ABC center-of-mass motion, we partition the remaining part of H_M into three parts:

$$H_M = K_{\mathbf{R}} + K_{\mathbf{r}} + W(\mathbf{R}, \mathbf{r}). \tag{2.30}$$

Here $\mathbf{R}$ is the radius vector separating A and the BC center of mass, $\mathbf{r}$ is the B–C separation; $W(\mathbf{R}, \mathbf{r})$ is the total potential energy of A, B, and C. The quantities

$$K_{\mathbf{R}} = \frac{-\hbar^2}{2\mu} \nabla_{\mathbf{R}}^2 \tag{2.31}$$

and

$$K_{\mathbf{r}} = \frac{-\hbar^2}{2m} \nabla_{\mathbf{r}}^2 \tag{2.32}$$

are the kinetic energy operators in $\mathbf{R}$ and $\mathbf{r}$ in the coordinate representation, with μ and m being the reduced masses,

$$\mu = m_{\mathrm{A}}(m_{\mathrm{B}} + m_{\mathrm{C}})/(m_{\mathrm{A}} + m_{\mathrm{B}} + m_{\mathrm{C}}), \qquad m = m_{\mathrm{B}}m_{\mathrm{C}}/(m_{\mathrm{B}} + m_{\mathrm{C}}). \qquad (2.33)$$

Denoting as $v(r)$ the asymptotic limit of $W(\mathbf{R}, \mathbf{r})$ as A departs from B–C,

$$v(r) = \lim_{R \to \infty} W(\mathbf{R}, \mathbf{r}), \qquad (2.34)$$

it is clear that the A–BC *interaction potential*, defined as

$$V(\mathbf{R}, \mathbf{r}) \equiv W(\mathbf{R}, \mathbf{r}) - v(r), \qquad (2.35)$$

vanishes as $R \to \infty$,

$$\lim_{R \to \infty} V(\mathbf{R}, \mathbf{r}) = 0. \qquad (2.36)$$

Defining the BC Hamiltonian as

$$h_{\mathbf{r}} \equiv K_{\mathbf{r}} + v(r), \qquad (2.37)$$

the triatomic Hamiltonian of Eq. (2.30) can now be broken into three different parts using Eq. (2.35),

$$H_M = K_{\mathbf{R}} + h_{\mathbf{r}} + V(\mathbf{R}, \mathbf{r}). \qquad (2.38)$$

We see that it is the interaction potential $V(\mathbf{R}, \mathbf{r})$ that couples the motion of the A atom to the motion of the BC diatomic. In its absence the two free fragments A and BC described by the free Hamiltonian

$$H_0 \equiv K_{\mathbf{R}} + h_{\mathbf{r}} \qquad (2.39)$$

move independently of one another. Because H_0 is a sum of two independent terms, its eigenstates, $|E, \mathbf{n}; 0\rangle$, satisfying

$$[E - H_0]|E, \mathbf{n}; 0\rangle = 0, \qquad (2.40)$$

are given as tensor products

$$|E, \mathbf{n}; 0\rangle = |e_{\mathbf{n}}\rangle|E - e_{\mathbf{n}}\rangle. \qquad (2.41)$$

Here $|e_{\mathbf{n}}\rangle$, the *internal* states, satisfy the eigenvalue relation,

$$[e_{\mathbf{n}} - h_{\mathbf{r}}]|e_{\mathbf{n}}\rangle = 0, \qquad (2.42)$$

with $e_{\mathbf{n}}$ being the *internal* (electronic, vibrational, rotational) energy of the BC diatomic. The state $|E - e_{\mathbf{n}}\rangle$, satisfying the eigenvalue relation,

$$[E - e_{\mathbf{n}} - K_{\mathbf{R}}]|E - e_{\mathbf{n}}\rangle = 0, \qquad (2.43)$$

describe the free (translational) motion of A relative to BC.

The $|e_n\rangle$ eigenstates of $h_{\mathbf{r}}$ are often called *channels* and a channel is said to be *open* if $E - e_{\mathbf{n}} > 0$; it is said to be *closed* if $E - e_{\mathbf{n}} < 0$. When a channel is open, the solution of Eq. (2.43), written in the coordinate representation,

$$\left[E - e_{\mathbf{n}} + \frac{\hbar^2}{2\mu}\nabla_{\mathbf{R}}^2\right]\langle \mathbf{R}|E - e_{\mathbf{n}}\rangle = 0, \qquad (2.44)$$

describes a plane wave of kinetic energy $E - e_{\mathbf{n}}$,

$$\langle \mathbf{R}|E - e_{\mathbf{n}}\rangle = \left[\frac{\mu k_{\mathbf{n}}}{\hbar^2(2\pi)^3}\right]^{1/2} \exp(i\mathbf{k}_{\mathbf{n}} \cdot \mathbf{R}), \qquad (2.45)$$

where

$$k_{\mathbf{n}} \equiv |\mathbf{k}_n| = \{2\mu(E - e_{\mathbf{n}})\}^{1/2}/\hbar \qquad (2.46)$$

is the wave vector of the free motion of A relative to the BC center of mass. Since the free solutions are continuous, they too satisfy the continuous spectrum normalization [Eq. (2.29)],

$$\langle E', \mathbf{m}; 0|E, \mathbf{n}; 0\rangle = \delta(E - E')\delta_{\mathbf{m},\mathbf{n}}. \qquad (2.47)$$

Note that if a channel is closed, $k_{\mathbf{n}}$ is imaginary [see Eq. (2.46)] and the $\langle \mathbf{R}|k_{\mathbf{n}}\rangle$ wave function is proportional to a decaying exponential $\exp(-k_{\mathbf{n}}R)$. Closed channels are therefore not observed in the $R \to \infty$ limit, i.e., they do not contribute to the scattering products.

The eigenstates $|E, \mathbf{n}\rangle$ of the fully interacting Hamiltonian H_M are related to $|E, \mathbf{n}; 0\rangle$ in the following way: We first write Eq. (2.27) as an inhomogeneous equation:

$$[E - H_0]|E, \mathbf{n}\rangle = V|E, \mathbf{n}\rangle, \qquad (2.48)$$

and now write the solution $|E, \mathbf{n}\rangle$ as a particular solution to Eq. (2.48), $|E, \mathbf{n}\rangle = [E - H_0]^{-1}V|E, \mathbf{n}\rangle$, plus any solution $|E, \mathbf{n}; 0\rangle$ to the homogeneous part of that equation, that is,

$$|E, \mathbf{n}\rangle = |E, \mathbf{n}; 0\rangle + [E - H_0]^{-1}V|E, \mathbf{n}\rangle. \qquad (2.49)$$

This is an integral equation. To see this, introduce the spectral resolution of an inverse of an operator,

$$[E - H_0]^{-1} = \int dE' \frac{|E', \mathbf{n}; 0\rangle \langle E', \mathbf{n}; 0|}{E - E'}, \tag{2.50}$$

and Eq. (2.49) becomes

$$|E, \mathbf{n}\rangle = |E, \mathbf{n}; 0\rangle + \int dE' \frac{|E', \mathbf{n}; 0\rangle \langle E', \mathbf{n}; 0|V|E, \mathbf{n}\rangle}{E - E'}. \tag{2.51}$$

However, the integral over E' is ill defined in the Riemann sense since the integrand diverges at $E' = E$. Hence, we calculate the integral as a Cauchy integral by defining its value as the limit of a series of well-defined Riemann integrals. That is, we generate a series of well-defined Riemann integrals by adding a small $i\epsilon$ imaginary part to E, thereby avoiding the divergence, calculate the integral, and finally let $\epsilon \to 0$. In what follows we usually omit writing the $\epsilon \to 0$ step, but it is always implied.

It turns out that the limiting value obtained in this way depends on the sign of ϵ. We therefore consider two cases, one in which we add $i\epsilon$ and one in which we subtract $i\epsilon$, with $\epsilon > 0$. The resulting two equations, which replace Eq. (2.49), are

$$|E, \mathbf{n}^{\pm}\rangle = |E, \mathbf{n}; 0\rangle + \lim_{\epsilon \to 0}[E \pm i\epsilon - H_0]^{-1}V|E, \mathbf{n}^{\pm}\rangle. \tag{2.52}$$

Each of these equations is known as the Lippmann–Schwinger equation. The plus (+) solutions are called the *outgoing* scattering states, and the minus (−) solutions are called the *incoming* scattering states. Though each such state is an independent solution of the full Schrödinger equation [Eq. (2.27)], the incoming and outgoing states are not orthogonal to one another, nor, significantly, do they satisfy the same boundary conditions.

We now use the Lippmann–Schwinger equation to explore the long-time behavior of the continuum part of the wave packet $\Psi'(t)$ that we have created with the laser pulse [Eq. (2.28)]. We can use either the outgoing or incoming states as the basis set of expanding $\Psi'(t)$. In what follows we shall expose the different boundary conditions and see which type of solution is best suited for which purpose. Substituting Eq. (2.52) in Eq. (2.28), we obtain that

$$|\Psi'(t)\rangle = \frac{2\pi i}{\hbar} \sum_{\mathbf{n}} \int dE \, \exp\left(\frac{-iEt}{\hbar}\right) \bar{\varepsilon}(\omega_{E,1}) \langle E, \mathbf{n}^{\pm}|\hat{\varepsilon} \cdot \mathbf{d}|E_1\rangle \{|E, \mathbf{n}; 0\rangle$$

$$+ [E \pm i\epsilon - H_0]^{-1}V|E, \mathbf{n}^{\pm}\rangle\}. \tag{2.53}$$

Using the spectral resolution of $[E \pm i\epsilon - H_0]^{-1}$ [Eq. (2.50)], we have from Eq. (2.53) that the amplitude for finding a free state $|E', \mathbf{m}; 0\rangle$ at time t is given as

$$\langle E', \mathbf{m}; 0|\Psi'(t)\rangle = \frac{2\pi i}{\hbar} \sum_{\mathbf{n}} \int dE \, \exp\left(\frac{-iEt}{\hbar}\right) \bar{\epsilon}(\omega_{E,1}) \langle E, \mathbf{n}^\pm | \hat{\varepsilon} \cdot \mathbf{d} | E_1 \rangle$$
$$\times \left\{ \langle E'\mathbf{m}; 0|E, \mathbf{n}; 0\rangle + [E \pm i\epsilon - E']^{-1} \langle E'\mathbf{m}; 0|V|E, \mathbf{n}^\pm\rangle \right\}.$$
$$(2.54)$$

Using the normalization of the free states [Eq. (2.47)], we have that

$$\langle E', \mathbf{m}; 0|\Psi'(t)\rangle = \frac{2\pi i}{\hbar} \exp\left(\frac{-iE't}{\hbar}\right) \bar{\epsilon}(\omega_{E',1}) \langle E', \mathbf{m}^\pm | \hat{\varepsilon} \cdot \mathbf{d} | E_1 \rangle$$
$$+ \frac{2\pi i}{\hbar} \sum_{\mathbf{n}} \int dE \, \exp\left(\frac{-iEt}{\hbar}\right) \bar{\epsilon}(\omega_{E,1}) \langle E, \mathbf{n}^\pm | \hat{\varepsilon} \cdot \mathbf{d} | E_1 \rangle [E \pm i\epsilon - E']^{-1}$$
$$\times \langle E'\mathbf{m}; 0|V|E, \mathbf{n}^\pm\rangle.$$
$$(2.55)$$

In the $t \to \infty$ limit, the integration over E can be performed analytically by contour integration (see Fig. 2.4). To see this we note that in that limit the integrand on a large semicircle in the lower part of the complex E plane is zero, since, for $E = Re^{i\theta}$, with $\theta < 0$,

$$\exp(-iEt/\hbar) = \exp(-iRe^{i\theta}t/\hbar) = \exp(-iR\cos\theta t/\hbar)\exp(R\sin\theta t/\hbar) \xrightarrow{t\to\infty} 0.$$
$$(2.56)$$

Hence the result of the real E integration remains unchanged by supplementing it with an integration along the above large semicircle in the lower half E plane. Since

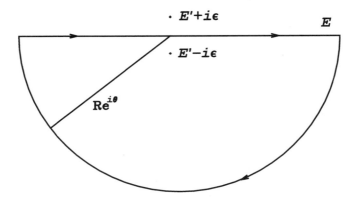

Figure 2.4 Complex energy plane contour integration.

in the $-i\epsilon$ case the integrand has a pole at $E = E' + i\epsilon$ residing outside the closed contour, the whole integral is zero. We obtain that

$$\lim_{t \to \infty} \langle E', \mathbf{m}; 0|\Psi'(t)\rangle = \frac{2\pi i}{\hbar} \bar{\varepsilon}(\omega_{E',1}) \exp\left(\frac{-iE't}{\hbar}\right) \langle E', \mathbf{m}^-|\hat{\varepsilon} \cdot \mathbf{d}|E_1\rangle. \qquad (2.57)$$

Thus, we see, via Eqs. (2.28) and (2.57) that the coefficients of expansion of the excited wave packet in terms of the $|E, \mathbf{m}^-\rangle$ states directly yield the probability amplitude for observing states $|E, \mathbf{m}; 0\rangle$ in the *distant future*.

If we expand $\Psi'(t)$ in the outgoing states instead of the incoming states, the closed contour integration encircles a pole at $E = E' - i\epsilon$. Hence the integration yields

$$\lim_{t \to \infty} \langle E', \mathbf{m}; 0|\Psi'(t)\rangle = \frac{2\pi i}{\hbar} \exp\left(\frac{-iE't}{\hbar}\right) \bar{\varepsilon}(\omega_{E',1}) \sum_{\mathbf{n}} S_{\mathbf{m},\mathbf{n}}(E') \langle E', \mathbf{n}^+|\hat{\varepsilon} \cdot \mathbf{d}|E_1\rangle,$$

$$(2.58)$$

where the $S_{\mathbf{m},\mathbf{n}}(E')$ matrix,

$$S_{\mathbf{m},\mathbf{n}}(E') \equiv \delta_{\mathbf{m},\mathbf{n}} - 2\pi i \langle E', \mathbf{m}; 0|V|E', \mathbf{n}^+\rangle \equiv \langle E', \mathbf{m}; 0|S|E', \mathbf{n}; 0\rangle, \qquad (2.59)$$

is called the *S-matrix* or scattering matrix. That is, the amplitude for transitions between asymptotic states is dictated by the S operator, defined by Eq. (2.59).

The form of Eq. (2.58) appears more complicated than that of Eq. (2.57) because the $\langle E, \mathbf{m}; 0|\Psi'(t)\rangle$ amplitude is composed of contributions from all degenerate $|E, \mathbf{n}; 0\rangle$ states. Why use the outgoing states at all then? The reason is that in ordinary scattering events (e.g., the collision of two particles) we use states whose past is well known to us. These are the outgoing states because when $t \to -\infty$ it is the contour on the semicircle in the upper half of the complex E plane that vanishes. Thus, supplementing the real E integration by such a contour keeps the $E = E' - i\epsilon$ pole out of the contour, and we obtain that

$$\lim_{t \to -\infty} \langle E', \mathbf{m}; 0|\Psi'(t)\rangle = \frac{2\pi i}{\hbar} \bar{\varepsilon}(\omega_{E',1}) \exp\left(\frac{-iE't}{\hbar}\right) \langle E', \mathbf{m}^+|\hat{\varepsilon} \cdot \mathbf{d}|E_1\rangle. \qquad (2.60)$$

That is, the $|E, \mathbf{m}^+\rangle$ states expose a simplified structure as $t \to -\infty$. By contrast, the $t \to -\infty$ limit appears more complicated when the wave packet is expanded in the incoming states because now the $E = E' + i\epsilon$ pole is enclosed within the integration contour, and we obtain that

$$\lim_{t \to -\infty} \langle E', \mathbf{m}; 0|\Psi'(t)\rangle = \frac{2\pi i}{\hbar} \exp\left(\frac{-iE't}{\hbar}\right) \bar{\varepsilon}(\omega_{E',1}) \sum_{\mathbf{n}} S_{\mathbf{m},\mathbf{n}}^-(E') \langle E', \mathbf{n}^-|\hat{\varepsilon} \cdot \mathbf{d}|E_1\rangle,$$

$$(2.61)$$

where the $S_{\mathbf{m,n}}^{-}(E')$ matrix is defined as

$$S_{\mathbf{m,n}}^{-}(E') \equiv \delta_{\mathbf{m,n}} + 2\pi i \langle E'\mathbf{m}; 0|V|E', \mathbf{n}^{-}\rangle. \qquad (2.62)$$

In the case of optical pulse excitation of molecules we use the incoming solutions $|E, \mathbf{n}^{-}\rangle$ because the nature of the system in the remote past is already known to us and we are interested in determining the fate of the system in the distant future.

The above results deal with photodissociation. A similar formulation can be applied to an inelastic scattering event, for example, the scattering of A + BC from an initial state $|E, \mathbf{m}; 0\rangle$ into a final state $|E', \mathbf{n}; 0\rangle$ of A + BC. This problem can be phrased as our asking for the amplitude of making a transition from a free state $|E, \mathbf{m}; 0\rangle$ at $t \to -\infty$ to free state $|E', \mathbf{n}; 0\rangle$ at $t \to \infty$. Since $|E, \mathbf{m}^{+}\rangle$ is known to have evolved from $|E, \mathbf{m}; 0\rangle$ and $|E', \mathbf{n}^{-}\rangle$ is known to evolve to state $|E, \mathbf{n}; 0\rangle$, the answer is given by the expansion coefficients of an outgoing state in terms of the incoming states. That is,

$$|E, \mathbf{m}^{+}\rangle = \sum_{\mathbf{n}} \int dE' |E', \mathbf{n}^{-}\rangle\langle E', \mathbf{n}^{-}|E, \mathbf{m}^{+}\rangle. \qquad (2.63)$$

It is easy to show by the same contour integration as above that

$$\langle E', \mathbf{n}^{-}|E, \mathbf{m}^{+}\rangle = \delta(E - E')S_{\mathbf{m,n}}(E), \qquad (2.64)$$

where $S_{\mathbf{m,n}}(E)$ is the S matrix of Eq. (2.59).

The above treatment assumes that the product comprises a single arrangement channel, that is, the formation of A + BC as the final product. The extension of this formalism to multiple product arrangements, for example, where A + BC and AB + C are products of ABC photodissociation, or where A + BC collide to form A + BC and AB + C, requires: (a) the addition of a channel label $q = 1, 2, \ldots$ to the descriptor of the state, so that $|E, \mathbf{m}, q; 0\rangle$ corresponds to arrangement q; and (b) rewriting Eqs. (2.30) to (2.52) to partition the Hamiltonian in a fashion appropriate to the final arrangement of interest. Thus, for example, for the AB + C arrangement, the vector $\mathbf{R}$ defines the AB to C distance, $\mathbf{r}$ defines the A–B separation, and so forth.

For example, and for later use, the amplitude for reactive scattering from A + BC (labeled q) to AB + C (labeled q') is given by Refs. [6–13] [see Eqs. (2.59) and (2.64)]:

$$S_{\mathbf{m},q':\mathbf{n},q}(E') \equiv \delta_{\mathbf{m,n}}\delta_{q,q'} - 2\pi i\langle E', \mathbf{m}, q; 0|V_q|E', \mathbf{n}, q'^{+}\rangle \equiv \langle E', \mathbf{m}, q'; 0|S|E', \mathbf{n}, q; 0\rangle,$$
$$(2.65)$$

where V_q is the part of the full potential that goes to zero as the distance between A and BC goes to infinity.

The boundary conditions that the state $|E, \mathbf{n}, q^-\rangle$ goes to $|E, \mathbf{n}, q; 0\rangle$ in the distant future and that $|E, \mathbf{n}, q^+\rangle$ originates from the $|E, \mathbf{n}, q; 0\rangle$ that existed in the remote past is summarized in the expression:

$$\lim_{t \to \mp\infty} e^{-i(E \pm i\epsilon)t/\hbar} |E, \mathbf{n}, q^\pm\rangle = e^{-i(E \pm i\epsilon)t/\hbar} |E, \mathbf{n}, q; 0\rangle. \tag{2.66}$$

2.3.2 Electronic States

Thus far, we have suppressed the eigenfunctions associated with the electronic state of the system. We continue to do so below, unless electronic state labels are necessary. A complete treatment, including the electronic states, would consider the total material Hamiltonian H_{MT} as a sum of two terms:

$$H_{MT} = H_M + H_{el} = (K + V_{NN}) + H_{el}. \tag{2.67}$$

Here H_M is the nuclear part, written as the sum of a kinetic term (K) and a nuclear–nuclear interaction V_{NN}, and H_{el} is the electronic part. Since H_{el} involves interactions between the electrons and nuclei, H_{MT} is not separable in the nuclear and electronic degrees of freedom. However, adopting the Born–Oppenheimer approximation substantially simplifies matters. In this approximation we: (a) define the Born–Oppenheimer potential seen by the nuclei in electronic state $|e\rangle$ as $W_e = \langle e|V_{NN} + H_{el}|e\rangle$ where

$$H_{el}|e\rangle = E_e|e\rangle \tag{2.68}$$

and (b) assume that

$$K|e\rangle = |e\rangle K. \tag{2.69}$$

Thus, as a consequence of Eqs. (2.68) and (2.69), we have that

$$\langle e|H_{MT}|e'\rangle = \delta_{ee'}\langle e|H_{MT}|e\rangle. \tag{2.70}$$

The total eigenfunctions are now of the form $|E_i\rangle|g\rangle$, $|E, \mathbf{n}^-\rangle|e\rangle$, and so forth, where $|g\rangle$ denotes the ground electronic state. If we consider optical transitions between the electronic states, then the matter–radiation Hamiltonian $H_{MR}(t)$ [Eq. (1.51)] should be replaced by

$$H'_{MR}(t) = \sum_{ee'} |e\rangle\langle e|H_{MR}(t)|e'\rangle\langle e'|, \tag{2.71}$$

where the sums are over all electronic states (including the ground electronic state) and where $H_{MR}(t)$ is given by Eq. (1.51). As a result, the matrix element $\langle E, \mathbf{n}^-|\hat{\varepsilon} \cdot \mathbf{d}|E_i\rangle$ would be replaced by $\langle E, \mathbf{n}^-|\langle e|\hat{\varepsilon} \cdot \mathbf{d}|g\rangle|E_i\rangle \equiv \langle E, \mathbf{n}^-|d_{e,g}|E_i\rangle$, where

$$d_{e,g} = \langle e|\hat{\varepsilon} \cdot \mathbf{d}|g\rangle \tag{2.72}$$

is an operator in the space of nuclear motion. Similarly, Eq. (2.28) becomes

$$|\Psi'(t)\rangle = \frac{2\pi i}{\hbar} \sum_{\mathbf{n}} \int dE \, \bar{\epsilon}(\omega_{E,1})|E, \mathbf{n}^-\rangle\langle E, \mathbf{n}^-|\mathbf{d}_{e,g}|E_1\rangle \exp\left(\frac{-iEt}{\hbar}\right). \qquad (2.73)$$

2.3.3 Energy-Resolved Quantities

We now focus attention on energy-resolved quantities, such as the cross section. Measurements of these quantities can be performed using cw excitation sources that excite only a single energy or by using pulsed sources and extracting energy-resolved information after the pulse is over.

The photodissociation probability into the state characterized by $\mathbf{n}$ at energy E, $P_{\mathbf{n}}(E|i)$, is given by the square of $A_{\mathbf{n}}(E|i)$, the photodissociation amplitude for observing the free state $\exp(-iEt/\hbar)|E, \mathbf{n}; 0\rangle$ in the long-time limit. That is,

$$P_{\mathbf{n}}(E|i) = |A_{\mathbf{n}}(E|i)|^2, \qquad (2.74)$$

with $A_{\mathbf{n}}(E|i)$ defined as

$$A_{\mathbf{n}}(E|i) = \lim_{t \to \infty} \exp\left(\frac{iEt}{\hbar}\right)\langle E, \mathbf{n}; 0|\Psi(t)\rangle. \qquad (2.75)$$

Because the bound and continuum wave functions usually belong to different electronic manifolds, they are orthogonal to one another, and the only term that contributes to $A_{\mathbf{n}}(E|i)$ derives from Ψ', the excited part of the wave packet. It follows from the boundary conditions on $|E, \mathbf{n}^-\rangle$ [Eq. (2.57)], that the $t \to \infty$ limit of Eq. (2.73) can be written as

$$|\Psi'(t \to \infty)\rangle = \frac{2\pi i}{\hbar} \sum_{\mathbf{n}} \int dE \, \bar{\epsilon}(\omega_{E,i})|E, \mathbf{n}; 0\rangle\langle E, \mathbf{n}^-|\mathbf{d}_{e,g}|E_i\rangle \exp\left(\frac{-iEt}{\hbar}\right). \qquad (2.76)$$

Hence, using the orthonormality of the $|E, \mathbf{n}; 0\rangle$ functions [Eq. (2.47)], we obtain from Eq. (2.75) that

$$A_{\mathbf{n}}(E|i) = \frac{2\pi i}{\hbar} \bar{\epsilon}(\omega_{E,i})\langle E, \mathbf{n}^-|\mathbf{d}_{e,g}|E_i\rangle. \qquad (2.77)$$

The photodissociation cross section $\sigma_{\mathbf{n}}(E|i)$ is defined as the photon energy absorbed in a transition to a final fragment state, divided by the incident intensity of light per unit energy. The energy absorbed is $\hbar\omega_{E,i}P_{\mathbf{n}}(E|i)$; the incident intensity (flux) per unit frequency, $I(\omega_{E,i})$, is the radiation energy density per unit frequency times the velocity of light, $|\epsilon(\omega_{E,i})|^2 c$. Hence the incident intensity per unit energy, $I(E)$, is $|\epsilon(\omega_{E,i})|^2 c/\hbar$. Thus, from Eqs. (2.74) and (2.75) we have that

$$\sigma_{\mathbf{n}}(E|i) = \frac{\hbar\omega_{E,i}P_{\mathbf{n}}(E|i)}{I(E)} = \frac{4\pi^2\omega_{E,i}}{c}|\langle E, \mathbf{n}^-|\mathbf{d}_{e,g}|E_i\rangle|^2. \qquad (2.78)$$

This formula forms the basis for many of the computations of detailed photodissociation cross sections and of angular distribution of photofragments reported in the literature [14–27].

Note, by comparing Eqs. (2.77) and (2.73), that the coefficient of the state $|E, \mathbf{n}^-\rangle$ at time $t = 0$ in Eq. (2.76) is exactly $A_\mathbf{n}(E|i)$, the long-time photodissociation amplitude. Thus, we obtain the crucial insight that *the probability of obtaining product in the state* $|E, \mathbf{n}; 0\rangle$ *is given solely by the probability of preparing the state* $|E, \mathbf{n}^-\rangle$ *at the time of preparation.*

Note also that the form of the pulse does not appear in the expression [Eq. (2.78)] for the cross section. This is because resolving the energy, embodied in the orthogonality expression [Eq. (2.47)], extracts a single frequency component of $\bar{\epsilon}(\omega)$, whose contribution is canceled in the division by the incident light intensity. Therefore, as shown in Appendix 2A, we can use any convenient pulse shape to compute energy-resolved quantities. This is not the case if we want to follow the real-time dynamics of the system, where the pulse shape is intimately linked with the observables. Indeed this link prevents a pulse-free definition of concepts such as the lifetime of a state. This issue is addressed in Appendix 2A.

APPENDIX 2A: MOLECULAR STATE LIFETIME IN PHOTODISSOCIATION

In this appendix we consider, in greater detail, the time dependence of photodissociation under various excitation conditions. To make a connection with wave packet methodologies, we first rewrite the preparation coefficients [Eq. (2.15)] for continuum states as the *finite-time* Fourier transform of the pulse:

$$\bar{\epsilon}(\omega_{E,i})c_E^+(\tau) + \bar{\epsilon}(-\omega_{E,i})c_E^-(\tau) = \int d\omega\, \bar{\epsilon}(\omega) \frac{\exp[i(\omega_{E,i} - \omega)t]}{i(\omega_{E,i} - \omega)}$$

$$= \int_{-\infty}^{t} dt'\, \varepsilon(z, t') \exp(i\omega_{E,i}t'). \qquad (2.79)$$

Using Eq. (2.79) in Eq. (2.73) we can write the excited portion of the wave packet, valid for any t, as

$$|\Psi'(t)\rangle = \frac{i}{\hbar} \sum_\mathbf{n} \int dE \int_{-\infty}^{t} dt'$$

$$\times \exp\left[\frac{i(E - E_i)t'}{\hbar}\right] \varepsilon(z, t') |E, \mathbf{n}^-\rangle \langle E, \mathbf{n}^- |d_{e,g}|E_i\rangle \exp\left(\frac{-iEt}{\hbar}\right). \qquad (2.80)$$

Using the spectral resolution of the evolution operator in the subspace spanned by the continuum wave functions,

$$\exp\left(\frac{-iH_c t}{\hbar}\right) = \sum_\mathbf{n} \int dE\, \exp\left(\frac{-iEt}{\hbar}\right) |E, \mathbf{n}^-\rangle \langle E, \mathbf{n}^- |, \qquad (2.81)$$

we can rewrite Eq. (2.80) as

$$|\Psi'(t)\rangle = \left(\frac{i}{\hbar}\right) \int_{-\infty}^{t} dt' \, \varepsilon(z, t') \exp\left[\frac{-iH_c(t - t')}{\hbar}\right] d_{e,g} |E_i(t')\rangle, \qquad (2.82)$$

where we have defined $|E_i(t')\rangle = \exp(-iE_i t'/\hbar)|E_i\rangle$.

Equation (2.82) unambiguously defines the prepared state, that is, the state is Eq. (2.82) for t greater than the pulse duration. One interesting case is the ultrashort pulse limit, $\varepsilon(t) \approx \epsilon_0 \delta(t)$, which is equivalent to choosing a completely white pulse, $\epsilon(\omega) = \epsilon_0/2\pi$. Substituting the ultrashort pulse form in Eq. (2.82) [or Eq. (2.28)], we obtain that

$$|\Psi'(t)\rangle = \frac{i\epsilon_0}{\hbar} \exp\left(\frac{-H_c t}{\hbar}\right) d_{e,g} |E_i\rangle. \qquad (2.83)$$

Equation (2.83) has been often used [28, 29] to describe photodissociation in the following way: At $t = 0$, the light pulse creates the wave packet $d_{e,g}|E_i\rangle$, which subsequently evolves under the action of $\exp(-iH_c t/\hbar)$. Strictly speaking, for Eq. (2.83) to hold, and hence for this qualitative interpretation to be valid, $\epsilon(\omega_{E,i})$ must vary more slowly with energy than any of the other variables in Eq. (2.28). In particular, it must vary more slowly than the energy (or frequency) dependence of the photodissociation amplitudes $\langle E, \mathbf{n}^-|d_{e,g}|E_i\rangle$.

An easy way to decide whether this is the case is to compare the bandwidth of the laser with the frequency width of the absorption spectrum of the molecule, where the latter is determined by $|\langle E, \mathbf{n}^-|d_{e,g}|E_i\rangle|^2$. Typically, for the case of direct dissociation, the absorption spectrum extends over a few thousands of reciprocal centimeters (cm^{-1}). In order to have a bandwidth broader than this, $\varepsilon(t)$ must, by Eq. (1.35), be as short as ≈ 1 to 5 femtoseconds (fs). Since most pulses used in real photodissociation experiments are much longer, Eq. (2.83) is not a valid description of many photodissociation experiments.

One can correct Eq. (2.83), by viewing "real-time" dynamics as composed of a superposition of wavelets created by a *sequence* of δ pulses [30a]–[36a]. This follows by writing the pulse $\varepsilon(t)$ in Eq. (2.82) as $\varepsilon(t) = \int dt' \, \varepsilon(t')\delta(t - t')$. Each $\delta(t - t')$ evolves according to Eq. (2.83) with a different starting time t'.

In most direct photodissociation cases the laser's bandwidth is much narrower than that of the absorption spectrum and the limit opposite to Eq. (2.83) is realized. Under such circumstances we can approximate, in Eq. (2.28), the narrow range of energies accessed by the laser by a *single* continuum energy level E_0, and write $|\Psi'(t)\rangle$ as

$$|\Psi'(t)\rangle \approx 2\pi i \sum_{\mathbf{n}} |E_0, \mathbf{n}^-\rangle\langle E_0, \mathbf{n}^-|d_{e,g}|E_i\rangle \left\{\int \frac{dE}{\hbar} \bar{\epsilon}(\omega_{E,i}) \exp\left(\frac{-iEt}{\hbar}\right)\right\}. \qquad (2.84)$$

We recognize the term in the curly bracket as $\mathbf{E}_+(t)$ [Eq. (1.35)]. Hence we see that under these circumstances the time dependence of the wave packet created by the laser is that of the laser pulse. In classical terms we would say that the dissociation is "faster" than the optical excitation process: Every photon absorbed leads to immediate dissociation, so that the only time evolution that can be observed is that of the laser pulse.

To obtain the lifetime of the state we integrate out over the spatial dependence of $\Psi'(\mathbf{R}, \mathbf{r}, t)$. This is most easily done by considering the *autocorrelation* function, defined by

$$F(t, t_0) \equiv \langle \Psi'(t_0) | \Psi'(t) \rangle. \tag{2.85}$$

It follows from Eq. (2.28) that $F(t, t_0)$ can be calculated as

$$F(t, t_0) = \frac{-1}{\hbar^2} \sum_{\mathbf{n}} \int dE \, c_E^+(\tau) c_E^{+*}(\tau_0) |\epsilon(\omega_{E,i}) \langle E, \mathbf{n}^- | d_{e,g} | E_i \rangle|^2 \exp\left[\frac{-iE(t - t_0)}{\hbar} \right], \tag{2.86}$$

where we have ignored the crw terms. If both t and t_0 refer to times after the pulse (i.e., $t, t_0 \gg 1/\Gamma$) we obtain, using Eq. (2.28), that

$$F(t, t_0) = -\left(\frac{2\pi}{\hbar} \right)^2 \sum_{\mathbf{n}} \int dE \, |\epsilon(\omega_{E,i}) \langle E, \mathbf{n}^- | d_{e,g} | E_i \rangle|^2 \exp\left[\frac{-iE(t - t_0)}{\hbar} \right], \tag{2.87}$$

We see that after the pulse is over the autocorrelation function is essentially the Fourier transform of the pulse-modulated absorption spectrum $I_\epsilon(E)$, where

$$I_\epsilon(E) \equiv \sum_{\mathbf{n}} |\epsilon(\omega_{E,i}) \langle E, \mathbf{n}^- | d_{e,g} | E_i \rangle|^2. \tag{2.88}$$

Quite clearly the decay of the autocorrelation function given in Eq. (2.87), and hence the "lifetime of the state," is a function of the pulse profile. There simply is no molecular lifetime measurement that is independent of the measurement apparatus. The only measurement that provides only molecular information is that where the laser width far exceeds the spectral bandwidth [Eq. (2.83)], giving the inverse of the spectral bandwidth.

As we shall see (see Section 3.5), this temporal behavior is separate from the issue of product selectivity (i.e., enhancing one product over another). In fact, in the weak-field regime, it is not possible to alter the yield of the $\mathbf{n}$ quantum numbers in a one-photon dissociation process by "shaping" (e.g., shortening) the pulse. To do this we need to either interfere two colors, use multiphoton processes, or work with strong pulses. This is discussed in further detail in the chapters that follow.

CHAPTER 3

WEAK-FIELD COHERENT CONTROL

Traditional molecular excitation and subsequent system evolution, discussed in Chapter 2, affords little opportunity for us to control the outcome of molecular events. According to perturbation theory, as developed in Section 2.3, the branching ratio (i.e., the relative probability to populate different product channels) at the end of the process depends entirely [see Eq. (2.77)] on the ratio between squares $|\langle E, \mathbf{n}^-|\hat{\varepsilon} \cdot \mathbf{d}|E_1\rangle|^2/|\langle E, \mathbf{m}^-|\hat{\varepsilon} \cdot \mathbf{d}|E_1\rangle|^2$ of the *purely material* transition dipole matrix elements. Thus, the electric field profile does not appear in the branching ratio expression. Since the electric field profile of the pulse is the means by which one could hope to influence and possibly control the outcome of the event, it seems that there is no way that we can change the natural branching ratio [37, 38].

This state of affairs holds true even if we consider pulsed excitation in the strong-field domain, provided that the excitation involves only a single precursor state $|E_1\rangle$. In order to see this [39], we generalize the perturbation theory expressions of Section 2.3.3 to the strong-field domain. We recall that the probability of populating a "free" state $|e_\mathbf{m}\rangle|\mathbf{k_m}\rangle$ at any given time is given as

$$P_m(E)(t) = |\langle e_\mathbf{m}|\langle \mathbf{k_m}|\Psi(t)\rangle|^2. \tag{3.1}$$

Using the expansion of the wave packet

$$|\Psi(t)\rangle = b_1(t)\exp{-\frac{iE_1 t}{\hbar}}|E_1\rangle + \sum_\mathbf{n}\int dE\, b_{E,\mathbf{n}}(t)\exp{-\frac{iEt}{\hbar}}|E, \mathbf{n}^-\rangle$$

we have that

$$
\begin{aligned}
\langle e_{\mathbf{m}} | \langle \mathbf{k_m} | \Psi(t) \rangle = {}& b_1(t) \langle e_{\mathbf{m}} | \langle \mathbf{k_m} | E_1 \rangle \exp\left(\frac{-iE_1 t}{\hbar}\right) \\
& + \sum_{\mathbf{n}} \int dE \, b_{E,\mathbf{n}}(t) \langle e_{\mathbf{m}} | \langle \mathbf{k_m} | E, \mathbf{n}^- \rangle \exp\left(\frac{-iEt}{\hbar}\right).
\end{aligned}
\tag{3.2}
$$

Assuming that $\langle e_{\mathbf{m}} | E_1 \rangle = 0$, (e.g., the two states belong to different electronic states), it follows from Eq. (3.2) that in the long-time limit

$$
P_m(E) = |\langle e_{\mathbf{m}} | \langle \mathbf{k_m} | \Psi(t \to \infty) \rangle|^2 = |b_{E,\mathbf{m}}(t \to \infty)|^2.
\tag{3.3}
$$

It follows from Eq. (2.3) (extended to continuum states) and Eq. (2.10) that we can write, in complete generality, that

$$
b_{E,\mathbf{n}}(t) = \frac{i}{\hbar} \langle E, \mathbf{n}^- | \hat{\varepsilon} \cdot \mathbf{d} | E_1 \rangle \int_{-\infty}^{t} dt' \, \varepsilon(t') \exp(-i\omega_{E,1} t) b_1(t').
\tag{3.4}
$$

Hence,

$$
P_n(E)/P_m(E) = |b_{E,\mathbf{n}}(\infty)/b_{E,\mathbf{m}}(\infty)|^2 = |\langle E, \mathbf{n}^- | \hat{\varepsilon} \cdot \mathbf{d} | E_1 \rangle / \langle E, \mathbf{m}^- | \hat{\varepsilon} \cdot \mathbf{d} | E_1 \rangle|^2.
\tag{3.5}
$$

Thus, the branching ratios at a fixed energy E are *independent of the laser power and pulse shape*. This result, which coincides with that of perturbation theory, holds true as long as there is only *one* initial state $|E_1\rangle$ that is excited to the continuum.

This argument motivates the idea that the way to control photodissociation is to use more than one initial state, or in greater generality, to use multiple excitation pathways. In this chapter we demonstrate that such a strategy allows us to actively influence and control which photodissociation product is formed. These ideas, which introduce the notion of *coherent control*, will be later shown to hold true for any dynamical process, not just for photodissociation.

3.1 PHOTODISSOCIATION FROM A SUPERPOSITION STATE

We introduce the basic principles of coherent control through a series of examples. We begin by extending the treatment of Chapter 2 to the photodissociation of a nonstationary *superposition* of bound states, $|\chi(0)\rangle = \sum_{j=1}^{N} a_j |E_j\rangle \exp(-iE_j t/\hbar)$. Numerous experimental techniques can be used to create such a state. Whatever the method of preparation, the amplitude and phase of the coefficients a_j are functions of the experimentally controllable parameters used in creating the superposition.

Repeating the full treatment of weak-field photodissociation given in Chapter 2, but now for an initial superposition state, gives the same result as replacing, in the final result [Eq. (2.25)], the single initial state $|E_1\rangle \exp(-iE_1 t/\hbar)$ by the superposition $\sum_{j=1}^{N} a_j |E_j\rangle \exp(-iE_j t/\hbar)$ and using the analogous expression for the continuum. Thus the system wave function at time t is given by

$$
|\Psi(t)\rangle = \sum_{j=1}^{N} a_j |E_j\rangle \exp\left(\frac{-iE_j t}{\hbar}\right) + \left(\frac{i}{\hbar}\right) \sum_{j=1}^{N} a_j \sum_{\mathbf{n}} \int dE [c_{E,j}^{+}(\tau)\bar{\epsilon}(\omega_{E,j})
$$

$$
+ c_{E,j}^{-}(\tau)\bar{\epsilon}(-\omega_{E,j})]\langle E, \mathbf{n}^{-}|d_{e,g}|E_j\rangle |E, \mathbf{n}^{-}\rangle \exp\left(\frac{-iEt}{\hbar}\right),
$$

(3.6)

where $\omega_{E,j} \equiv (E - E_j)/\hbar$. Here the definition of the c_E preparation coefficients [Eq. (2.15)] has been extended to include many states, that is,

$$
c_{E,j}^{\pm}(t) \equiv \frac{1}{\bar{\epsilon}(\pm\omega_{E,j})} \int_0^{\infty} d\omega \, \bar{\epsilon}(\omega) \frac{\exp[i(\omega_{E,j} \mp \omega)t]}{i(\omega_{E,j} \mp \omega)}.
$$

(3.7)

As in Eq. (2.22), $c_{E,j}^{+} \to 2\pi$ and $c_{E,j}^{-} \to 0$ for $t \gg 1/\Gamma$ so that after the pulse Eq. (3.6) assumes the form

$$
|\Psi(t)\rangle = \sum_{j=1}^{N} a_j |E_j\rangle \exp\left(\frac{-iE_j t}{\hbar}\right) + \left(\frac{2\pi i}{\hbar}\right)
$$

$$
\times \sum_{j=1}^{N} a_j \sum_{\mathbf{n}} \int dE \, \bar{\epsilon}(\omega_{E,j})\langle E, \mathbf{n}^{-}|d_{e,g}|E_j\rangle |E, \mathbf{n}^{-}\rangle \exp\left(\frac{-iEt}{\hbar}\right).
$$

(3.8)

The probability $P_{\mathbf{n}}(E)$ of being in the final state $|E, \mathbf{n}; 0\rangle$ is

$$
P_{\mathbf{n}}(E) = |A_{\mathbf{n}}(E)|^2,
$$

(3.9)

where the probability amplitude $A_{\mathbf{n}}(E)$ is given by [using Eqs. (2.47) and (2.75)]

$$
A_{\mathbf{n}}(E) = \lim_{t \to \infty} \exp\left(\frac{iEt}{\hbar}\right) \langle E, \mathbf{n}; 0|\Psi(t)\rangle = \frac{2\pi i}{\hbar} \sum_{j=1}^{N} a_j \bar{\epsilon}(\omega_{E,j})\langle E, \mathbf{n}^{-} d_{e,g}|E_j\rangle. \quad (3.10)
$$

Of particular interest is the probability of being in a complete subspace of states, denoted by the label q; that is, all $\mathbf{m}$ associated with a fixed q, where $\mathbf{n} = (\mathbf{m}, q)$. Of greatest concern in chemistry is the case where q labels the chemical identity (i.e., the arrangement channel) of the product of a chemical reaction; hence below we often explicitly refer to q in this manner. However, it should be clear that the theory applies to any other q chosen from the set of $\mathbf{n}$ quantum numbers.

The probability of forming product with a particular value of q is given by

$$P_q(E) = \sum_{\mathbf{m}} P_{\mathbf{m},q}(E) = \sum_{\mathbf{m}} |A_{\mathbf{m},q}(E)|^2. \tag{3.11}$$

Inserting $A_{\mathbf{m},q}(E)$ from Eq. (3.10) gives

$$P_q(E) = \left(\frac{2\pi}{\hbar}\right)^2 \sum_{i,j=1}^{N} \left[a_i a_j^* \bar{\epsilon}(\omega_{E,i}) \bar{\epsilon}^*(\omega_{E,j})\right] d_q(ji), \tag{3.12}$$

where

$$d_q(ji) = \sum_{\mathbf{m}} \langle E_j | d_{g,e} | E, \mathbf{m}, q^- \rangle \langle E, \mathbf{m}, q^- | d_{e,g} | E_i \rangle, \tag{3.13}$$

and $d_{g,e} = d_{e,g}^*$. The branching ratio between two channels at energy E, $R_{q,q'}(E)$, which we control below, is then

$$R_{q,q'}(E) = P_q(E)/P_{q'}(E). \tag{3.14}$$

Consider then the nature of $P_q(E)$ [Eq. (3.12)]. The diagonal terms ($i = j$) give the standard probability, at energy E, of photodissociation out of a bound state $|E_j\rangle$ to produce a product in channel q. The off-diagonal terms ($i \neq j$) correspond to interference terms between these photodissociation routes. These interference terms describe the constructive enhancement, or destructive cancellation, of product formation in subspace q. Equation (3.12) is important *in practice* because the interference terms have coefficients $[a_i a_j^* \bar{\epsilon}(\omega_{E,i}) \bar{\epsilon}^*(\omega_{E,j})]$ whose magnitude and sign depend upon *experimentally controllable* parameters. Thus the experimentalist can manipulate laboratory parameters and, in doing so, alter the interference term and hence control the reaction product yield. This control approach can be extended to the domain of moderately strong fields, as shown in Chapter 10, Eq. (10.20).

It is important to note, for experimental implementation, that the interference term contains a z dependence through $\bar{\epsilon}(\omega_{E,i}) \bar{\epsilon}^*(\omega_{E,j})$. Thus, it is necessary to design the step that prepares the a_i coefficients in such a way as to cancel this spatial dependence. Failure to do so results in a spatially dependent interference term and hence in the cancellation of control when spatially averaged. Several methods in which the spatial dependence are eliminated are described in Sections 3.3.2 and 3.5.

Equation (3.12) displays another important feature. That is, the entire control map $P_q(E)$ or $R_{q,q'}(E)$ as a function of the control parameters, is a function of very few molecular parameters, that is, the $d_q(ji)$. As a consequence, the experimentalist need only determine these few parameters in order to produce the entire control map. This statement constitutes the weak-field version of *adaptive feedback control* [40–43], discussed in the context of strong-field optimal control in Chapter 4. That is, in the general strong-field regime, a numerical nonlinear search procedure must be performed (see Chapter 4), to achieve a desired optimization. However, in the weak-field regime discussed here, because of the simple bilinear dependence of each $P_q(E)$ on the $a_j \bar{\epsilon}(\omega_{E,j})$ experimental parameters, we need only carry out N^2 measurements for each q to determine all the $d_q(ji)$ coefficients. Once these coefficients are known, the bilinear $P_q(E)$ function can be analytically evaluated to give any desired branching ratio between, and including, the extrema of $R_{q,q'}(E)$.

3.1.1 Bichromatic Control

Experimentally attaining control via Eq. (3.12) requires a light source containing N frequencies ω_i, $(i = 1, \ldots, N)$. Both pulsed excitation with a source whose frequency width encompasses these frequencies, as well as excitation with N continuous wave (cw) lasers of frequencies $\omega_i = \omega_{E,i}$, $(i = 1, \ldots, N)$ are possible approaches, as depicted in Figure 3.1. Here we focus on $N = 2$, that is, the effect of two cw lasers on a system in a superposition of two states (Fig. 3.1c), a scenario we call *bichromatic control*.

cw light at frequency ω is described by

$$\epsilon(\omega) = \hat{\epsilon}[\epsilon(\omega)\delta(\omega - \omega_i) + \epsilon(-\omega)\delta(\omega + \omega_i)]. \tag{3.15}$$

Using Eqs. (1.35), (1.36), (1.37), and (2.9) we obtain

$$\begin{aligned} \mathbf{E}(z, t) &= 2\hat{\epsilon}|\epsilon(\omega_i)| \cos[\omega_i \tau - \phi(\omega_i)] \\ &= 2\hat{\epsilon} \, \text{Re}[\bar{\epsilon}(\omega_i) \exp(-i\omega_i t)], \end{aligned} \tag{3.16}$$

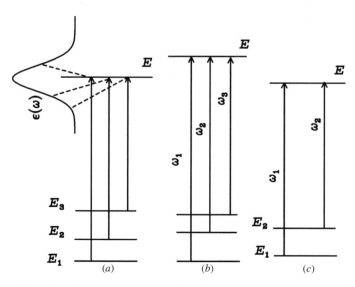

Figure 3.1   Photodissociation of a superposition of N levels using (a) a pulsed light source ($N = 3$ is shown); (b) N cw lasers ($N = 3$ is shown), and (c) $N = 2$ with two cw lasers.

where we have dropped the subscript R from the notation for the electric field of the radiation since no confusion can arise here. Consider then two parallel cw fields of frequencies ω_1 and ω_2 incident on a molecule. The light–molecule interaction [Eq. (1.51)] is then

$$H_{MR}(t) = -\sum_{i=1}^{2} 2\mathbf{d} \cdot \hat{\boldsymbol{\epsilon}} \, \text{Re}[\bar{\epsilon}(\omega_i)\exp(-i\omega_i t)].$$ (3.17)

Tuning the ω_1 and ω_2 frequencies such that $\omega_2 - \omega_1 = (E_1 - E_2)/\hbar$, we have that $P_q(E)$ of Eq. (3.12), at energy $E = E_1 + \hbar\omega_1 = E_2 + \hbar\omega_2$, only has two contributions, corresponding to the excitations shown in Figure 3.1c. The quantities $P_q(E = E_1 + \hbar\omega_1)$ and $R_{q,q'}(E)$ are therefore given by [44–48]

$$\left(\frac{\hbar}{2\pi}\right)^2 P_q(E = E_1 + \hbar\omega_1) = |a_1|^2 |\bar{\epsilon}(\omega_1)|^2 d_q(11) + |a_2|^2 |\bar{\epsilon}(\omega_2)|^2 d_q(22)$$

$$+ 2\,\text{Re}[a_1 a_2^* \bar{\epsilon}(\omega_1)\bar{\epsilon}^*(\omega_2) d_q(12)],$$ (3.18)

$$R_{q,q'}(E = E_1 + \hbar\omega_1) = \frac{|d_q(11) + x^2|d_q(22)| + 2x\cos(\theta_1 - \theta_2 + \alpha_q(12))|d_q(12)|}{|d_{q'}(11)| + x^2|d_{q'}(22)| + 2x\cos(\theta_1 - \theta_2 + \alpha_{q'}(12))|d_{q'}(12)|},$$ (3.19)

where $\alpha_q(ij)$ and θ_j are defined via

$$d_q(ij) = |d_q(ij)|\exp(i\alpha_q(ij)),$$

$$x = \frac{|\bar{\epsilon}(\omega_2)a_2|}{|\bar{\epsilon}(\omega_1)a_1|},$$ (3.20)

$$\tan\theta_j = \frac{\text{Im}[\bar{\epsilon}(\omega_j)a_j]}{\text{Re}[\bar{\epsilon}(\omega_j)a_j]}.$$

For convenience we have introduced the control variables

$$\Delta\theta = \theta_1 - \theta_2 \quad \text{and} \quad s = x^2/(x^2 + 1).$$

The range $0 \leq s \leq 1$ covers all possible values of relative laser intensities. Varying $\Delta\theta$ or s, changes the interference term and thus gives us control over the dissociation probabilities. These changes may be accomplished either by varying the coefficients of the initial superposition state, $\{a_j\}$, or by changing the intensity and relative phases of the dissociation lasers. Note, in particular, that varying $\Delta\theta$ corresponds to just varying a phase. The dependence of the yield on $\Delta\theta$ hence emphasizes the quantum-interference-based nature of the control.

As an example of this approach we consider control over the relative probability of forming $^2P_{3/2}$ vs. $^2P_{1/2}$ atomic iodine, denoted I and I*, in the dissociation of methyl iodide in the regime of 266 nm,

$$CH_3 + I^*(^2P_{1/2}) \leftarrow CH_3I \rightarrow CH_3 + I(^2P_{3/2}). \tag{3.21}$$

This reaction is an example of electronic branching of photodissociation products. The results reported below are for a nonrotating two-dimensional collinear model [49, 50] in which the H_3 center of mass, the C, and the I atoms are assumed to lie on a line. Results for a rotating collinear model are discussed in the next section.

Typical results for the control of the I vs. I* channel are shown in Figure 3.2 as contour plots of the yield of $CH_3 + I$ as a function of the control parameters. Two cases are shown: photodissociation out of the two superposition states $|\chi(0)\rangle = a_1|E_1\rangle + a_2|E_2\rangle$ (Fig. 3.2a) and $|\chi(0)\rangle = a_1|E_1\rangle + a_3|E_3\rangle$ (Fig. 3.2b). Here $|E_1\rangle$ is the ground state and $|E_2\rangle$ and $|E_3\rangle$ correspond to states with one and

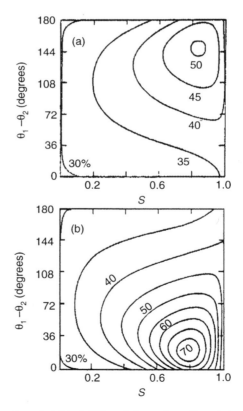

Figure 3.2   Contour plot of yield of $CH_3 + I$ from photodissociation of CH_3I from super-position of states at $\omega_1 = 37593.9\,\text{cm}^{-1}$. (a) $|\chi(0)\rangle = a_1|E_1\rangle + a_2|E_2\rangle$, (b) $|\chi(0)\rangle = a_1|E_1\rangle + a_3|E_3\rangle$. (Taken from Fig. 1, Ref. [44].)

two quanta of excitation in the C–I bond. The results show a large range of possible control. For example, in Figure 3.2b, the yield changes from 30 to 70% as one varies s at small $\theta_1 - \theta_2$. In addition, a comparison of the two figures shows that the topology of the control plot depends strongly on the states that comprise the superposition state.

3.1.2 Energy Averaging and Satellite Contributions

In general, experiments measure energy-averaged quantities such as

$$P_q = \int dE\, P_q(E),$$
$$R_{q,q'} = P_q / P_{q'},$$

(3.22)

since products are not distinguished on the basis of total energy. As such, it is necessary to compute photodissociation to all energies.

For the case considered above, two states irradiated with two cw fields of frequencies ω_1 and ω_2, $P_q(E)$ [Eq. (3.12)] is nonzero at three energies: $E = E_1 + \hbar\omega_1 = E_2 + \hbar\omega_2$, $E' = E_1 + \hbar\omega_2$, and $E'' = E_2 + \hbar\omega_1$.

The contribution from the first of these energies $P_q(E = E_1 + \hbar\omega_1)$ is given in Eq. (3.18) and shown as the left two arrows of Figure 3.3. The remaining contributions, shown on the right-hand side of Figure 3.3 are

$$P_q(E' = E_1 + \hbar\omega_2) = \left(\frac{2\pi}{\hbar}\right)^2 |a_1 \bar{\epsilon}(\omega_2)|^2 d_q(11),$$
$$P_q(E'' = E_2 + \hbar\omega_1) = \left(\frac{2\pi}{\hbar}\right)^2 |a_1 \bar{\epsilon}(\omega_1)|^2 d_q(22).$$

(3.23)

Thus, the overall P_q for $N = 2$ is given by

$$P_q = P_q(E = E_1 + \hbar\omega_1) + P_q(E' = E_1 + \hbar\omega_2) + P_q(E'' = E_2 + \hbar\omega_1). \quad (3.24)$$

The latter two terms correspond to traditional photodissociation terms without associated interference contributions and provide *uncontrollable* photodissociation terms that we call "satellites." In this, and all coherent control scenarios discussed below, it is important to attempt to reduce the relative magnitude of the satellite terms in order to increase overall controllability.

We make the general observation that interference between terms of different energies contain oscillatory $\exp[i(E_1 - E_2)t/\hbar]$ terms that average out to zero with time. (This is not to say that the oscillatory interference term cannot be put to good use. See, for example, our proposal for generating THz radiation [52] using such oscillatory terms, and a related experiment [52a]).

Results demonstrating control over electronic branching ratios in the photodissociation of CH_3I, including the satellite terms [51] are shown in Figure 3.4. In this

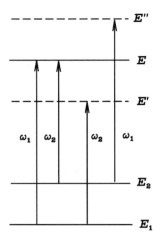

Figure 3.3 Contributions for two levels photodissociated by two frequencies. Interference terms contribute at total energy $= E$. Satellite terms correspond to total energies $= E'$ and E''.

computation CH_3I was treated as a rotating pseudo-triatomic molecule [49, 50]. The bound states $|E_i\rangle$ in this case are described by the vibrational, rotational, and magnetic numbers (v, J, M_J). Figure 3.4 shows contour plots of the $CH_3 + I^*$ product in the photodissociation of CH_3I at two different total energies, obtained by using $\omega_1 = 39,639\,cm^{-1}$ and $\omega_1 = 42,367\,cm^{-1}$, where the initial state is a superposition of $(v_1, J_1) = (0, 2)$ and $(v_2, J_2) = (1, 2)$. The M_J magnetic quantum

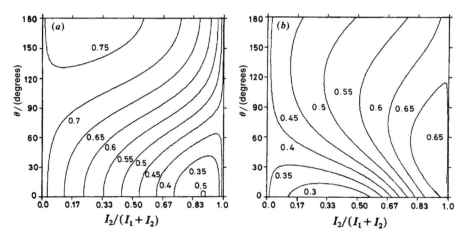

Figure 3.4 Contour plot of yield of I^* (i.e., fraction of I^* as product) in the photodissociation of CH_3I starting from an M-averaged initial state. (a) $\omega_1 = 39,639\,cm^{-1}$, (b) $\omega_1 = 42,367\,cm^{-1}$. In both cases $\omega_2 = (E_1 - E_2)/\hbar + \omega_1$. Here $\theta \equiv \theta_1 - \theta_2$ and $I_2/[I_1 + I_2]$ is equivalent to S in previous figures. (Taken from Fig. 4, Ref. [51].)

number is averaged over, and all satellite terms are included. The results show that the I* yield varies over the same large range as that seen in the nonrotating model study. In addition, a comparison of Figures 3.4*a* with 3.4*b*, which correspond to results at different excitation frequencies, shows that there is considerable dependence of the control contour topology on total energy.

Note that the entire theory needs to be modified only slightly to accommodate control of scattering into different angles, that is, the differential cross section, into channel q. A specific example of this type of control is discussed in Section 3.4 where we apply this approach to manipulating electric currents in semiconductors.

3.2 PRINCIPLE OF COHERENT CONTROL

Control of the type discussed above, in which quantum interference effects are used to constructively or destructively alter product properties, is called coherent control (CC). Photodissociation of a superposition state, the scenario described above, will be seen to be just one particular implementation of a general principle of coherent control: *Coherently driving a state with phase coherence through multiple, coherent, indistinguishable, optical excitation routes to the same final state allows for the possibility of control.* This procedure has a well-known analogy, the interference between paths as a beam of either particles or of light passes through a double slit. In that case interference between two coherent beams leads to spatial patterns of enhanced or reduced probabilities on an observation screen. In the case of coherent control the overall coherence of a pure state plus laser source allows for the constructive or destructive manipulation of probabilities in product channels. Active control results because the excitation process explicitly imparts experimentally controllable phase and amplitude information to the molecule.

A few comments are in order:

As mentioned above, in general, control can only arise from energetically degenerate states. Another way of seeing this is to note that products of states of different energies E and E' appearing in the square of the wave packet of Eq. (3.8) cannot contribute to any measurement where the total energy is resolved. Such a measurement, which filters out all the wave packet components save those belonging to a given value of E, eliminates all the $E \neq E'$ products. Alternatively, we may note that two states of different energy are in principle distinguishable. Hence they cannot interfere with one another.

Weak-field coherent control (CC) scenarios lead to simple analytic expressions for reaction probabilities in terms of a few molecular parameters and a few control parameters. Hence, as noted above, the entire dependence of product probabilities on the control parameters can be easily generated experimentally once the molecular terms are determined from a fit of the control expression to a small number of experimentally determined yields.

Numerous other scenarios can be designed that rely upon the essential coherent control principle. Several are discussed in the following sections. As discussed in Appendix B, there is an alternative approach to the control of chemical reactions, called *mode-selective chemistry*, which does not rely upon quantum interferences. When applied to photodissociation, this approach would entail attempting to excite specific bonds in the molecule (e.g., the A–B bond in the A–B–C molecule) in order to produce a specific product (e.g., the A + B–C product in the given example). Mode-selective chemistry, though very useful under favorable circumstances [53–60], is of limited scope because in most cases the chemical bond we wish to excite is strongly coupled to other bonds (i.e., the "local mode" corresponding to excitation of one bond is not an eigenstate of the system Hamiltonian). As a consequence, most excitations result in the production of a highly delocalized wave packet that entails excitation of many bonds. This phenomenon, to be discussed more fully in Appendix B, is called intramolecular vibrational redistribution (IVR).

3.3 INTERFERENCE BETWEEN N-PHOTON AND M-PHOTON ROUTES

Another important example of coherent control introduces the possibility of quantum interference that arises through competitive optical routes in the excitation of a single bound state to an energy E. Specifically, we consider the photodissociation of a single state via two simultaneous pathways, an N-photon and an M-photon dissociation route. As will become evident in our discussion of selection rules (Section 3.3.2), the N vs. M scenarios are of two types, N and M of the same parity (e.g., both N and M odd or both even) or of opposite parity (one of N, M being odd and the other being even). The latter allows for control over the photodissociation differential cross sections (i.e., scattering into different angles) only, whereas the former allows for control over both the integral and differential cross sections. For simplicity we focus on the two lowest order cases $(N, M) = (1, 2)$ and $(N, M) = (1, 3)$. We begin by deriving the expressions for one-, two-, and three-photon absorption, which will serve as input into these control scenarios.

3.3.1 Multiphoton Absorption

In Section 1.1 we developed the theory of the absorption of a single photon using first-order perturbation theory. We now consider processes involving the absorption of many photons. Such multiphoton processes can be treated using high-order perturbation solutions to the time-dependent Schrödinger equation [Eq. (1.50)].

The computation becomes somewhat simpler if we work in the interaction representation [61]. Defining

$$
\begin{aligned}
V^I(t) &= \exp(iH_M t/\hbar)H_{\mathrm{MR}}(t)\exp(-iH_M t/\hbar), \\
|\psi^I(t)\rangle &= \exp(iH_M t/\hbar)|\Psi(t)\rangle,
\end{aligned}
\tag{3.25}
$$

we obtain, upon substitution into Eq. (1.50), that

$$
i\hbar\frac{\partial|\psi^I(t)\rangle}{\partial t} = V^I(t)|\psi^I(t)\rangle.
\tag{3.26}
$$

The solution to Eq. (3.26), where $|\psi^I(-\infty)\rangle = |E_i\rangle$ is given by

$$
|\psi^I(t)\rangle = |E_i\rangle - \frac{i}{\hbar}\int_{-\infty}^{t} dt_1\, V^I(t_1)|\psi^I(t_1)\rangle.
\tag{3.27}
$$

Iterating this equation gives the series solution

$$
|\psi^I(t)\rangle = \left[1 + \sum_{n=1}^{\infty}\mathcal{V}^{(n)}(t)\right]|E_i\rangle,
\tag{3.28}
$$

with

$$
\begin{aligned}
\mathcal{V}^{(1)}(t) &= -\frac{i}{\hbar}\int_{-\infty}^{t} dt_1\, V^I(t_1), \\
\mathcal{V}^{(2)}(t) &= \left(-\frac{i}{\hbar}\right)^2 \int_{-\infty}^{t} dt_1\, V^I(t_1)\int_{-\infty}^{t_1} dt_2\, V^I(t_2), \\
&\;\;\vdots \\
\mathcal{V}^{(n)}(t) &= \left(-\frac{i}{\hbar}\right)^n \int_{-\infty}^{t} dt_1\, V^I(t_1)\int_{-\infty}^{t_1} dt_2\, V^I(t_2)\cdots\int_{-\infty}^{t_{n-1}} dt_n\, V^I(t_n).
\end{aligned}
\tag{3.29}
$$

Using Eq. (3.25), the photodissociation probability amplitude becomes

$$
\begin{aligned}
\lim_{t\to\infty}\langle E, \mathbf{m}, q^-|\Psi(t)\rangle &= \lim_{t\to\infty}\exp\left(\frac{-iEt}{\hbar}\right)\langle E, \mathbf{m}, q^-|\psi^I(t)\rangle \\
&= \lim_{t\to\infty}\exp\left(\frac{-iEt}{\hbar}\right)\langle E, \mathbf{m}, q^-|\sum_{n=1}^{\infty}\mathcal{V}^{(n)}(t)|E_i\rangle,
\end{aligned}
\tag{3.30}
$$

where we have used the facts that $|E, \mathbf{m}, q^-\rangle$ are eigenstates of H_M, and that $\langle E_i|E, \mathbf{m}, q^-\rangle = 0$. Henceforth we drop the term $\exp(-iEt/\hbar)$ since it provides an overall, and therefore irrelevant, phase factor.

Multiphoton excitation using light pulses that are moderately strong may be treated directly using Eq. (1.51). Here we focus solely on cw excitation with an adiabatically switched interaction potential,

$$H_{MR}(t) = \lim_{\epsilon \to 0} -\mathbf{d} \cdot \mathbf{E}(z, t)e^{-\epsilon|t|} = \lim_{\epsilon \to 0} -2\hat{\epsilon} \cdot \mathbf{d} \, \mathrm{Re}[\bar{\epsilon}(\omega)e^{-i\omega t}e^{-\epsilon|t|}]. \tag{3.31}$$

The adiabatic switching is introduced via the slowly varying $e^{-\epsilon|t|}$ term that guarantees that the interaction vanishes as $t \to \pm\infty$. It is the exact time-dependent analog of the procedure used in the $\pm iE$ derivation of the Lippmann–Schwinger equation [Eq. (2.52)] in the energy domain.

Inserting Eq. (3.31) into Eq. (3.29), and carrying out the integration, gives the first-order probability amplitude in the long-time limit:

$$\langle E, \mathbf{m}, q^- | \mathcal{V}^{(1)}(t \to \infty) | E_i \rangle = \frac{i}{\hbar} \langle E, \mathbf{m}, q^- | \int_{-\infty}^{\infty} dt_1 \exp\left(\frac{iH_M t_1}{\hbar}\right) 2\hat{\epsilon} \cdot \mathbf{d} \, \mathrm{Re}[\bar{\epsilon}(\omega)e^{-i\omega t_1}]$$

$$\times \exp\left(\frac{-iH_M t_1}{\hbar}\right) | E_i \rangle. \tag{3.32}$$

Remembering that $|E_i\rangle$ and $|E, \mathbf{m}, q^-\rangle$ are eigenstates of H_M, we have that

$$\langle E, \mathbf{m}, q^- | \mathcal{V}^{(1)}(t \to \infty) | E_i \rangle = \frac{i}{\hbar} \langle E, \mathbf{m}, q^- | \int_{-\infty}^{\infty} dt_1 \, \hat{\epsilon} \cdot \mathbf{d}[\bar{\epsilon}(\omega)e^{i(E-E_i-\hbar\omega)t_1/\hbar}$$

$$+ \bar{\epsilon}^*(\omega)e^{i(E-E_i+\hbar\omega)t_1/\hbar}]|E_i\rangle. \tag{3.33}$$

After integration over t_1, we obtain, using the well-known equality $\int \exp(i\omega t) \, dt = 2\pi\delta(\omega)$, that

$$\langle E, \mathbf{m}, q^- | \mathcal{V}^{(1)}(t \to \infty) | E_i \rangle = 2\pi i \langle E, \mathbf{m}, q^- | \hat{\epsilon} \cdot \mathbf{d} | E_i \rangle [\bar{\epsilon}(\omega)\delta(E - \hbar\omega - E_i)$$

$$+ \bar{\epsilon}^*(\omega)\delta(E + \hbar\omega - E_i)]. \tag{3.34}$$

When $E > E_i$ only the first term in the square brackets can be nonzero, and we obtain that

$$\langle E, \mathbf{m}, q^- | \mathcal{V}^{(1)}(t \to \infty) | E_i \rangle = 2\pi i \bar{\epsilon}(\omega)\delta(E - \hbar\omega - E_i)\langle E, \mathbf{m}, q^- | \hat{\epsilon} \cdot \mathbf{d} | E_i \rangle. \tag{3.35}$$

This term is the probability amplitude for the one-photon absorption. It is the cw analog of the expression obtained in Section 2.3 for a general pulse. The second term in the square brackets of Eq. (3.34) contributes when $E < E_i$, in which case the expression describes the process of photoemission.

In a similar way we can obtain an expression for the matrix element resulting from the second-order perturbation term:

$$\langle E, \mathbf{m}, q^- | \mathcal{V}^{(2)}(t \to \infty) | E_i \rangle = \left(\frac{-4}{\hbar^2} \right) \int_{-\infty}^{\infty} dt_1 \langle E, \mathbf{m}, q^- | e^{iH_M t_1/\hbar} e^{-\epsilon|t_1|}$$

$$\times \hat{\varepsilon} \cdot \mathbf{d} \operatorname{Re}[\bar{\epsilon}(\omega)e^{-i\omega t_1}] e^{-iH_M t_1/\hbar} \int_{-\infty}^{t_1} dt_2 \ e^{iH_M t_2/\hbar} \hat{\varepsilon} \cdot \mathbf{d} \operatorname{Re}[\bar{\epsilon}(\omega)e^{-i\omega t_2}]$$

$$\times e^{-\epsilon|t_2|} e^{-iH_M t_2/\hbar} | E_i \rangle \tag{3.36}$$

We can time integrate the exponential operators as if their arguments were numbers, provided that we maintain the proper ordering of all noncommuting operators. Using the fact that $|E_i\rangle$ is an eigenstate of H_M, we obtain that the integral over t_2 is given as

$$\int_{-\infty}^{t_1} dt_2 \ e^{iH_M t_2/\hbar} \hat{\varepsilon} \cdot \mathbf{d} \operatorname{Re}[\bar{\epsilon}(\omega)e^{-i\omega t_2}] e^{-\epsilon|t_2|} e^{-iH_M t_2/\hbar} | E_i \rangle$$

$$= \left(\frac{i\hbar}{2} \right) \{ e^{-i(E_i - i\epsilon - H_M + \hbar\omega)t_1/\hbar} [E_i - i\epsilon - H_M + \hbar\omega]^{-1} \bar{\epsilon}(\omega)\hat{\varepsilon} \cdot \mathbf{d} | E_i \rangle$$

$$+ e^{-i(E_i - i\epsilon - H_M - \hbar\omega)t_1/\hbar} [E_i - i\epsilon - H_M - \hbar\omega]^{-1} \bar{\epsilon}^*(\omega)\hat{\varepsilon} \cdot \mathbf{d} | E_i \rangle \}, \tag{3.37}$$

where the contribution of the lower limit has vanished due to the $e^{-\epsilon|t_2|}$ term.

When the above expression is substituted in Eq. (3.36), and using the fact that $\langle E, \mathbf{m}, q^- \rangle$ is an eigenstate of H_M, we obtain that

$$\langle E, \mathbf{m}, q^- | \mathcal{V}^{(2)}(t \to \infty) | E_i \rangle = \left(\frac{-i}{\hbar} \right) \int_{-\infty}^{\infty} dt_1 \langle E, \mathbf{m}, q^- | [\bar{\epsilon}^2(\omega)e^{-i(E_i - i\epsilon + 2\hbar\omega - E)t_1/\hbar}$$

$$+ |\bar{\epsilon}(\omega)|^2 e^{-i(E_i - i\epsilon - E)t_1/\hbar}] \hat{\varepsilon} \cdot \mathbf{d} [E_i - i\epsilon - H_M + \hbar\omega]^{-1} \hat{\varepsilon} \cdot \mathbf{d} | E_i \rangle$$

$$+ \langle E, \mathbf{m}, q^- | [|\bar{\epsilon}(\omega)|^2 e^{-i(E_i - i\epsilon - E)t_1/\hbar} + (\bar{\epsilon}^*)^2(\omega)e^{-i(E_i - i\epsilon - E - 2\hbar\omega)t_1/\hbar}]$$

$$\times \hat{\varepsilon} \cdot \mathbf{d} [E_i - i\epsilon - H_M - \hbar\omega]^{-1} \hat{\varepsilon} \cdot \mathbf{d} | E_i \rangle. \tag{3.38}$$

Using the equality $\int_{-\infty}^{\infty} \exp(i\omega t) = 2\pi\delta(\omega)$ again gives

$$\langle E, \mathbf{m}, q^- | \mathcal{V}^{(2)}(t \to \infty) | E_i \rangle = -2\pi\{\langle E, \mathbf{m}, q^- | [\bar{\epsilon}^2(\omega)\delta(E_i + 2\hbar\omega - E)$$

$$+ |\bar{\epsilon}(\omega)|^2 \delta(E_i - E)]\hat{\epsilon} \cdot \mathbf{d}[E_i - i\epsilon - H_M + \hbar\omega]^{-1}\hat{\epsilon} \cdot \mathbf{d}|E_i\rangle$$

$$+ \langle E, \mathbf{m}, q^- | [|\bar{\epsilon}(\omega)|^2 \delta(E_i - E) + (\bar{\epsilon}^*)^2(\omega)\delta(E_i - 2\hbar\omega - E)]$$

$$\times \hat{\epsilon} \cdot \mathbf{d}[E_i - i\epsilon - H_M - \hbar\omega]^{-1}\hat{\epsilon} \cdot \mathbf{d}|E_i\rangle\}. \tag{3.39}$$

When $E > E_i$, the only second-order perturbation theory contribution comes from the first term of Eq. (3.39). Thus,

$$\langle E, \mathbf{m}, q^- | \mathcal{V}^{(2)}(t \to \infty) | E_i = -2\pi i \bar{\epsilon}^2(\omega)\delta(E_i + 2\hbar\omega - E)$$

$$\times \langle E, \mathbf{m}, q^- | \hat{\epsilon} \cdot \mathbf{d}[E_i - i\epsilon + \hbar\omega - H_M]^{-1}\hat{\epsilon} \cdot \mathbf{d}|E_i\rangle. \tag{3.40}$$

The $\delta(E_i + 2\hbar\omega - E)$ term clearly identifies this second-order perturbation theory term as a process in which the molecule undergoes a transition from level E_i to level E by absorbing two photons of frequency ω. [The second and third terms of Eq. (3.39), surviving when $E = E_i$, represent (elastic) light scattering. The fourth term, surviving when $E < E_i$, represents two-photon stimulated emission.]

In a similar way we can show that the probability amplitude for three-photon absorption, obtained from the third-order perturbation theory term, is given as

$$\langle E, \mathbf{m}, q^- | \mathcal{V}^{(3)}(t \to \infty) | E_i \rangle = 2\pi i \bar{\epsilon}^3(\omega)\delta(E_i + 3\hbar\omega - E)$$

$$\times \langle E, \mathbf{m}, q^- | \hat{\epsilon} \cdot \mathbf{d}[E_i - i\epsilon + 2\hbar\omega - H_M]^{-1}\hat{\epsilon} \cdot \mathbf{d}[E_i - i\epsilon + \hbar\omega - H_M]^{-1}\hat{\epsilon} \cdot \mathbf{d}|E_i\rangle. \tag{3.41}$$

For notational simplicity we have neglected the electronic degrees of freedom in deriving Eq. (3.41). To properly incorporate them we can repeat this derivation with $H_{MR}(t)$ replaced by $H'_{MR}(t)$ [Eq. (2.71)], with H_M replaced by H_{MT} [Eq. (2.67)] and with eigenstates including the electronic components. Adopting the Born–Oppenheimer approximation and its consequences [Eqs. (2.68) to (2.70)] and noting that at energies corresponding to visible and ultraviolet (UV) light the relevant dipole matrix elements within electronic states are small compared to those between different electronic states, gives, instead of Eqs. (3.35), (3.40) and (3.41),

$$\langle E, \mathbf{m}, q^- | \mathcal{V}^{(1)}(t \to \infty) | E_i \rangle = 2\pi i \bar{\epsilon}(\omega) \delta(E - \hbar\omega - E_i) \langle E, \mathbf{m}, q^- | d_{e,g} | E_i \rangle, \qquad (3.42)$$

$$\langle E, \mathbf{m}, q^- | \mathcal{V}^{(2)}(t \to \infty) | E_i \rangle = -2\pi i \bar{\epsilon}^2(\omega) \delta(E - 2\hbar\omega - E_i)$$

$$\times \sum_{e'} \langle E, \mathbf{m}, q^- | d_{e,e'} [\hbar\omega + E_i - i\epsilon - H_{e'}]^{-1} d_{e',g} | E_i \rangle$$

$$\equiv -2\pi i \bar{\epsilon}^2(\omega) \delta(E - 2\hbar\omega - E_i) \langle E, \mathbf{m}, q^- | D | E_i \rangle, \qquad (3.43)$$

$$\langle E, \mathbf{m}, q^- | \mathcal{V}^{(3)}(t \to \infty) | E_i \rangle = 2\pi i \bar{\epsilon}^3(\omega) \delta(E - 3\hbar\omega - E_i)$$

$$\times \sum_{e'e''} \langle E, \mathbf{m}, q^- | d_{e,e''} [2\hbar\omega + E_i - i\epsilon - H_{e''}]^{-1}$$

$$\times d_{e'',e'} [\hbar\omega + E_i - i\epsilon - H_{e'}]^{-1} d_{e',g} | E_i \rangle$$

$$\equiv 2\pi i \bar{\epsilon}^3(\omega) \delta(E - 3\hbar\omega - E_i) \langle E, \mathbf{m}, q^- | T | E_i \rangle. \qquad (3.44)$$

Here the matrix elements $H_{e'} \equiv \langle e' | H_{\mathrm{MT}} | e' \rangle$ are operators with respect to nuclear wave functions in the ground and excited electronic states and $d_{e,g} \equiv \hat{\epsilon} \cdot \mathbf{d}_{e,g}$, and so forth. The two-and three-photon absorption operators D and T are defined by the above identities.

3.3.2 One- vs. Three-Photon Interference

Given the above results, we can consider [62] now a molecule initially in state $|E_i\rangle$ subjected to two co-propagating cw fields of frequencies ω_1 and ω_3, with $\omega_3 = 3\omega_1$. The interaction potential is given by

$$H_{\mathrm{MR}}(t) = -2\mathbf{d} \cdot \mathrm{Re}[\hat{\epsilon}_3 \bar{\epsilon}_3 \exp(-i\omega_3 t) + \hat{\epsilon}_1 \bar{\epsilon}_1 \exp(-i\omega_1 t)], \qquad (3.45)$$

where $\bar{\epsilon}_i = \bar{\epsilon}(\omega_i)$.

We assume the following physics: (a) the dipole transitions within electronic states are negligible compared to those between electronic states; (b) the fields are sufficiently weak to allow the use of perturbation theory, and (c) $E_i + 2\hbar\omega_1$ is below the dissociation threshold, with dissociation occurring from the excited electronic state. In accord with Eq. (3.44) the lowest order expression for the one-photon or three-photon dissociation amplitude $A_{\mathbf{m},q}(E = E_i + \hbar\omega_3)$ is

$$A_{\mathbf{m},q}(E = E_i + \hbar\omega_3) = \frac{2\pi i}{\hbar} [\delta(\omega_3 - \omega_{E,i}) \bar{\epsilon}_3 \langle E, \mathbf{m}, q^- | d_{e,g} | E_i \rangle$$

$$+ \delta(3\omega_1 - \omega_{E,i}) \bar{\epsilon}_1^3 \langle E, \mathbf{m}, q^- | T_{e,g} | E_i \rangle],$$

where $T_{e,g}$ is the three-photon transition operator, given according to Eq. (3.44) by

$$T_{e,g} = \sum_{e'e''} d_{e,e'} (E_i - H_{e'} + 2\hbar\omega_1)^{-1} d_{e',e''} (E_i - H_{e''} + \hbar\omega_1)^{-1} d_{e'',g}. \qquad (3.47)$$

The probability to produce fragments q at a fixed energy E is therefore

$$P_q(E) = \sum_{\mathbf{m}} |A_{\mathbf{m},q}(E_i + \hbar\omega_3)|^2 = P_q^{(1)}(E) + P_q^{(3)}(E) + P_q^{(13)}(E), \tag{3.48}$$

where the one-photon photodissociation probability is

$$P_q^{(1)}(E) = \left(\frac{2\pi}{\hbar}\right)^2 |\bar{\epsilon}_3|^2 \sum_{\mathbf{m}} |\langle E, \mathbf{m}, q^- |d_{e,g}|E_i\rangle|^2, \tag{3.49}$$

the three-photon photodissociation probability is

$$P_q^{(3)}(E) = \left(\frac{2\pi}{\hbar}\right)^2 |\bar{\epsilon}_1|^6 \sum_{\mathbf{m}} |\langle E, \mathbf{m}, q^- |T_{e,g}|E_i\rangle|^2, \tag{3.50}$$

and the combined one-photon three-photon interference term is

$$P_q^{(13)}(E) = \left(\frac{2\pi}{\hbar}\right)^2 |\bar{\epsilon}_3\bar{\epsilon}_1^3 \sum_{\mathbf{m}} \langle E_i|T_{g,e}|E, \mathbf{m}, q^-\rangle\langle E, \mathbf{m}, q^- |d_{e,g}|E_i\rangle + \text{c.c.}], \tag{3.51}$$

where c.c. is complex conjugate. As in our discussion of the photodissociation of a superposition state (Section 3.1), we define a "molecular" interference amplitude $|F_q^{(13)}|$ and a "molecular" phase $\alpha_q(13)$ as

$$|F_q(13)| \exp[i\alpha_q(13)] = \sum_{\mathbf{m}} \langle E_i|T_{g,e}|E, \mathbf{m}, q^-\rangle\langle E, \mathbf{m}, q^- |d_{e,g}|E_i\rangle. \tag{3.52}$$

Recognizing that $\bar{\epsilon}_i$ is a complex number, $\bar{\epsilon}_i = |\bar{\epsilon}_i|e^{i\phi_i}$, we can write the above interference term as

$$P_q^{(13)}(E) = -2\left(\frac{2\pi}{\hbar}\right)^2 |\bar{\epsilon}_3\bar{\epsilon}_1^3| \cos(\phi_3 - 3\phi_1 + \alpha_q(13)|F_q(13)|. \tag{3.53}$$

The branching ratio $R_{qq'}(E)$ for channels q and q', [see Eq. (3.14)] can now be written as

$$R_{qq'}(E) = \frac{F_q(11) - 2x \cos[\phi_3 - 3\phi_1 + \alpha_q(13)]\epsilon_0^2|F_q(13)| + x^2\epsilon_0^4 F_q(33)}{F_{q'}(11) - 2x \cos[\phi_3 - 3\phi_1 + \alpha_{q'}(13)]\epsilon_0^2|F_{q'}(13)| + x^2\epsilon_0^4 F_{q'}(33)}, \tag{3.54}$$

where

$$F_q(11) = \left(\frac{\hbar}{\pi|\bar{\epsilon}_3|}\right)^2 P_q^{(1)}(E); \quad F_q(33) = \left(\frac{\hbar}{\pi|\bar{\epsilon}_1|^3}\right)^2 P_q^{(3)}(E); \quad \text{and} \quad x = \frac{|\bar{\epsilon}_1|^3}{\epsilon_0^2|\bar{\epsilon}_3|}, \tag{3.55}$$

where ϵ_0 is defined as a single unit of electric field; x is therefore a dimensionless parameter.

The numerator and denominator of Eq. (3.54) each display the canonical form for coherent control, that is, a form similar to Eq. (3.19) in which there are independent contributions from more than one route, modulated by an interference term. Since the interference term is controllable through variation of the (x and $\phi_3 - 3\phi_1$) laboratory parameters, so too is the branching ratio $R_{qq'}(E)$. Thus, the principle upon which this control scenario is based is the same as that in Section 3.1, but the interference is introduced in an entirely different way.

Three Dimensional Formalism With the qualitative principle of interfering pathways exposed, it remains to demonstrate the quantitative extent to which the one- vs. three-photon scenario alters the yield ratio in a realistic system. To this end we consider the photodissociation of IBr,

$$I + Br \leftarrow IBr \rightarrow I + Br^*, \qquad (3.56)$$

where $Br = Br(^2P_{3/2})$ and $Br^* = Br(^2P_{1/2})$; the IBr potential curves used in the calculation are shown in Figure 3.5. Details of the computation are discussed below to demonstrate the role of selection rules in the one- vs. three-photon scenario and the extent of achievable control.

Because we want to consider the one- vs. three-photon control of IBr in three dimensions [63], we replace the notation $|E_i\rangle$ for the initial state by $|E_i, J_i, M_i\rangle$, where E_i is, as before, the energy of the state, J_i is its angular momentum, and M_i is the angular momentum projection along the z axis. Where no confusion arises, we continue to use $|E_i\rangle$ for simplicity.

The collection of final channel quantum numbers **n** in the photodissociation amplitude $\langle E, \mathbf{n}, q^- | d_{e,g} | E_i, J_i, M_i \rangle$ must now include the scattering angles $\hat{\mathbf{k}} = (\phi_k, \theta_k)$ and, in the case of IBr, the quantum number of primary interest $q = 1, 2$ labels the $Br(^2P_{3/2})$ or $Br^*(^2P_{1/2})$ electronic states. Further, the continuum states $|E, \mathbf{n}, q^-\rangle$ are conveniently treated by partial wave expansion in states of total angular momentum J and associated projection M_J along the z axis.

The required $P^{(1)}$, $P^{(3)}$, and $P^{(13)}$ terms of Eq. (3.48) are conveniently expressed in terms of $d_q(E_j, J_j, M_j; E_i, J_i, M_i; E)$, the angle-averaged products of two amplitudes,

$$d_q(E_j, J_j, M_j; E_i, J_i, M_i; E) = \int d\hat{\mathbf{k}} \langle E_j, J_j, M_j | d_{g,e} | E, \hat{\mathbf{k}}, q^- \rangle \langle E, \hat{\mathbf{k}}, q^- | d_{e,g} | E_i, J_i, M_i \rangle.$$

$$(3.57)$$

Evaluating this integral [63] shows that the $d_q(E_j, J_j, M_j; E_i, J_i, M_i; E)$ is proportional to δ_{M_i, M_j} and contains nonzero contributions only from terms where $|E_j, J_j, M_j\rangle$ and $|E_i, J_i, M_i\rangle$ are excited to continuum states of the same J.

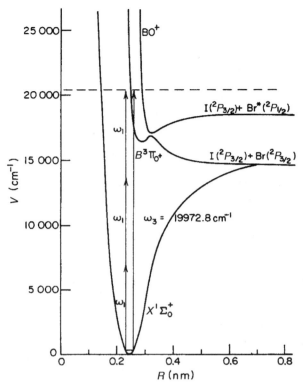

Figure 3.5 IBr potential energy curves relevant to one-photon vs. three-photon dissociation. (Taken from Fig. 1, Ref. [63].)

The probability $P^{(1)}$ [Eq. (3.49] is given in terms of d_q by

$$P_q^{(1)}(E, E_i, J_i, M_i) = \left(\frac{\pi}{\hbar}\right)^2 \bar{\epsilon}_3^2 d_q(E_i, J_i, M_i; E_i, J_i, M_i; E). \tag{3.58}$$

Assuming that only two electronic states $|g\rangle$ and $|e\rangle$ are of importance, the terms $P^{(13)}$ and $P^{(3)}$ can also be written in terms of the d_q. To do so we express $\langle E, \mathbf{n}, q^- | T_{e,g} | E_i \rangle$ of Eq. (3.52) for two electronic states in terms of $\langle E, \mathbf{n}, q^- | \hat{\varepsilon} \cdot \mathbf{d} | E_i \rangle$ by inserting appropriate resolutions of the identity,

$$\langle E, \mathbf{n}, q^- | T_{e,g} | E_i \rangle$$
$$= \sum_{j, \mathbf{n}', q'} \int dE' \frac{\langle E, \mathbf{n}, q^- | d_{e,g} | E_j \rangle \langle E_j | d_{g,e} | E', \mathbf{n}', q'^- \rangle \langle E', \mathbf{n}', q'^- | d_{e,g} | E_i \rangle}{(E_j - E_i - 2\hbar\omega_1)(E' - E_i - \hbar\omega_1)}. \tag{3.59}$$

Here, as noted above, $|E_i\rangle$ denotes $|E_i, J_i, M_i\rangle$ and the j summation indicates a sum over all bound states $|E_j\rangle$ of the ground $X^1\Sigma_0^+$ potential surface. Computationally, of all the bound eigenstates of $X^1\Sigma_0^+$, the contribution to $\langle E, \mathbf{n}, q^- |T|E_i\rangle$ is dominated by those states with energy E_j that nearly satisfy the two-photon resonance condition $E_j \approx E_i + 2\hbar\omega_1$.

From Eq. (3.59) and the definition of d_q, $|F_q^{(13)}| \exp(i\delta_q^{(13)})$ in the cross term $P^{(13)}$ [Eq. (3.53)] is given by

$$|F_q^{(13)}| \exp(i\delta_q^{(13)}) =$$

$$\sum_{E_j, J_j, q'} \int dE' \frac{d_q(E_j, J_j, M_i; E_i, J_i, M_i; E)d_{q'}^*(E_j, J_j, M_i; E_i, J_i, M_i; E')}{(E_j - E_i - 2\hbar\omega_1)(E' - E_i - \hbar\omega_1)}, \quad (3.60)$$

where the J_j and the E_j summation indicates a summation over all bound eigenstates of the $X^1\Sigma_0^+$ state. Recall [Eq. (3.57)] that the term $d_q(E_j, J_j, M_i; E_i, J_i, M_i; E)$ arose from two dipole matrix elements, one coupling $|E_i, J_i, M_i\rangle$ to the continuum in the one-photon route and one coupling $|E_j, J_j, M_j\rangle$ to the continuum as the third step in the three-photon route. Thus, in accord with the discussion above, the angular momentum J of the continuum accessed by these two routes must be equal for $F_q^{(13)}$ to be nonzero.

Finally, using Eqs. (3.55), (3.50), and (3.59), the probability of three-photon photodissociation is given by

$$P_q^{(3)}(E, E_i, J_i, M_i) = \sum_{E_j, E_l, J_j, J_l, \bar{q}, q'} \int dE' \int d\bar{E}$$

$$\times \frac{d_q(E_l, J_l, M_i; E_j, J_j, M_i; E)d_{q'}(E_j, J_j, M_i; E_i, J_i, M_i; E')d_{\bar{q}}^*(E_l, J_l, M_i; E_i, J_i, M_i; \bar{E})}{(E_j - E_i - 2\hbar\omega_1)(E_l - E_i - 2\hbar\omega_1)(E' - E_i - \hbar\omega_1)(\bar{E} - E_i - \hbar\omega_1)}.$$

$$(3.61)$$

Given these results, the branching ratio $(R_{qq'})$ [Eq. (3.54)] can be easily written in terms of d_q using Eqs. (3.55), (3.58), (3.60), and (3.61).

Equations (3.50) and (3.59) are quite complex, and simple qualitative rules for tabulating the terms have been developed [63]. Figures 3.6 and 3.7 provide a means of demonstrating these rules and exposing features of the angular momentum as it affects control.

Consider, as an example, the case of $J_i = M_i = 0$. The angular momentum values associated with the one- and three-photon absorption routes are summarized in Figure 3.6*a*. In the figure, the solid horizontal line represents angular momentum states of the ground potential surface $(X^1\Sigma_0^+)$ and the dashed horizontal line represents those of the excited potential surfaces $(B^3\Pi_{0^+}$ and $BO^+)$. The succession of transitions follows from the usual dipole selection rules for linearly polarized light $(\Delta J = \pm 1, \Delta M = 0)$ for transitions between $\Omega = 0$ states (where Ω labels the total electronic angular momentum along the molecular axis). Note that there are two J states that arise from the three-photon absorption, that is, $J = 1$ and $J = 3$. Only the

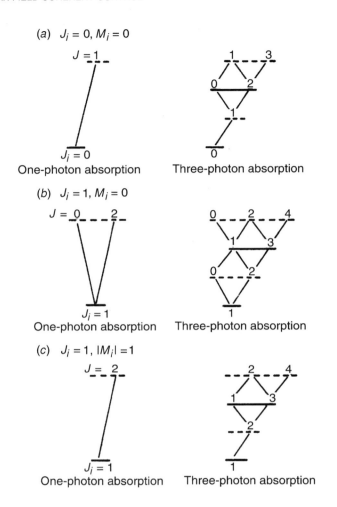

Figure 3.6 Angular momentum levels available for one- and three-photon absorption from (a) $J_i = 0$ and (b) $J_i = 1$, $M_i = 0$ and (c) $J_i = 1$, $|M_i| = 1$. (Taken from Fig. 2, Ref. [63].)

$J = 1$ term interferes with the one-photon route, which also generates an excited state with $J = 1$. That is, there is no interference term involving the $J = 3$ route, which is therefore an uncontrollable satellite.

To use these diagrams to tabulate the dipole matrix elements that contribute to $P^{(1)}$, $P^{(3)}$, and $P^{(13)}$ consider, for example, contributions to $P^{(3)}$ [Eq. (3.61)] for the case of $J_i = M_i = 0$. The relevant diagram is then that shown on the right-hand side of Figure 3.6a. Detailed consideration of Eq. (3.61) shows that the following diagrammatic method gives the appropriate dipole contributions to $P^{(3)}$. One starts at the bottom of the diagram (here $J = 0$) and proceeds to a state at the top of the

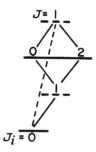

Figure 3.7   Angular momentum levels (for the $J_i = 0$ case) connected by both one- and three-photon routes to the continuum J state, which is shared by both types of absorption. (Taken from Fig. 3, Ref. [63].)

diagram, and then back down to the initial state. Each pair of levels encountered in this route contributes one dipole matrix element to the product in Eq. (3.61). A sample case would be the route $0 \to 1 \to 2 \to 1 \to 0 \to 1 \to 0$, which contributes a term of the form [where the value of the angular momentum is indicated and where the electronic states are distinguished as being unprimed (ground state) or primed (excited state)]: $\langle 0|\hat{\varepsilon} \cdot \mathbf{d}|1'\rangle \langle 1'|\hat{\varepsilon} \cdot \mathbf{d}|2\rangle \langle 2|\hat{\varepsilon} \cdot \mathbf{d}|1'\rangle \langle 1'|\hat{\varepsilon} \cdot \mathbf{d}|0\rangle \langle 0|\hat{\varepsilon} \cdot \mathbf{d}|1'\rangle \langle 1'|\hat{\varepsilon} \cdot \mathbf{d}|0\rangle$. Equation (3.61) would include all terms, of which this is one, with products of six dipole transition matrix elements that can arise from this figure in the manner prescribed. Similarly, contributions to $P^{(1)}$ arise from the only diagram contributing to the left-hand side of Figure 3.6*a*, that is, $|\langle 0|\hat{\varepsilon} \cdot \mathbf{d}|1'\rangle|^2$.

Finally, contributions to the interference term $P^{(13)}$ can be obtained from Figure 3.7, which results from superimposing the two angular momentum ladders in Figure 3.6*a* for those continuum angular momentum states that contribute to the interference term. To obtain the contributions to $P^{(13)}$, one constructs all matrix element products that involve *four* dipole transitions in making a complete circuit from ground state $J = 0$, through the continuum level, and back down to the ground state.

Computational results were obtained [63] using the one-photon vs. three-photon scenario for IBr photodissociation [Eq. (3.56)]. Two different cases were examined, those corresponding to fixed initial M_i values and those corresponding to averaging over a random distribution of M_i, for fixed J_i. Results typical of those obtained are shown in Figures 3.8 and 3.9, where we provide a contour plot of the yield of Br*($^2P_{1/2}$) for the case of excitation from $J_i = 1$, $M_i = 0$ and $J_i = 42$ while averaging over M_i, as a function of laser control parameters [(relative intensity $S = x^2/(1 + x^2)$ and relative phase)].

The range of control in each case is impressive, with essentially no loss of control due to the M_i averaging. Note also the similarity in behavior of the lower J_i and higher J_i cases, although at a different range of S. A related high-field study of two-photon vs. four-photon control in the photodissociation of Cl_2 has been carried out by Bandrauk and co-workers [64, 65]. Strong-field extensions of the one- vs. three-photon scenario have also been discussed [66–69].

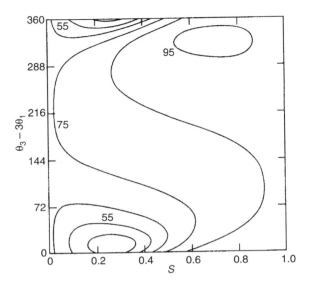

Figure 3.8   Contour plot of the yield of Br*($^2P_{1/2}$) (percentage of Br* as product) in the photodissociation of IBr from an initial bound state in $X^1\Sigma_0^+$ with $v = 0, J_i = 1, M_i = 0$. Results arise from simultaneous (ω_1, ω_3) excitation $(\omega_3 = 3\omega_1)$ with $\omega_1 = 6657.5\,\mathrm{cm}^{-1}$. Abscissa is labeled by the amplitude parameter $s = x^2/(1 + x^2)$ and ordinate by the relative phase parameter $\theta_3 - 3\theta_1$, equivalent to $\phi_3 - 3\phi_1$ of the text. (Taken from Fig. 4, Ref. [63].)

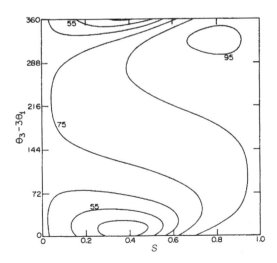

Figure 3.9   As in Figure 3.8 but for $v = 0, J_i = 42, \omega_1 = 6635.0\,\mathrm{cm}^{-1}$. Results are M averaged, with $\epsilon_0 = 0.125$ a.u. Here $\theta_3 - 3\theta_1$ is equivalent to $\phi_3 - 3\phi_1$ in the text. (Taken from Fig. 5, Ref. [63].)

The three-photon vs. one-photon scenario has been experimentally realized by Chen et al. in atoms [70], and by Gordon and co-workers [71–75] in a series of experiments on HCl and CO.

In the case of HCl, shown in Figure 3.10 the molecule was excited to an intermediate $^3\Sigma^-(\Omega^+)$ vib-rotational resonance, using a combination of three ω_1 ($\lambda_1 = 336$ nm) photons and one ω_3 ($\lambda_3 = 112$ nm) photon. The ω_3 beam was generated from an ω_1 beam by tripling in a Kr gas cell. Ionization of the intermediate state takes place by absorption of one additional ω_1 photon. The relative phase of the light fields was varied by passing the ω_1 and ω_3 beams through a second Ar or H_2 ("tuning") gas cell of variable pressure.

The HCl experiments verified the predictions of coherent control theory concerning the sinusoidal dependence of the ionization rates on the relative phase of the two

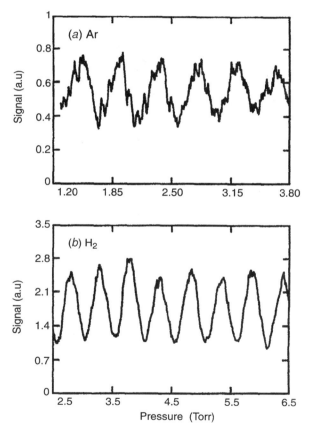

Figure 3.10   Ionization signal for the HCl $R(2)$ transition as a function of pressure in the tuning cell, using either (*a*) Ar or (*b*) H_2 to control the relative phases of ω_1 and ω_3. (Taken from Fig. 2, Ref. [71].)

exciting lasers. The HCl experiment also verified the prediction of the dependence of the strength of the sinusoidal modulation of the ionization current on the relative laser field intensities. Similar demonstrations for ammonia, trimethylamine, triethyl-amine, cyclooctatetraene, and 1,1-dimethylhydrazine by Wang, Bersohn and co-workers [76] have been reported.

Kleiman, Gordon and co-workers [73] have also demonstrated control of ion-ization in H_2S in a jet with a large distribution of j states. Although in this case both a dissociation and an ionization channel are possible, that is, $H_2S^+ \leftarrow H_2S \rightarrow H + HS$, no discrimination between the possible outcomes of the photoexcitation has been observed: The signals of all final channels oscillate in phase as the relative phase $\phi_3 - 3\phi_1$ is varied.

By contrast, in the $HI^+ \leftarrow HI \rightarrow H + I$ case, control over the production of different channels, specifically the HI^+ vs. the $H + I$ channels, has been observed [74, 75]. The three-photon and one-photon excitation routes possible in this system are shown in Figure 3.11 for both the ionization process and the dissociation process.

The experimental results, shown in Figure 3.12 are highly significant as, contrary to the H_2S case, the modulations in the I^+ signal are seen to be *out of phase* with those of the HI^+ signal. Thus, control over different reaction products has been demonstrated. That is, by changing $\phi_3 - 3\phi_1$, the phase difference between the ω_3 and the ω_1 laser fields, through the change in the pressure of the H_2 gas in the tuning cell, different I^+/HI^+ ratios are attained.

The quantitative nature of the observed control depends upon the values of $F_q^{(13)}$ and the *molecular phase*, α_q. In particular, the value of $\alpha_q - \alpha_{q'}$ dictates the shift between the peaks in $P_q(E)$ and $P_{q'}(E)$. For example, a molecular case where $\alpha_q - \alpha_{q'} \approx 0$ (e.g., in the $H_2S^+ \leftarrow H_2S \rightarrow H + HS$ case discussed above) shows less discrimination between channels than does a molecular case where $\alpha_q - \alpha_{q'} = \pi$. Hence the relationship between the nature of the dynamics and the α_q values is of interest, a topic studied in detail by Gordon, Seideman and co-workers [77] and discussed in further detail in Chapter 6.

Note that a crucial aspect of the experiments is that the two co-propagating ω_1 and ω_3 beams satisfy the *phase-matching* condition $\mathbf{k}_3 = 3\mathbf{k}_1$. As a result, Eq. (3.54) no longer depends upon the spatial coordinate z (via the $\bar{\epsilon}_i$) and the interference term is independent of the position in space. Finally we note that N-photon vs. M-photon control was demonstrated earlier in high field experiments in Bucksbaum's group [77a]. These results show that it is possible to control the integral cross section into channel q via one-photon vs. three-photon absorption. A similar result obtains for any N-photon vs. M-photon absorption scenario where N and M are of the same parity. In addition, these scenarios allow for control over differential cross sections as well. To see this, consider rewriting Eqs. (3.49) to (3.52) so that it applies to the probability of observing the product in channel q, but at a fixed scattering angle. Then the sum over the channel indices $\mathbf{m}$ no longer includes an integral over scattering angles. The resultant cross term $P^{(13)}$ is nonzero so that varying properties of the lasers will indeed alter the differential cross section into channel q.

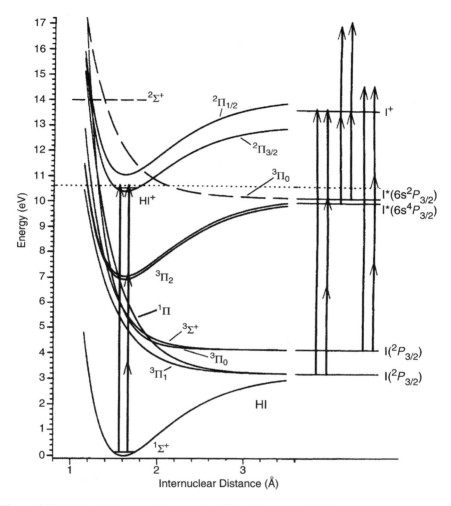

Figure 3.11 Potential energy diagram for HI, with arrows showing the one- and three-photon paths whose interference is used to control the ratio of products formed in branching reactions HI $\rightarrow$ HI$^+$ + e and HI $\rightarrow$ H + I. (Taken from Fig. 2, Ref. [74].)

3.3.3 One- vs. Two-Photon Interference: Symmetry Breaking

Although scenarios for interference between an N-photon route and an M-photon route, where N, M are of the same parity, allow for control over both the differential and integral photodissociation cross sections, this is not the case when N and M are of different parity. In this case only control over the differential cross section is possible. However, the control is such that it leads to the breaking of the usual backward–forward symmetry. This is but one example of the breaking of symmetry

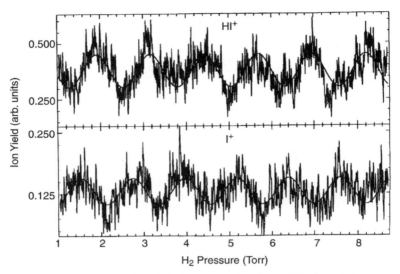

Figure 3.12 Modulation of HI$^+$ and I$^+$ signals as a function of difference between one- and three-photon phases (proportional to the H$_2$ pressure in the cell used to phase shift the beams). (Taken from Fig. 3, Ref. [74].)

afforded via coherent control techniques. A more spectacular example, that of chiral (asymmetric) synthesis, is presented in Chapter 8.

To understand why control over the total cross section is lost and how the backward–forward symmetry is broken, we analyze in some detail the simplest case in this class, namely the interference between a one-photon and a two-photon absorption process [78]. Consider irradiating a molecule by a field composed of two modes, ω_2 and ω_1, with $\omega_2 = 2\omega_1$, for which the light-matter interaction is

$$H_{\mathrm{MR}}(t) = -2\mathbf{d} \cdot \mathrm{Re}[\hat{\varepsilon}_2 \bar{\epsilon}_2 \exp(-i\omega_2 t) + \hat{\varepsilon}_1 \bar{\epsilon}_1 \exp(-i\omega_1 t)]. \tag{3.62}$$

It follows from Eqs. (3.30) and (3.40) that the amplitude for the combined one-photon, two-photon absorption process is

$$A_{q,\mathbf{m}}(E = E_i + \hbar\omega_2) = \left(\frac{2\pi i}{\hbar}\right)\delta(\omega_2 - \omega_{E,i})[\bar{\epsilon}_2 \langle E, \mathbf{m}, q^- | \mathrm{d}_{e,g} | E_i \rangle$$

$$+ \bar{\epsilon}_1^2 \langle E, \mathbf{m}, q^- | D_{e,g} | E_i \rangle], \tag{3.63}$$

where $\langle E_i | D_{e,g} | E, \hat{\mathbf{k}}, q^- \rangle$ is the two-photon dissociation amplitude, defined in Eq. (3.44). Here we have implicitly assumed that $E_i + \hbar\omega_1$ is below the threshold for photodissociation.

Suppressing for the moment all channel indices **m** (which can be readily included), save for the final direction $\hat{\mathbf{k}}$, we square the amplitude to obtain $P_q(E, \hat{\mathbf{k}})$, the probability of photodissociation into channel q at angles $\hat{\mathbf{k}} \equiv (\theta_k, \phi_k)$,

$$P_q(E, \hat{\mathbf{k}}) = |A_q(\hat{\mathbf{k}}, E_i + \hbar\omega_2)|^2 = P_q^{(1)}(E, \hat{\mathbf{k}}) + P_q^{(12)}(E, \hat{\mathbf{k}}) + P_q^{(2)}(E, \hat{\mathbf{k}}), \quad (3.64)$$

where

$$P_q^{(1)}(E, \hat{\mathbf{k}}) = \left(\frac{2\pi}{\hbar}\right)^2 |\bar{\epsilon}_2|^2 |\langle E, \hat{\mathbf{k}}, q^- | \mathrm{d}_{e,g} | E_i\rangle|^2,$$

$$P_q^{(2)}(E, \hat{\mathbf{k}}) = \left(\frac{2\pi}{\hbar}\right)^2 |\bar{\epsilon}_1|^4 |\langle E, \hat{\mathbf{k}}, q^- | D_{e,g} | E_i\rangle|^2, \quad (3.65)$$

$$P_q^{(12)}(E, \hat{\mathbf{k}}) = -2\left(\frac{2\pi}{\hbar}\right)^2 |\bar{\epsilon}_2 \bar{\epsilon}_1^2| \cos[\phi_2 - 2\phi_1 + \alpha_q^{(12)}(\hat{\mathbf{k}})]|F_q^{(12)}(\hat{\mathbf{k}})|,$$

with the amplitude $F_q^{(12)}(\hat{\mathbf{k}})$ and phase $\delta_q^{(12)}(\hat{\mathbf{k}})$ defined by

$$|F_q^{(12)}(\hat{\mathbf{k}})| \exp[i\alpha_q^{(12)}(\hat{\mathbf{k}})] = \langle E_i | D | E, \hat{\mathbf{k}}, q^-\rangle \langle E, \hat{\mathbf{k}}, q^- | \mathrm{d}_{e,g} | E_i\rangle. \quad (3.66)$$

The interference term $P_q^{(12)}(E, \hat{\mathbf{k}})$ is generally nonzero, so that control over the differential cross section is possible.

Consider, however, the integral cross section into channel q, that is,

$$P_q(E) = \int d\hat{\mathbf{k}}\, P_q(E, \hat{\mathbf{k}}), \quad (3.67)$$

and focus on the contribution from $P_q^{(12)}(E, \hat{\mathbf{k}})$. That is, consider

$$
\begin{aligned}
P_q^{(12)}(E) &= \int d\hat{\mathbf{k}} |F_q^{(12)}(\hat{\mathbf{k}})| \exp[i\delta_q^{(12)}(\hat{\mathbf{k}})] \\
&= \int d\hat{\mathbf{k}} \langle E_i, J_i, M_i | D_{g,e} | E, \hat{\mathbf{k}}, q^-\rangle \langle E, \hat{\mathbf{k}}, q^- | \mathrm{d}_{e,g} | E_i, J_i, M_i\rangle,
\end{aligned} \quad (3.68)
$$

where we have explicitly inserted the angular momentum characteristics of the initial state. Using the definition of D and inserting unity in terms of the states $|E_j, J_j, M_j\rangle$ of the intermediate electronic states gives

$$
\begin{aligned}
P_q^{(12)}(E) = \sum_{j,e'} \int d\hat{\mathbf{k}} [\hbar\omega_1 + E_i - E_j]^{-1} \\
\times \langle E_i, J_i, M_i | \mathrm{d}_{g,e'} | E_j, J_j, M_j\rangle \langle E_j, J_j, M_j | \mathrm{d}_{e',e} | E, \hat{\mathbf{k}}, q^-\rangle \\
\times \langle E, \hat{\mathbf{k}}, q^- | \mathrm{d}_{e,g} | E_i, J_i, M_i\rangle.
\end{aligned} \quad (3.69)
$$

For convenience consider the case of diatomic dissociation. Examination of the selection rules shows that when the transition-dipole operators $\mathbf{d}_{eg}$ and $\mathbf{d}_{e'e}$ are parallel to the nuclear axis, the two-photon amplitude is nonzero only if $J_j - J_i = \pm 2, 0$. By contrast, in that case the one-photon matrix element $\langle E_i, J_i, M_i | \mathbf{d}_{g,e'} | E_j, J_j, M_j \rangle$ is nonzero only if $J_j - J_i = \pm 1$. Since these two conditions are contradictory, $P_q^{(12)}(E)$ is zero. Hence coherent control over integral cross sections is not possible using the one- vs. two-photon scenario.

This result holds true even when the transition-dipole operators are perpendicular to the nuclear axis. Thus, lack of control over the integral cross section in the one vs. two scenario will also occur in polyatomic molecules and for any N- vs. M-photon process where N and M are of different parities. The loss of integral control emanates from the fact that the total parity of any molecular wave function is reversed each time a photon is absorbed, since the parity of each photon is negative and the total parity of the photon + molecule system must be conserved. Thus, the parity of a molecular state resulting from a given initial state absorbing an odd number (N) of photons is opposite that resulting from the absorption of an even number (M) of photons by the same initial state. The integrated interference term, which reflects the overlap integral between such states, is zero.

However, these features do not prevent control over the differential cross sections (i.e., scattering into fixed angles) for N and M of different parity because no integration over angles is required. In fact, because the continuum state $|E, \hat{\mathbf{k}}^-\rangle$ accessed via multiphoton pathways of opposite parity has contributions from angular momentum states of opposite parity, the probability of seeing products in a given direction $\hat{\mathbf{k}}$ is not the same as the probability of observing products in the opposite direction $-\hat{\mathbf{k}}$. That is, the forward–backward symmetry has been broken. As shown below, this is but one example of symmetry breaking induced by many coherent control scenarios. One particularly interesting example is control over right- vs. left-handed enantiomers, discussed in detail in Chapter 8.

The experimental implementation of the one-plus-two photon absorption scenario has taken a variety of forms [79–84]. Several theoretical studies in addition to Ref. [78] have also analyzed this phenomenon [85–87].

For example, experiments in Corkum's laboratory [82] have been carried out on one- vs. two-photon absorption in crafted quantum wells, depicted schematically in Figure 3.13. As shown in Figure 3.14 by varying $\phi_2 - 2\phi_1$, the relative phase between the second harmonic and twice that of the fundamental frequency (at 10.6 µm), the experimentalists were able to direct the electric current to move in either the forward or backward direction, or to generate a current that was equally probable in both directions.

Related results were obtained with molecules. For example, following the theoretical predictions of Charron et al. [68] on the photodissociation of H_2^+, displayed in Figure 3.15, Sheehy et al. [83] performed an experiment (shown schematically in Fig. 3.16) to control product directionality in HD^+ dissociation to $H + D^+$ and $H^+ + D$. Here a combination of a one-photon process, induced by the second harmonic, and a two-photon process, induced by the fundamental frequency, were used to excite the molecule to a repulsive $2p\sigma$ state, yielding either the $H + D^+$ or

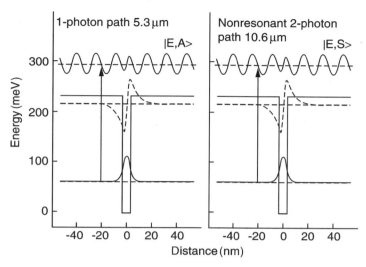

Figure 3.13   Energy band diagram of a 55-A GaAs/Ga$_{0.74}$Al$_{0.26}$ As quantum well and wave functions of the states implied in a 5.3-μm single-photon pathway and a 10.6-μm two-photon process. Neither dephasing nor reflections of the electronic waves on the neighbor quantum wells are considered in this simplified figure. (Taken from Fig. 1, Ref. [82].)

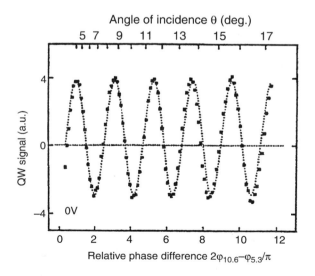

Figure 3.14   The integrated quantum well response versus the relative laser phase. Dashed line: sinusoidal fit. (Taken from Fig. 4, Ref. [82].)

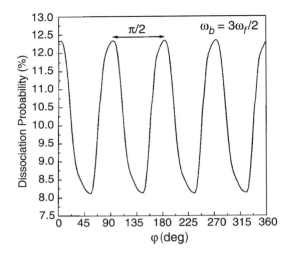

Figure 3.15   Computed H^+ current resulting from the photodissociation of H_2^+ as a function of $\varphi = \phi_2 - 2\phi_1$, the difference between the second-harmonic phase and twice the phase of the fundamental photon. (Taken from Fig. 6, Ref. [68].)

the $D + H^+$ products. The results of the experiment are shown in Figure 3.17. We see that the angle at which the ions appear can be varied by changing the $\phi_2 - 2\phi_1$ relative phase.

It is interesting to note (see Fig. 3.17b) that the ratio between the H^+ and D^+ ions does not vary with the relative phase. This is partly in agreement with the analysis

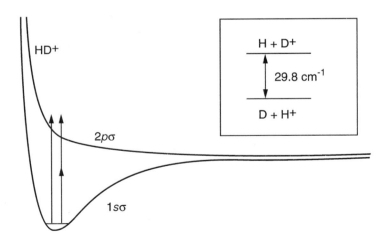

Figure 3.16   Potential curves for the $1s\sigma$ and $2p\sigma$ states of HD^+. In the homonuclear case (H_2^+), the two states are asymptotically degenerate; the degeneracy is lifted in the heteronuclear case by $29.8\,\text{cm}^{-1}$ (inset). (Taken from Fig. 1, Ref. [83].)

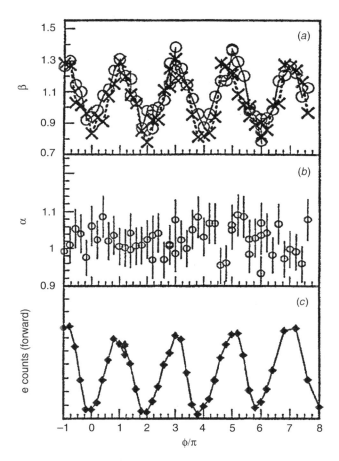

Figure 3.17 (*a*) Forward/backward yield ratios of H^+, $\beta_{H^+} = H_f^+/H_b^+$ (circles), and of D^+, $\beta_{D^+} = D_f^+/D_b^+$ (crosses), of protons and deuterons in the dissociation of HD^+ vs. $\phi = \phi_2 - 2\phi_1$, the difference between the phase of the second-harmonic phase and twice the phase of the fundamental photon. (*b*) Plot of ratio of isotopes, α vs. ϕ. The uncertainty is indicated by error bars. (*c*) Yield of photoelectron arising from the $Kr \xrightarrow{\hbar\omega_3, 3\hbar\omega_1} Kr^+ + e^-$ photoionization moving toward the detector vs. ϕ. The modulations were used to calibrate ϕ. (Taken from Fig. 3, Ref. [83].)

presented above that, within lowest order perturbation theory, the ratio between integral cross sections of different channels cannot be controlled by an *N*- vs. *M*-photon scenario when *N* and *M* possess different parities. This means that, within the confines of perturbation theory, when we average over all angles we should find a phase-independent H^+/D^+ branching ratio. This argument does not, however, explain why this lack of discrimination should hold for each and every angle, i.e., other experimental results show that the H^+/D^+ ratio is independent of the dissociation angle. Moreover, an argument based on low-order perturbation theory is not expected to hold in the long-wavelength regime where multiple photon transitions are involved, and isotopic discrimination is therefore expected to occur [68].

We conclude that in the short-wavelength regime what is being affected in the dissociation of HD^+ is the motion of the (lone) *electron*. The electron is seen to direct itself toward the forward or backward directions in the *laboratory frame* as the $\phi_2 - 2\phi_1$ relative phase is varied. Since the experiment monitored only dissociative events where the electron is still bound to the molecule, the electron simply "rides" on whatever ion happens to be pointing in its preferred laboratory direction. If, while the molecule is rotating and dissociating, the electron finds the D^+ nucleus pointing in its preferred direction, it attaches itself to the deuteron, and the neutral D atom will emerge in that direction (with the H^+ ion emerging in the opposite direction). The situation is reversed if the proton happens to be moving in the direction preferred by the electron.

These conclusions, that even if ionization does not occur, it is often the electron that is being controlled rather than the nuclei, follow the work of Aubanel and Bandrauk [88] who have shown such electronic control in the photodissociation of Cl_2. The case for electronic control is naturally stronger when the lasers are intense enough to ionize the molecule. In that case the interference between the one-photon and two-photon processes has been shown to affect the ionization yield [69].

Additional theoretical and experimental work on the control of *atomic* phenomena in $\omega + 2\omega$ and $\omega + 3\omega$ scenario has been reviewed in detail by Ehlotzky [89].

3.4 POLARIZATION CONTROL OF DIFFERENTIAL CROSS SECTIONS

Rather than attempting coherent control with two different frequencies, it appears that using two different polarizations of the same frequency would be much easier to implement experimentally. It turns out that this scenario is akin to the one- vs. two-photon control in the sense that integral control is not possible. Although differential cross sections can be controlled [90], one cannot break the forward–backward symmetry in this case.

To see this, we consider the photodissociation of a *single* bound state $|E_1\rangle$ by a single cw source with arbitrary polarization vector $\hat{\varepsilon}$,

$$\hat{\varepsilon} = \eta_1 \exp(i\alpha_1)\hat{\varepsilon}_1 + \eta_2 \exp(i\alpha_2)\hat{\varepsilon}_2, \tag{3.70}$$

where $\hat{\varepsilon}_1$ and $\hat{\varepsilon}_2$ are two orthonormal vectors (see Fig. 3.18).

We can regard the two components $\hat{\varepsilon}_1$ and $\hat{\varepsilon}_2$ as inducing two independent excitation routes. Choosing $\hat{\varepsilon}_1$ and $\hat{\varepsilon}_2$ parallel and perpendicular to the quantization (z) axis, respectively, the differential cross section is composed of three terms; one corresponds to photodissociation of $|E_1\rangle$ by the $\hat{\varepsilon}_1$ component, one by the $\hat{\varepsilon}_2$ component, and one being the cross term between these two contributions. Excitation by the parallel component allows $\Delta M_J = 0$ transitions, while excitation by the

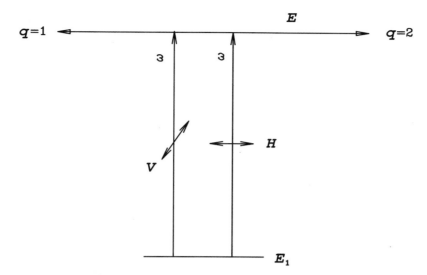

Figure 3.18 Photodissociation by cw source of arbitrary polarization. Two components of the polarization vector are shown.

perpendicular $\hat{\varepsilon}_2$ component allows $\Delta M_J = \pm 1$ transitions. The interference term is therefore comprised of a product of two bound-continuum matrix elements, where the two continua differ in M_J by ± 1. If this cross term is nonzero, then control over the differential cross section is possible. However, producing the *integral* cross section necessitates integrating the differential cross section over $\hat{\mathbf{k}}$ and, under these circumstances, the cross term vanishes.

Contrary to the one- vs. two-photon case, the states comprising the $|E, \hat{\mathbf{k}}^-\rangle$ state are of the same parity. Thus, in the differential cross section, the backward–forward symmetry is not broken. Rather, control is manifest in our ability to sharpen or broaden the angular distribution about a given recoil direction.

3.5 PUMP–DUMP CONTROL: FEW-LEVEL EXCITATION

Control of the dynamics via a pump–dump scenario was first introduced by Tannor and Rice [91, 92] using insight afforded by localized wave packets [93], an approach that is associated with many-level excitation, and which is discussed in Section 4.1. Here we consider the few-levels case shown qualitatively in Figure 3.19. It can be regarded as a useful extension of the scenario outlined in Section 3.1 in which the initial superposition of bound states is prepared with one laser pulse and subsequently dissociated by another. It is also related to a two- vs. two-photon scenario (Section 6.1) since, as shown in Figure 3.20, pump–dump excitation contains two- vs. two-photon pathways to a collection of energies E.

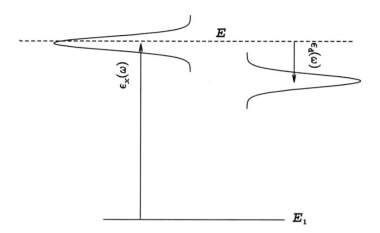

Figure 3.19   Pump–dump control scenario.

Consider then a system irradiated by two pulses, termed the pump-and-dump pulse. These pulses are assumed to be temporally separated by a time delay Δ_d. The analysis below shows that under these circumstances control over the photodissociation yields is obtained by varying the central frequency of the pump pulse and the time delay between the two pulses.

Consider a molecule, initially ($t = 0$) in an eigenstate $|E_1\rangle$ of the molecular Hamiltonian H_M that is subjected to two transform-limited light pulses. The electric

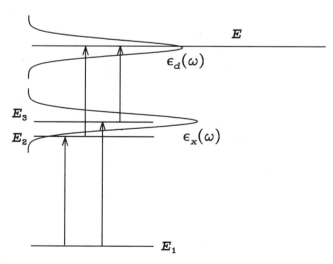

Figure 3.20   Interfering two two-photon pathways to energy E contained in pump–dump control scheme of Figure 3.19. The ϵ_x, ϵ_d labels indicate whether the excitation or dissociation laser is causing the indicated transition.

field consists of two temporally separated pulses $\mathbf{E}_x(\tau)$, $\mathbf{E}_d(\tau)$ [where τ is the retarded time $(t - z/c)$]. For both pulses the electric field is of the form $\mathbf{E}(\tau) = 2\hat{\varepsilon}\varepsilon(\tau)\cos(\omega_0\tau)$, which is a parameterization of Eq. (1.35). Here ω_0 is the carrier frequency and $\varepsilon(\tau)$ describes the pulse envelope. Thus, the molecule is subjected to the field,

$$\mathbf{E}(\tau) = \mathbf{E}_x(\tau) + \mathbf{E}_d(\tau) \tag{3.71}$$

For convenience we use Gaussian pulses that peak at $t = t_x$ and t_d, respectively. In particular, the excitation pulse is of the form

$$\mathbf{E}_x(\tau) = \hat{\varepsilon}_x \epsilon_x \exp[-i(\omega_x\tau + \delta_x)]\exp[-(\tau - t_x)^2/\tau_x^2]/2, \tag{3.72}$$

where the Gaussian pulse is spread with width τ_x about time t_x and carries an overall phase δ_x. The associated frequency profile is given by the Fourier transform of Eq. (3.72):

$$\epsilon_x(\omega) = (\sqrt{\pi}/2)\epsilon_x\tau_x \exp[-i(\omega_x - \omega)t_x]\exp[-\tau_x^2(\omega_x - \omega)^2/4]\exp(-i\delta_x). \tag{3.73}$$

Further, we define $\bar{\epsilon}_x(\omega)$ as in Eq. (2.9), with $\phi(\omega) = (\omega - \omega_x)t_x - \delta_x$.

The analogous quantities for the dissociation laser, $\mathbf{E}_d(\tau)$, $\epsilon_d(\omega)$, and $\bar{\epsilon}_d(\omega)$ are defined similarly, with the parameters t_d and ω_d replacing t_x and ω_x, and so forth. The pump pulse $\mathbf{E}_x(\tau)$ induces a transition to a linear combination of the eigenstates $|E_i\rangle$ of the excited electronic state. The pump pulse may be chosen to encompass any number of states. Here we choose the pump pulse sufficiently narrow in frequency to excite only two of these states, $|E_2\rangle$ and $|E_3\rangle$. The dump pulse $\mathbf{E}_d(\tau)$ dissociates the molecule by further exciting it to the continuous part of the spectrum. Both fields are chosen sufficiently weak for perturbation theory to be valid.

Since the two pulses are temporally distinct, it is convenient to deal with their effects consecutively. After the first pulse is over, the superposition state prepared by the $\mathbf{E}_x(\tau)$ pulse, whose width is chosen to encompass just the two levels $|E_2\rangle$ and $|E_3\rangle$, is given in first-order perturbation theory as

$$|\phi(t)\rangle = |E_1\rangle e^{-iE_1t/\hbar} + b_2|E_2\rangle e^{-iE_2t/\hbar} + b_3|E_3\rangle e^{-iE_3t/\hbar}, \tag{3.74}$$

where [Eq. (2.23)]

$$b_k = (2\pi i/\hbar)\langle E_k|\hat{\varepsilon}_x \cdot \mathbf{d}|E_1\rangle\bar{\epsilon}_x(\omega_{k,1}), \qquad k = 2, 3, \tag{3.75}$$

with $\omega_{k,1} \equiv (E_k - E_1)/\hbar$.

After a delay time of $\Delta_d \equiv t_d - t_x$ the system is subjected to the $\mathbf{E}_d(\tau)$ pulse. It follows from Eq. (3.74) that after this delay time each preparation coefficient has picked up an extra factor of $e^{-iE_k\Delta_d/\hbar}$, $k = 2, 3$. Hence, the phase of b_2 relative to b_3 at that time increases by $[-(E_2 - E_3)\Delta_d/\hbar = \omega_{3,2}\Delta_d]$. Thus, the natural two-state time evolution controls the relative phase of the two terms, replacing the externally controlled relative laser phase of the two-frequency control scenario of Section 3.1.

After the action and subsequent decay of the $\mathbf{E}_d(\tau)$ pulse, the system wave function is:

$$|\psi(t)\rangle = |\phi(t)\rangle + \sum_{n,q} \int dE\, b_{E,\mathbf{m},q}(t)|E, \mathbf{m}, q^-\rangle e^{-iEt/\hbar}. \tag{3.76}$$

In accord with Eqs. (2.74) to (2.77), the probability of observing the q product at total energy E in the remote future is therefore

$$P_q(E) = \sum_{\mathbf{m}} |b_{E,\mathbf{m},q}(t=\infty)|^2$$

$$= \left(\frac{2\pi}{\hbar^2}\right)^2 \sum_{\mathbf{m}} | \sum_{k=2,3} b_k \langle E, \mathbf{m}, q^-|\mathrm{d}_{e,g}|E_k\rangle \bar{\epsilon}_d(\omega_{EE_k})|^2, \tag{3.77}$$

where $\omega_{EE_k} = (E - E_k)/\hbar$, b_k is given by Eq. (3.75), and where $\bar{\epsilon}_d(\omega)$ is given via an expression analogous to Eq. (3.73).

Expanding the square and using the Gaussian pulse shape [Eqs. (3.72) and (3.73)] gives

$$P_q(E) = \left(\frac{2\pi}{\hbar^2}\right)^2 [|b_2|^2 \mathrm{d}_q(22)\bar{\epsilon}_2^2 + |b_3|^2 \mathrm{d}_q(33)\bar{\epsilon}_3^2$$

$$+ 2|b_2 b_3^* \bar{\epsilon}_2 \bar{\epsilon}_3^* \mathrm{d}_q(32)| \cos(\omega_{3,2}\Delta_d + \alpha_q(32) + \chi)], \tag{3.78}$$

where $\bar{\epsilon}_i = |\bar{\epsilon}_d(\omega_{EE_i})|$, $\omega_{3,2} = (E_3 - E_2)/\hbar$, and the phases χ, $\alpha_q(32)$ are defined via

$$\langle E_1|\mathrm{d}_{e,g}|E_g\rangle\langle E_g|\mathrm{d}_{g,e}|E_2\rangle \equiv |\langle E_1|\mathrm{d}_{e,g}|E_g\rangle\langle E_g|\mathrm{d}_{g,e}|E_2\rangle|e^{i\chi},$$

$$\mathrm{d}_q(ki) \equiv |\mathrm{d}_q(ki)|e^{i\alpha_q(ki)} = \sum_{\mathbf{m}} \langle E_k|\mathrm{d}_{g,e}|E, \mathbf{m}, q^-\rangle\langle E, \mathbf{m}, q^-|\mathrm{d}_{e,g}|E_i\rangle. \tag{3.79}$$

Integrating Eq. (3.78) over E to encompass the full width of the second pulse yields the final expressions for the quantities we wish to control: P_q, the probability of forming channel q, and $R_{q,q'}$, the ratio of product probabilities into q vs. q' [see Eq. (3.22)].

Examination of Eq. (3.78) makes clear that $R_{q,q'}$ can be varied by changing the delay time $\Delta_d = (t_d - t_x)$ or the ratio $x = |b_2/b_3|$; the latter is most conveniently done by detuning the initial excitation pulse. Note that, once again, as in the scenarios above, the spatial (z) dependence of P_q vanishes due to cancellation between the excitation and dump steps. In addition, the phases δ_x, δ_d do not appear in the final $R_{q,q'}$ expression, so that the relative phases of the two pulses do not affect the result.

This approach was applied to realistic systems such as the control of the Br to Br* branching ratio in the photodissociation of IBr [94], and the control of Li$_2$ photodissociation [95], discussed later. To gain insight into the control afforded by this scenario, we also applied it to an artificial model of the photodissociation of a hypothetical collinear DH_2 complex [96]:

$$H + HD \leftarrow DH_2 \rightarrow D + H_2. \tag{3.80}$$

The first pulse is used to excite a pair of states in an electronic state supporting bound states, and the second pulse is used to dissociate the system by deexciting it back to the ground state, above the dissociation threshold.

The model potentials used [96] are shown in Figure 3.21 and typical control results are shown in Figure 3.22. Specifically, Figure 3.22 shows contours of equal DH yield as a function of $E_x - E_{AV}$ and Δ_d. Here $E_x - E_{AV}$ measures the deviation of the central excitation energy of the pump pulse from the mean energy E_{AV} of the pair of bound states which it excites. The DH yield is shown to vary significantly, from 16 to 72%, as the control parameters are varied. This is an extreme range of control, especially if one considers that the product channels only differ by a mass factor.

It is highly instructive to examine the nature of the superposition state prepared in the initial excitation [Eq. (3.74)] and its time evolution during the delay between pulses. An example is shown in Figure 3.23 where we plot the wave function for the collinear model of model DH_2 photodissociation. Specifically, the axes are the reaction coordinate S and the coordinate x orthogonal to it. The wave function is shown evolving over one half of its total possible period. An examination of Figure 3.23 in conjunction with Figure 3.22 shows that deexciting this superposition state at the time of panel (b) would yield a substantially different product yield than deexciting at the time of panel (e). However, Figure 3.23 shows that there is clearly no particular preference of the wave function for either large positive or large negative S at these particular times, which would be the case if the reaction control were a result of some spatial characteristics of the wave function. Rather, the results make clear that the essential control characteristics of the wave function are encrypted in the quantum amplitude and phase of the created superposition state.

The controlled photodissociation of Li_2 into different atomic states provides another example of pump–dump control [95]. At the energy of interest three product pairs, Li(2s) + Li(2p), Li(2s) + Li(3p), and Li(2s) + Li(3s) may be produced. The photodissociation computations were carried out using realistic potential curves [97] shown in Figure 3.24. A schematic description of a suggested multipulse scheme (a preliminary experiment along these lines has been reported by Papanikolas et al. [98]) is also shown in this figure. In this case, a cw laser prepares a single vib-rotational state of the $A(1^1\Sigma_u^+)$ electronic state, chosen as $v = 14, J = 22$, which serves as the starting point for the subsequent pump–pump control. The pump pulse "lifts" the nuclear wave function in this electronic state to the $E(3^1\Sigma_g^+)$ state, forming a coherent superposition of ro-vibrational states. The subsequent pulse excites the system above the dissociation energy to form the products.

Figures 3.25 and 3.26 show the control resulting from the use of a narrow-band pump laser to prepare a superposition of the $(v = 14, J = 21)$ and $(v = 14, J = 23)$ vib-rotational states in the excited E electronic manifold. The pump pulse center was

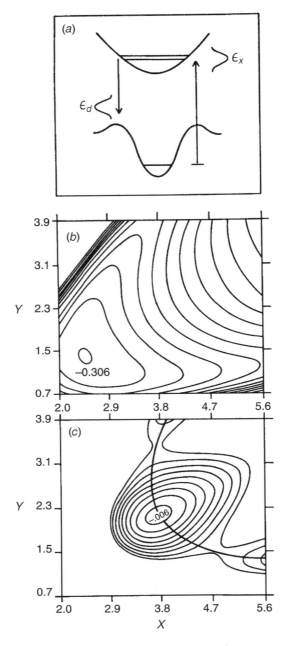

Figure 3.21 (a) Schematic diagram of a pump–dump scheme to control the model DH + H ← DH$_2$ → D + H$_2$ branching photodissociation reaction. Here ϵ_x is the excitation pulse and ϵ_d is the dump pulse. (b) Ground potential surface. Contour lines are spaced by 0.02 a.u., increasing outwards from the indicated minimum. (c) Excited potential surface. Contour lines are spaced by 0.0098 a.u., increasing outwards from the indicated minimum. Reaction coordinate S is shown as a thick line that is chosen here to coincide with the minimum energy path connecting the DH + H and the D + H$_2$ products. (Taken from Fig. 1, Ref. [96].)

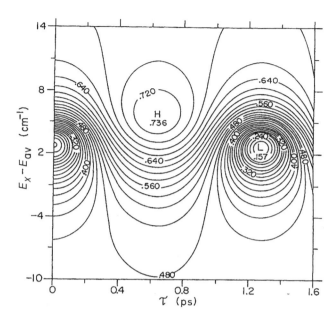

Figure 3.22   Contour plot of the DH yield as a function of the detuning of the exciting pulse $E_x - E_{AV}$, and the delay variable $\tau \equiv \Delta_d$. The actual delay is $(8.44 + 2.11n)$ picoseconds $+ \tau$, where n is an arbitrary positive integer chosen high enough to eliminate any overlap between the pulses. Here the initially created superposition state is between levels 56 and 57 ($E_1 = 0.323849$ a.u., $E_2 = 0.323968$ a.u.) of the excited surface. The letters H and L denote the positions of the absolute maxima and minima, whose magnitudes are explicitly shown. (Taken from Fig. 6, Ref. [48].)

tuned to the midfrequency between these two states; that is, $\lambda_1 = 805.6$ nm, $\lambda_2 = 1045$ nm, with $\Delta_{1\omega} = 36$ cm^{-1}, and $\Delta_{2\omega} = 45$ cm^{-1}. Here Δ_{iw} is a measure of the frequency width of the ith pulse, that is, $\Delta_{iw} = 4\sqrt{\ln 2}\tau_i$ [see Eq. (3.73)]. The yield of the third channel, Li(2s) + Li(3s), was found to vary from 0.46 to 1.03% and to be relatively invariant to changes in Δ_d. For this reason its behavior is ignored below.

Figure 3.25 shows the percent yield of forming Li(2s) + Li(2p) (solid curve) and Li(2s) + Li(3p) (dashed curve) as a function of delay time Δ_d. The results clearly show an extensive variation of the relative yield of the different products as a function of τ with the Li(2s) + Li(2p) yield varying from 82.2 to 20.4%, and the Li(2s) + Li(3p) yield varying from 17.2 to 78.7%. This corresponds to the change in the yield ratio $R = P(q = 2p)/P(q = 3p)$ from 4.8 to 0.26. The yield of the Li(2p) product is seen to change more dramatically with Δ_d, a consequence of the fact that the $d_q(12)$ matrix element associated with the Li(2p) product is approximately three times larger than that associated with the Li(3p) product.

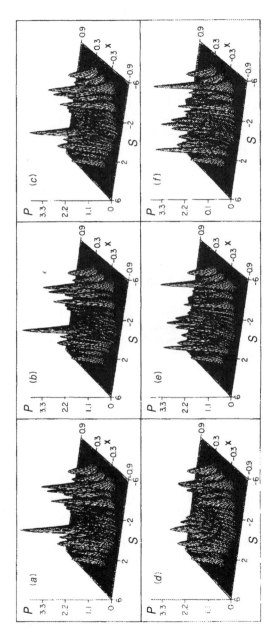

Figure 3.23 Time evolution of the square of the wave function for a superposition state comprised of levels 56 and 57. Probability is shown as a function of S and its orthogonal coordinate x at times (*a*) 0 ps, (*b*) 0.0825 ps, (*c*) 0.165 ps, (*d*) 0.33 ps, (*e*) 0.495 ps, (*f*) 0.66 ps, which correspond to fractions of the period $2\pi/\omega_{2,1}$. (Taken from Fig. 6, Ref. [96].)

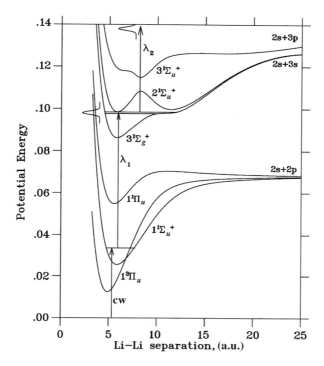

Figure 3.24 Schematic of the pulse sequencing control scheme used in this work. Realistic Li$_2$ potentials used in this work are shown. (From Fig. 1, Ref. [95].)

In accord with Eq. (3.78), the product probability is seen to be periodic, with an approximate period of $T = 2\pi/\omega_{2,1} = 1773$ fs, which corresponds to the rotational spacing $\omega_{2,1} \equiv (E_{v=14,J=23} - E_{v=14,J=23})/\hbar$ of 18.8 cm^{-1}.

The product ratios can also be controlled by shifting the central frequency of the pump laser, thus altering the b_j coefficients of the superposition state [Eq. (3.74)]. Sample results are shown in Figure 3.26 where we display a contour plot of the Li(2s) + Li(2p) product yield as a function of $\tau \equiv \Delta_d$ and of the detuning of ω_1 from $\omega_{AV} = \dfrac{1}{2\hbar}(E_{v=14,J=21} + E_{v=14,J=23} - 2E_1)$, the halfway transition frequency to the two vib-rotational levels. In this case $\Delta_{1\omega} = 18$ cm^{-1} and $\Delta_{2\omega} = 45$ cm^{-1} with the remaining parameters of the second laser chosen as in Figure 3.25. In Figure 3.26, λ_1 is varied from 806.2 nm to 805 nm, corresponding to the centering of the first pulse on levels $E_{v=14,J=21}$ and $E_{v=14,J=23}$, respectively. The results indicate a large range of control, from 81 to 20% as one varies τ at fixed pulse detuning. Substantial control is also attained by changing ω_1 at fixed τ (e.g., at 890 fs). The results clearly show, however, that the highest level of control is attained when the energy of the first pulse is centered close to E_{AV} ($\lambda_1 = 805.6$ nm). This is because, under these circumstances, the bound-bound dipole matrix elements $\langle E_j|\hat{\varepsilon} \cdot \mathbf{d}|E_s\rangle$,

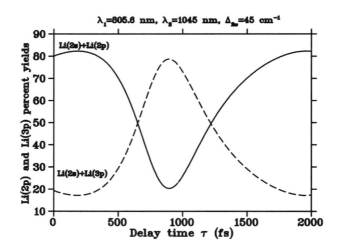

Figure 3.25 Yields of Li(2p) (solid curve) and Li(3p) (dashed curve) as a function of the delay time $\tau \equiv \Delta_d$. Here the initially created superposition state is composed of the $(v = 14, J = 21)$ and $(v = 14, J = 23)$ states in the E electronic manifold. (From Fig. 2, Ref. [95].)

$j = 1, 2$ contributing via Eq. (3.75) have very similar values for the $(v = 14, J = 21)$ and $(v = 14, J = 23)$ levels.

Varying other parameters, such as the width of the pulses, also has substantial effect on product control. For example, the effect of exciting more vib-rotational levels in the E electronic state by using a broader pump pulse is shown in Figure 3.27, where $\Delta_{1\omega} = 60 \text{ cm}^{-1}$ and $\Delta_{2\omega} = 100 \text{ cm}^{-1}$. The superposition state prepared by the first pulse consists of the $v = 14, 15$ and $J = 21, 23$ levels, where the pulse is centered at $\lambda_1 = 803.88$ nm, corresponding to the frequency halfway between the $(v = 14, J = 21)$ and the $(v = 15, J = 23)$ levels. The resultant behavior of P_q is more complicated than the previous cases since more than two terms are included in Eq. (3.78). Interestingly, we find that overall control is reduced with increasing $\Delta_{1\omega}$. That is, in this case the yield in the Li(2s) + Li(2p) channel only varies from 51.4 to 2.6%, a much smaller control range than in the comparable (narrower pump pulse) case. Adding additional levels to the initial superposition state by pumping with a wider pulse ($\Delta_{1\omega} = 100 \text{ cm}^{-1}$) resulted in slightly more complicated behavior of P_q as a function of τ, and an even further small reduction in the extent of control. Hence, faster pulses, corresponding to larger $\Delta_{1\omega}$ decrease the observed control.

The pump–dump scheme described above has also been applied to the control of the

$$\text{D} + \text{OH} \leftarrow \text{HOD} \rightarrow \text{H} + \text{OD}$$

dissociation reaction, proceeding via the B^1A' excited state of HOD. In this case both asymptotic channels have identical potential energy surfaces so that control over the

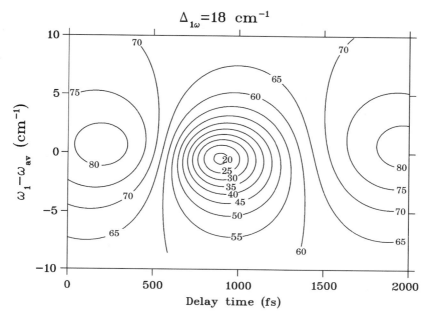

Figure 3.26 Contour plot of the Li(2p) yield as a function of the detuning of the pump pulse $\omega_1 - \omega_{AV}$ and the delay time $\tau \equiv \Delta_d$. Initially created superposition state is between levels $v = 14$, $J = 21$ and $J = 23$. Laser parameters are given in text. (From Fig 4, Ref. [95].)

relative yield is challenging. To consider the extent of possible control, we excite an initial superposition of the $(0, 2, 0)$ and $(1, 0, 0)$ states of ground state HOD [$(0, 2, 0)$ denotes two quanta in the bend mode and $(1, 0, 0)$ denotes one quantum of excitation of the OD stretch]. A subsequent pulse dissociates HOD via the B^1A' continuum. A typical result is displayed in Figure 3.28 which shows contours of constant percentage yield of H + OD [i.e., 100 P(OD + H)/[P(OD + H) + P(OH + D)] as a function of the time delay Δ_d and of the detuning of the pump laser pulse $E_x - E_{AV}$. Features of this result are of note. First, significant variations of yield ratio accompany changes in $(E_x - E_{AV})$. Second, the dependence of the yield ratio on the time delay is weak. The former feature merely reflects a natural preference, on the part of either of the two excited states $|E_2\rangle$, $|E_3\rangle$ to favor production of OD over OH. Changing $E_x - E_{AV}$ changes the relative contribution of each of these two states thereby changing the yield ratio. Thus, changes in yield ratio with changes in $E_x - E_{AV}$ are not due to coherent control. Rather, quantum interference effects are reflected in variations of the yield ratio with changes in Δ_d. The fact that this is weak is indeed a reflection of the similarity of the two product channels.

The approach discussed thus far relies heavily on the interference generated between a very small number of energy levels. An alternative fundamental perspective on control, based on localized wave packets, which are a superposition of many levels, was originally introduced by Tannor and Rice [91, 92] and Tannor et al.

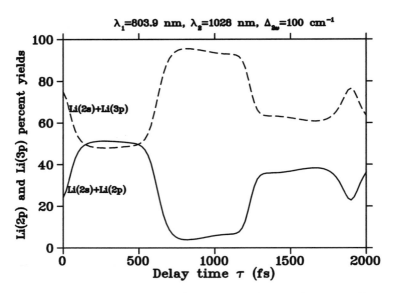

Figure 3.27 Control results with broader first pulse. Yield ratios of Li(2p) (solid curve) and Li(3p) (dashed curve) as a function of $\tau \equiv \Delta_d$ for the integrated probability $P_q = \int_0^\infty dE\, P_q(E)$ with wavelengths $\lambda_1 = 803.9$ nm and $\lambda_2 = 1028$ nm and pulse widths $\Delta_{1\omega} = 60\,\mathrm{cm}^{-1}$ and $\Delta_{2\omega} = 100\,\mathrm{cm}^{-1}$. Superposition state consists of $v = 14$, $J = 21$ and $J = 23$, and $v = 15$, $J = 21$ and $J = 23$ levels. (From Fig 8, Ref. [95].)

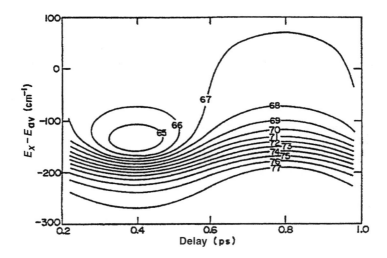

Figure 3.28 Percentage yield of the $H + OD$ channel in the photodissociation of the $DOH(0, 2, 0 + 1, 0, 0)$ superposition state. The excitation pulse bandwidth is $50\,\mathrm{cm}^{-1}$, the dissociation pulse bandwidth is $50\,\mathrm{cm}^{-1}$, and the enter frequency is $71,600\,\mathrm{cm}^{-1}$. Ordinate is the detuning of the excitation pulse ω_x from the energy center of the $(0, 2, 0)$ and $(1, 0, 0)$ states. (Taken from Fig. 3, Ref. [99].)

[100]. Their approach, which founded the fundamentals of pump–dump control and of optimal control, is the subject of Chapter 4.

APPENDIX 3A: MODE-SELECTIVE CHEMISTRY

Coherent control is quite distinct from the mode-selective approach to controlling chemistry, an approach which was advocated for some time [53–60]. Examining the difference between mode selectivity and coherent control, as done in this section, affords considerable insight into both.

The essence of the mode-selective approach is concisely stated in the Pimentel and Coonrod report of the National Academy of Sciences [101] as "an effort to excite a particular degree of freedom in a molecule so that the molecules react as if this particular degree of freedom was at a very high temperature whereas the rest of the molecular degrees of freedom are cold." For example, in the case of $A + BC \leftarrow ABC \rightarrow AB + C$ one might excite the AB bond in an attempt to enhance production of $A + BC$. In addition, the Pimentel and Coonrod report identifies that problem that is widely regarded as the major difficulty to overcome in order to achieve control via mode selectivity. That is [101], "apparently the problem is that vibrational redistribution (IVR) takes place within these molecules," presumably preventing maintenance of the selective control based upon the idea of initially localizing energy in some part of the molecule. For this reason, advocates of this approach argued for faster excitation methods. Below we show that rapid laser excitation is unnecessary and that the mode-selective approach affords little new in controlling reactions of isolated molecules, assuming that one uses one-photon processes.

Consider the general case of a pure state wave packet prepared by any of a variety of schemes that are only required to be turned off at long times. The prepared time-dependent wave packet is

$$|\Psi(t)\rangle = \int dE \sum_{\mathbf{n},q} b_{E,\mathbf{n},q} |E, \mathbf{n}, q^-\rangle e^{-iEt/\hbar}. \tag{3.81}$$

Some mode-selective chemistry advocates argued that it is the time dependence of the wave packet, that is, the phase relationships between energy levels, that affords the possibility of effectively controlling reactions. Here we show that this is not the case.

The probability $P_{\mathbf{n}',q'}(E')$ of forming the product with energy E', quantum numbers $\mathbf{n}'$, and product arrangement q' [Eq. (2.74)] is

$$P_{\mathbf{n}',q'}(E') = |\lim_{t \to \infty} e^{iE't/\hbar} \langle E', \mathbf{n}', q'; 0|\psi(t)\rangle|^2 = |b_{E',\mathbf{n}',q'}|^2. \tag{3.82}$$

Thus, the probability of forming the product state $P_{\mathbf{n},q}(E)$ is given by the sum over the squares of the $b_{E,\mathbf{n},q}$ coefficients, a quantity that is *totally independent of any coherence established between nondegenerate energy levels in the initially created state.* Indeed as shown below, the product yield is exactly the same as that which would be obtained if the initially created state were a totally time-independent energy-incoherent mixture,

$$\rho(0) = \int dE \sum_{\mathbf{n},q} \sum_{\mathbf{n}',q'} b_{E,\mathbf{n},q} b^*_{E,\mathbf{n}',q'} |E, \mathbf{n}, q^-\rangle\langle E, \mathbf{n}', q'^-| \qquad (3.83)$$

with the associated probability

$$P_{\mathbf{n},q}(E) = \langle E, \mathbf{n}, q; 0|\rho|E, \mathbf{n}, q; 0\rangle. \qquad (3.84)$$

This being the case, the coherence among energy levels, hence the time dependence of the initial state, has no influence whatsoever on the product yield. This implies, for example, that in the weak-field regime, the product yield obtainable with a subfemtosecond laser pulse with frequency spectrum $I(\omega)$ is exactly equal to the sum of a set of, for example, *microsecond* pulses with an appropriate set of frequencies and intensities that have the same cumulative $I(\omega)$. Shorter pulses alter the product probabilities only to the extent that they excite a larger number of states as the frequency spectrum broadens with diminishing pulse duration. Clearly, an observation of increased yield of a particular product upon use of a shorter pulse is then due solely to the fact that the shorter pulse encountered some states with a preference for a particular product.

One further point is worth noting. The mode selective approach seeks to excite modes that enhance production of a desired product. The associated assumption is that the correct modes are simple and intuitive, for example, such as exciting local bond modes (e.g., the AB bond in ABC) to selectively dissociate the molecule. However, our discussion (Sections 3.1 and 3.3) makes clear that the proper modes to excite to produce product in channel q, for example, are the $|E, \mathbf{n}, q^-\rangle$ states, since these states are guaranteed to correlate with the q product. In some instances the appropriate $|E, \mathbf{n}, q^-\rangle$ may indeed correspond to bond excitation. This is the case with hydrogen-containing molecules such as HOOH [54], C_2HD [59], and HOD [102], where excitation of the OH bond is known to enhance the breaking of the OH in preference to the OD bond and vice versa [57, 102]. However, in general, the structure of $|E, \mathbf{n}, q^-\rangle$ is considerably different than that of simple bond excitation.

The delta function energy normalization [e.g., Eq. (2.47)] of the scattering states $|E, \mathbf{n}, q^-\rangle$ and $|E, \mathbf{n}, q; 0\rangle$ necessitates that care be taken in deriving Eq. (3.84). The nuances are discussed below. For notational simplicity we drop the $(\mathbf{n}, q)$ labels, which, as discrete indices, can readily be included later.

Consider the stationary density matrix

$$\rho = \int dE \, |E^-\rangle\langle E^-| g(E), \qquad (3.85)$$

where $g(E)$ is a square integrable function and $|E^-\rangle$ is a scattering state at energy E. Equation (3.85) is the analog of Eq. (3.83) where we suppress the $(\mathbf{n}, q)$ variables. In light of the delta function normalization $\langle E^- | E_1^- \rangle = \delta(E - E_1)$, and $\mathrm{Tr}(\rho) = \infty$, the operator ρ is then said to be "non-trace-class" [103] and, as a consequence, standard operations, such as taking the average values, are meaningless.

To bypass these problems, we initially introduce broadening in energy, replacing the diagonal density matrix [Eq. (3.85)] by the nondiagonal

$$\rho(0) = \lim_{\epsilon \to 0} \int dE_0 \int \int dE_1 \, dE_2 \, f_\epsilon(E_1, E_0) f_\epsilon(E_2, E_0) |E_1^-\rangle \langle E_2^- | g(E_1) g^*(E_2), \quad (3.86)$$

where the argument of $\rho(0)$ denotes time zero. Here $f_\epsilon(E_1, E_0)$ is a function with nonzero values within the E_1 range $(E_0 - \epsilon, E_0 + \epsilon)$. We choose $f_\epsilon(x, x')$ such that

$$\lim_{\epsilon \to 0} |f_\epsilon(x, x')|^2 = \delta(x - x'). \quad (3.87)$$

In this limit, Eq. (3.86) is essentially a stationary mixture of projectors $|E_0^-\rangle \langle E_0^- |$. Note that a small degree of coherence between energy levels has been introduced into Eq. (3.86) to account for the continuous character of E.

Consider now $\rho(t \to \infty)$, given $\rho(0)$ [Eq. (3.86)]. At long times $\rho(t \to \infty)$ is of the same form as Eq. (3.86), but with $|E_1^-\rangle \langle E_2^-|$ being $|E_1\rangle \langle E_2|$ states, that is, products of eigenstates of the asymptotic Hamiltonian. Specifically, if $|E_1\rangle \langle E_2|$ denote such asymptotic states, then at long times

$$\rho(t) = \lim_{\epsilon \to 0} \int dE_0 \int \int dE_1 \, dE_2 \, f_\epsilon(E_1, E_0) f_\epsilon(E_2, E_0) |E_1\rangle \langle E_2| g(E_1) g^*(E_2) e^{-i(E_1 - E_2)t/\hbar}.$$

$$(3.88)$$

Then the probability of observing product states within the range $E - \Delta$ to $E + \Delta$ is

$$\lim_{\epsilon \to 0} \int_{E-\Delta}^{E+\Delta} d\bar{E} \langle \bar{E} | \rho | \bar{E} \rangle = \lim_{\epsilon \to 0} \int_{E-\Delta}^{E+\Delta} d\bar{E} \int dE_0 |f_\epsilon(\bar{E}, E_0)|^2 |g(\bar{E})|^2 = \int_{E-\Delta}^{E+\Delta} dE_0 |g(E_0)|^2.$$

$$(3.89)$$

Here, the first equality arises from using the delta function normalization of $|E\rangle$ and the second equality from Eq. (3.87).

Equation (3.89) is the relevant result, indicating that the product probability is independent of any coherence established initially between nondegenerate states. It reiterates the discussion above and emphasizes that product control resides in control over the wave functions $|E, \mathbf{n}, q^-\rangle$ that exist at fixed energies.

CHAPTER 4

OPTIMAL CONTROL THEORY

4.1 PUMP–DUMP EXCITATION WITH MANY LEVELS: TANNOR–RICE SCHEME

The approach discussed thus far relies heavily on the interference generated between a small number of energy levels. An alternative perspective on control, based on localized wave packets, was pioneered by Tannor and Rice [91, 92] and Tannor and co-workers [100]. As an example of their approach, consider the potential energy profile shown in Figure 4.1, where the product A + BC emerges on the potential surface when leaving through one exit and AB + C emerges from the other. An initial wave packet, localized on the ground electronic state, is laser-excited to the upper electronic state where it evolves. A second laser then causes stimulated emission back to the ground electronic state. Assuming only small spreading of the wave packet, a properly timed second laser pulse induces stimulated emission to either deposit the excited wave function preferentially in the A + BC or AB + C exit channel region, enhancing production of that product. In addition to timing, Tannor and Rice introduced the idea of optimizing the shape of the laser pulses using the calculus of variations so as to alter the final state, hence controlling the outcome of the dynamics. This approach was the underlying basis for developments in optimal control theory, discussed in Section 4.2.

Assuming that the states are spatially localized implies that they are comprised of a large number of energy eigenstates. Further, since excitations to produce superpositions of large numbers of states require broadband coherent laser excitation, the required lasers are in the femtosecond domain. Finally, spatial localization of wave packets suggests behavior near the classical limit. In this regime, the formalism of Section 3.5 becomes unreasonably complicated, involving interferences between a

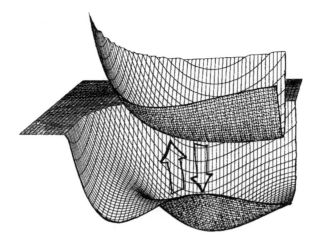

Figure 4.1 Schematic illustration of the original Tannor–Rice scenario for enhancing the yield in a given arrangement channel. The case of enhancing the right channel is shown. (From Fig. 1, Ref. [104].)

myriad of energy eigenstates, and the direct wave packet propagation approach based on perturbation theory (Section 3.3.1) now proves more useful.

According to this approach, the amplitude $A_{\mathbf{m},q}(E)$ for photodissociation into the final state $|E, \mathbf{m}, q; 0\rangle|g\rangle$ at time t, having started in state $|E_i\rangle|g\rangle$, is using second-order perturbation theory with respect to the light–matter interaction [Eq. (3.30) restricted to $n = 2$ at finite time]:

$$
\begin{aligned}
A_{\mathbf{m},q}(E) &= \langle E, \mathbf{m}, q; 0|\langle g|\Psi(t)\rangle = \exp(-iEt/\hbar)\langle E, \mathbf{m}, q^-|\langle g|\psi^I(t)\rangle \\
&= \exp(-iEt/\hbar)\langle E, \mathbf{m}, q^-|\langle g|\mathcal{V}^{(2)}(t)|g\rangle|E_i\rangle.
\end{aligned}
\tag{4.1}
$$

Here we have explicitly included the electronic state labels ($|g\rangle$ is the ground electronic state) and have chosen the electronic ground state energy as the zero of energy. Note that $\mathcal{V}^{(1)}$ is neglected since it does not contribute significantly to transitions between states of the ground electronic state $|g\rangle$. The $\mathcal{V}^{(2)}$ operator is given by an equation analogous to Eq. (3.29) in which the electronic state terms are explicitly included. That is, H_M is replaced by H_{MT} [Eq. (2.67)] and V^I [Eq. (3.25)] is written in terms of $H'_{MR}(t)$ [Eq. (2.71)]. We obtain that

$$
\begin{aligned}
A_{\mathbf{m},q}(E) &= \left(\frac{-i}{\hbar}\right)^2 \langle E, \mathbf{m}, q^-|\langle g|e^{-iEt/\hbar} \int_{-\infty}^{t} dt_1\, e^{iH_{MT}t_1/\hbar} H'_{MR}(t_1)e^{-iH_{MT}t_1/\hbar} \\
&\quad \times \int_{-\infty}^{t_1} dt_2\, e^{iH_{MT}t_2/\hbar} H'_{MR}(t_2)e^{-iH_{MT}t_2/\hbar}|g\rangle|E_i\rangle.
\end{aligned}
\tag{4.2}
$$

Using Eqs. (2.67) and (2.71) allows us to rewrite the photodissociation amplitude as

$$A_{\mathbf{m},q}(E) = \left(\frac{-i}{\hbar}\right)^2 \sum_e \langle E, \mathbf{m}, q^- | \int_{-\infty}^{t} dt_1 \; e^{-iE(t-t_1)/\hbar} \langle g|H_{MR}(t_1)|e\rangle$$

$$\times \int_{-\infty}^{t_1} dt_2 \; e^{-iH_e(t_1-t_2)/\hbar} \langle e|H_{MR}(t_2)|g\rangle e^{-E_i t_2/\hbar}|E_i\rangle, \tag{4.3}$$

where $H_e \equiv \langle e|H_{MT}|e\rangle$ is an operator on the nuclear coordinates describing the nuclear motion on electronic state $|e\rangle$. Likewise, $\langle g|H_{MT}|g\rangle$ describes the nuclear motion on $|g\rangle$. Given that $H_{MR}(t) = -\varepsilon(z, t)\hat{\varepsilon} \cdot \mathbf{d}$ [Eq. (2.10)], and assuming that the laser is resonant with only one electronic state, we obtain that

$$A_{\mathbf{m},q}(E) = \left(\frac{-i}{\hbar}\right)^2 \langle E, \mathbf{m}, q^- | \int_{-\infty}^{t} dt_1 \; e^{-iE(t-t_1)/\hbar} \mathbf{d}_{g,e}\varepsilon(z, t_1)$$

$$\times \int_{-\infty}^{t_1} dt_2 \; e^{-iH_e(t_1-t_2)/\hbar} \mathbf{d}_{e,g}\varepsilon(z, t_2)e^{-E_i t_2/\hbar}|E_i\rangle. \tag{4.4}$$

Equation (4.4) has the qualitative interpretation that $|E_i\rangle$ evolves on the ground state surface until time $t = t_2$ when it makes a transition to the excited electronic surface. It then propagates for a time t_2 to t_1 at which time it makes a transition back to the ground electronic state. The wave function then evolves on the ground state surface until time t when we measure its overlap with the product state. Since the fields are spread out over time, we have to integrate over both t_1 and t_2.

As in Section 3.5, $\varepsilon(z, t)$ is often comprised of two temporally distinct pulses where the timing between the two pulses serves as a control parameter. It is interesting to examine, for example, the limiting case where $\varepsilon(z, t)$ comprises two delta function pulses. In this case the pulses have an infinitely wide profile in frequency space and hence encompass a complete set of levels. This is the extreme opposite of the case discussed in Section 3.5 where the laser pulse only encompasses two levels. Here, neglecting the spatial dependence, the field is of the form

$$\varepsilon(z, t) = \mathcal{E}_x \exp(i\omega_x t + i\phi_x)\delta(t - t_x) + \mathcal{E}_d \exp(i\omega_d t + i\phi_d)\delta(t - t_d) \tag{4.5}$$

with $t_d > t_x$. Noting that

$$\int_{-\infty}^{t} dt_1 \int_{-\infty}^{t_1} dt_2 \, f(t_1, t_2)\delta(t_1 - t_d)\delta(t_2 - t_x) = f(t_d, t_x), \tag{4.6}$$

and inserting Eq. (4.5) in Eq. (4.4), we obtain for $-\infty < t_x < t$ and $t_x < t_d < t$ that the dissociation probability is

$$P_q(E) = \sum_{\mathbf{m}} |A_{\mathbf{m},q}(E)|^2 = (\hbar)^{-4}\mathcal{E}_x^2\mathcal{E}_d^2 \sum_{\mathbf{m}} |\langle E, \mathbf{m}, q^- |\mathbf{d}_{g,e}e^{-iH_e(t_d-t_x)/\hbar}\mathbf{d}_{e,g}|E_i\rangle|^2. \tag{4.7}$$

Thus, in this case, the ratio of product into various arrangement channels is controlled entirely by the time delay $(t_d - t_x)$ between pulses.

This example also makes clear that ϕ_x and ϕ_d, the phases of the two pulses, do not enter the final expression, a result that holds true even for pulses of finite time duration. As a consequence, we are at liberty to choose arbitrary values for these quantities. We say that the phases of the pump and dump pulses need not be "locked" relative to one another to achieve this type of control. However, when pulses of finite durations are used, all other phases [such as those associated with each pulse envelope, $\varepsilon_x(t)$ and $\varepsilon_d(t)$] must be defined.

Sample results for pulses of finite duration are shown in Figure 4.2 where the magnitude of the excited state wave function before the second pulse, and of the ground state wave function after the second pulse, are shown. In this case the delay time of 1900 a.u. (1 a.u. of time $= 2.41888 \times 10^{-17}$ s) corresponds to the time it

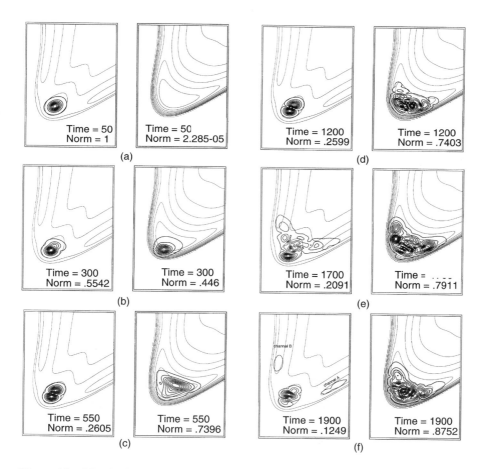

(a)

Time = 50
Norm = 1

Time = 50
Norm = 2.285-05

(b)

Time = 300
Norm = .5542

Time = 300
Norm = .446

(c)

Time = 550
Norm = .2605

Time = 550
Norm = .7396

(d)

Time = 1200
Norm = .2599

Time = 1200
Norm = .7403

(e)

Time = 1700
Norm = .2091

Time = ...
Norm = .7911

(f)

Time = 1900
Norm = .1249

Time = 1900
Norm = .8752

Figure 4.2 Magnitude of the wave function for a model collinear problem. The left panel in each figure shows results on the ground potential surface, and the right panel shows results on excited potential surface. Times shown are at various intervals from $t = 50$ a.u. to $t = 1900$ a.u. Note only a slight increase in probability density in outgoing channel A on the ground potential surface at last time shown. (From Fig. 2, Ref. [104].)

would take a compact excited state wave packet to move to the channel A configuration, assuming that the excited state wave packet does not spread appreciably. As
shown in Figure 4.2 this in fact is not the case, as the excited state wave packet
spreads considerably while losing its original shape. Thus, a sizable portion of the
probability is seen to leak into the "unwanted" channel B.

The deficiency of the simple model, which assumes that the excited state wave
packet remains a compact object, can be corrected, as discussed below, by optimally
shaping the pump and dump pulses. Alternatively, as shown in Figure 4.3, one can
simply scan the delay times while leaving the pulses with their original shapes. We
see that the timing between the pulses strongly affects the relative yield of each of
the two products. In this way it is possible to find the "optimal" delay time to
enhance a given channel, a topic to be discussed at great length below.

Numerous computational studies of many-level pump–dump scenarios have been
considered by Rice, Tannor, and Kosloff and are discussed in detail in Refs. [105]
and [93]. Further, a number of experiments have been carried out demonstrating
control of molecular processes using pulsed light sources. In addition, there are a
host of experiments in the literature [106] that use a first pulse to initiate a dynamical
process and a second pulse to interrogate the process. By the formalism above, these
too are pulsed control experiments. Often, however, they only involved a single
product arrangement channel so that our primary challenge, that is, enhancing one
channel over another, is not addressed.

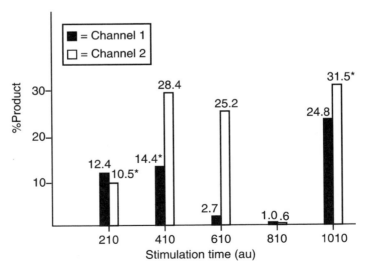

Figure 4.3   Branching ratio between two product channels in a collinear model as a function
of the time of the second pulse. Note strong dependence of branching ratio on pulse time.
(From Fig. 5, *Adv. Chem. Phys.*, **101**, 213 (1997).

The initial experiment demonstrating control in accordance with the Tannor–Rice scenario is due to Gerber and co-workers [107, 108] in which control was demonstrated over the two-channel ionization:

$$\mathrm{Na_2^+ + e^- \leftarrow Na_2 \rightarrow Na^+ + Na + e^-.} \tag{4.8}$$

Specifically, as shown in Figure 4.4, $\mathrm{Na_2}$ is pumped from the ground electronic state to the $2^1\Pi_g$ state in a two-photon process by an initial pulse. The wave packet propagates on this potential curve until an additional pulse carries it to the ionized state. The Franck–Condon factors favor production of the $\mathrm{Na_2^+}$ product if excitation is when the packet is at the inner turning point of the $2^1\Pi_g$ curve, whereas the excited $\mathrm{Na_2}$ decays to the $\mathrm{Na^+ + Na + e^-}$ product if excitation is at the outer turning point. The experimental results on the ratio of the $\mathrm{Na^+}$ and $\mathrm{Na_2^+}$ signals, as a

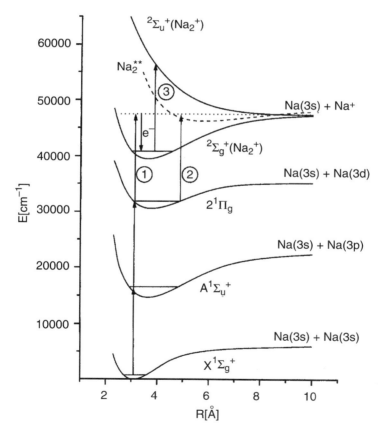

Figure 4.4 Potential energy surfaces and excitation scheme involved in Tannor–Rice controlled $\mathrm{Na_2}$ ionization. (Taken from Fig. 3, Ref. [108].)

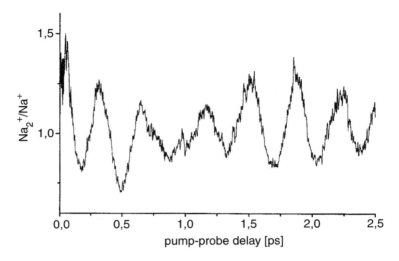

Figure 4.5   Ratio of Na^+ to Na_2^+ product as a function of the time delay between pulses. (Taken from Fig. 5, Ref. [108].)

function of the delay time between the two 80-fs long pulses, are shown in Figure 4.5. Clearly, the ratio is a strong function of the delay time.

A number of other early experiments confirming the Tannor–Rice scenario are discussed in detail in the monograph of Rice and Zhao [105].

4.2   OPTIMAL CONTROL THEORY

Optimal control theory (OCT) [91–93, 100, 104, 105, 109–132] is a generalization of the original ideas of Tannor–Rice [91, 100]. It was developed by Rabitz et al. [109–111, 116–118], by Tannor, Kosloff, Rice et al. [92, 104, 119, 123, 132]; and by Jakubetz, Manz et al. [113, 115, 124, 125].

Tannor and Rice's original idea, discussed above, had considerable obvious appeal. However, it soon became apparent that for most realistic surfaces, intuitive ideas do not necessarily tell us what pulses do the best job. For example, the wave packet, once created, does not remain localized, but quickly disperses. An example of this can be seen in Figure 4.2 where the wave packet in the excited state is seen to spread, resulting in less than ideal channel A/channel B branching ratios. As a consequence, intuition (which was based on simulations [91, 92, 100]) was abandoned in favor of a more systematic pulse optimization scheme.

In what follows we outline the general principles of OCT. Specific cases are described in Chapter 13.

4.2.1 General Principles of Optimal Control Theory

Optimal control theory aims to maximize or minimize certain transition probabilities, called *objectives*, such as the production of a specified wave function Φ at a specified time t_f, given a wave function $\Psi(t_0)$ at time t_0. The general principles of OCT are best understood via a case study due to Rice and coworkers [104, 119], illustrated in Figure 4.2, in which the objective is to concentrate the wave function Φ in one of the exit channels of a bifurcating chemical reaction:

$$\text{(channel B)} \qquad A + BC \leftarrow ABC \rightarrow AB + C, \qquad \text{(channel A)} \qquad (4.9)$$

Mathematically speaking, our objective is to maximize

$$J = \langle \Psi(t_f)|P|\Psi(t_f)\rangle, \qquad (4.10)$$

where $P \equiv |\Phi\rangle\langle\Phi|$ is a projection operator onto the product state of interest.

The maximization of J is subject to a set of physical and practical constraints that make the optimization problem meaningful. For example, we usually wish to optimize J using a laser pulse of a fixed total energy I. This results in the (practical) constraint equation (known as a "penalty") where

$$\int_{t_0}^{t_f} dt\, |\varepsilon(t)|^2 - I = 0. \qquad (4.11)$$

Dynamics is introduced as an additional constraint. In our case this is the requirement that Ψ be a solution of the time-dependent Schrödinger equation,

$$(i\hbar \partial/\partial t - H(t))|\Psi\rangle = 0. \qquad (4.12)$$

For a problem involving a ground and excited electronic state, it is convenient to rewrite the Schrödinger equation in matrix notation. Specifically [91, 92, 94, 96, 100, 104], $|\Psi\rangle$ is replaced by $|\boldsymbol{\Psi}\rangle$, a two-component state vector:

$$|\boldsymbol{\Psi}\rangle \equiv \begin{pmatrix} |\psi_e(t)\rangle \\ |\psi_g(t)\rangle \end{pmatrix}, \qquad (4.13)$$

where $|\psi_g(t)\rangle$ and $|\psi_e(t)\rangle$ are (time-dependent) wave functions on the ground and excited electronic surfaces, respectively. The Hamiltonian $\underline{\underline{H}}(t)$ is now a 2×2 matrix (for a full justification of this treatment see Section 12.53):

$$\underline{\underline{H}}(t) \equiv \begin{pmatrix} H_e & -d_{g,e}\varepsilon^*(t) \\ -d_{e,g}\varepsilon(t) & H_g \end{pmatrix}, \qquad (4.14)$$

where $d_{g,e}$, $d_{e,g}$ is the transition dipole moment introduced in Eq. (2.72).

The problem of maximizing the objective J subject to the above constraints can be transformed into an unconstrained problem by using Lagrange multipliers. According to this standard procedure, we multiply Eq. (4.11) by an unknown number λ and Eq. (4.12) by an unknown two-component state vector:

$$|\chi(t)\rangle \equiv \begin{pmatrix} |\chi_e(t)\rangle \\ |\chi_g(t)\rangle \end{pmatrix}, \tag{4.15}$$

and add the result to Eq. (4.10). Note that the latter is an extension of the simple Lagrange multiplier procedure to constraints on continuous functions. The resulting unconstrained objective

$$\bar{J} = \langle \Psi(t_f)| \cdot \mathsf{P} \cdot |\Psi(t_f)\rangle + \lambda \left[\int_{t_0}^{t_f} |\varepsilon(t)|^2 - I \right]$$
$$+ i \int_{t_0}^{t_f} dt \left\{ \langle \chi| \cdot \left[\frac{i\hbar\partial}{\partial t} - \underline{\underline{H}} \right] \cdot |\Psi\rangle + \text{c.c.} \right\}, \tag{4.16}$$

is maximized by imposing the $\delta \bar{J} = 0$ (with respect to changes in $|\Psi\rangle$) condition. When this is done, it follows from Eq. (4.16) that the Lagrange multipliers $|\chi\rangle$ satisfy [104]

$$i\hbar \frac{\partial |\chi\rangle}{\partial t} = \underline{\underline{H}} \cdot |\chi\rangle, \tag{4.17}$$

with the boundary condition,

$$|\chi(t_f)\rangle = \mathsf{P} \cdot |\Psi(t_f)\rangle. \tag{4.18}$$

The optimum field, defined via the requirement $\delta \bar{J} = 0$ is then related to the above solutions and to λ as

$$\varepsilon(t) = O(t)/\lambda, \tag{4.19}$$

where

$$O(t) = i[\langle \chi_e|d_{e,g}|\psi_g\rangle - \langle \psi_e|d_{g,e}|\chi_g\rangle]. \tag{4.20}$$

Once $O(t)$ is known, λ can be obtained by substituting Eq. (4.19) in Eq. (4.11) to obtain

$$\lambda = \pm \left(\frac{1}{I} \int_{t_0}^{t_f} dt |O(t)|^2 \right)^{1/2}. \tag{4.21}$$

Likewise, $\varepsilon(t)$ can be written explicitly, using Eq. (4.19) and Eq. (4.21), as

$$\varepsilon(t) = O(t)\left(\frac{1}{I}\int_{t_0}^{t_f} dt|O(t)|^2\right)^{-1/2}. \tag{4.22}$$

The above set of equations can be solved by an interative procedure that starts by guessing some $\varepsilon(t)$ function. One then determines $\underline{\underline{H}}(t)$ using Eq. (4.14), and $\boldsymbol{\Psi}(t)$ by propagating Eq. (4.13), using Eq. (4.12), from t_0 to t_f. The final value of $\boldsymbol{\chi}$, which by Eq. (4.18) is just $\mathsf{P}\cdot|\boldsymbol{\Psi}(t_f)\rangle$, is used to obtain $\boldsymbol{\chi}(t)$ from Eq. (4.17) for all $t < t_f$. An improved guess for $\varepsilon(t)$ is obtained from Eqs. (4.20) and (4.22), and the whole procedure is repeated until convergence.

An example of the success of this procedure is given in Figure 4.6 where the probability density for the model studied in Figure 4.2 is shown for a field optimized to deposit the system in channel A. We see that at the time shown [1950 a.u. ($=47.2$ fs)] essentially all the probability density in the ground state (whose absolute norm is still quite small $= 0.0046$) is deposited in channel A, while none exists in channel B. This result should be contrasted with Figure 4.2 where no pulse shaping was carried out and a nonoptimal delay between the pump and dump pulse was applied.

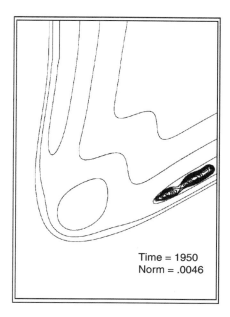

Time = 1950
Norm = .0046

Figure 4.6 Magnitude of the wave function for the same model collinear problem as in Fig. 4.2, but for field optimized to deposit excited state wave packet in channel A. (From Fig. 3, Ref. [104].)

This procedure was applied to a number of cases by Kosloff et al. [104] who studied a model for the theoretical

$$H + HD \leftarrow HHD \rightarrow H_2 + D$$

dissociation reaction, Amstrup et al. [120] who studied the control of the HgAr and I_2 photodissociation, Hartke et al. [115] who simulated the dynamics of control of the Br*/Br branching ratio in the photodissociation of Br_2, and Jin et al. [123] who studied the control of population inversion between two displaced harmonic oscillators. In all cases it was possible to show theoretically appreciable increase in the yield of the desired product by optimizing the pulse shape. Although the increase in yield relative to the zero-order guess is dramatic, the final selectivity is often similar to that obtained in the linear regime by the coherent control (CC) procedure. This has been shown explicitly in the HHD case for the Tannor–Rice potential [96].

Both CC and OCT attempt to extremize the same transition probabilities. At a first glance the two approaches appear to be different. CC is based on using multiple interfering pathways, and relies heavily on the (linear) superposition principle of quantum mechanics, whereas OCT is a (nonlinear) theory in which one attempts to perform a completely general optimization of some external inputs in order to reach the desired objectives. However, in similar problems the control fields calculated by the two methods are often quite similar. This can be traced back to the finding, discussed in Section 3.2, that the presence of interfering pathways is essential to quantum control, whether included explicitly in the algorithm (as in CC) or not (as in OCT). In the absence of interference the degree of control attainable is greatly reduced.

General global optimization is a difficult procedure [104], involving search in a complex function space (see, however, Ref. [123] for a description of an improved search routine using conjugate-gradient methods). The search is difficult because each step necessitates the solution of the time-dependent Schrödinger equation, which although done very efficiently [133–135], is still time consuming. An additional complication arises because of uncertainties in the parameters of the system to be optimized, the most obvious such parameter being the Hamiltonian itself (interaction potentials are only rarely known to sufficient accuracy).

There are now direct experimental confirmations [41–43], to be discussed in Section 13.3, of the effect of the pulse shaping. In addition, the role of the phase between two pulses, predicted in both OCT [104, 119] and in CC studies [94, 96, 136], has been confirmed experimentally by Fleming et al. [137, 138], Girard et al. [139–142], Kinrot et al. [143], and Warmuth et al. [144] in the so-called wave packet interferometry experiments [145]. For example, Fleming et al. [137, 138] and Warmuth et al. [144] describe experiments where the fluorescence from I_2 in the B state is influenced by constructive or destructive interference between two wave packets induced by two-phase related excitation pulses. This study relates to a large volume of work on wave packet interferometry in atoms [146–148], as well as to various femtochemistry experiments, where similar effects were seen in absorption [106].

Within OCT the very existence of a unique solution is not guaranteed. This point has been investigated in the context of square integrable functions by Rabitz's group

[111] and for unbounded systems [121, 122] where the existence of an optimal solution under certain conditions was established. Yao et al. [118] have shown that nonlinearities may give rise to multiple solutions, each producing exactly the same physical effect on the molecule. This result can be viewed from two alternate perspectives. From a positive perspective, this implies that the experimenter can obtain the optimal result using equipment that may be readily available in his laboratory. From a negative perspective, this result makes it very difficult to *understand* the physics behind optimization since many different electric fields give essentially the same final result.

Theoretical strategies for designing optimal pulses that are least sensitive to errors in the molecular Hamiltonians have also been proposed [117]. Basically one confines the search to pulses that prohibit the wave packet from making excursions into regions of uncertainties [117]. Such pulses may not always yield the best results as far as objectives. In fact, studies [104, 119] show great sensitivity of the outcome to even slight changes in pulse shapes. A "phase-space" representation in frequency and time of optimal pulses shows surprisingly small difference between pulses designed to select the AA + B channel vs. the AB + A channel. This means that one has to tailor these pulses very carefully in order to achieve the desired product, a task now achievable due to the progress made in pulse-shaping techniques [149–155]. This problem can be made easier by smoothing the pulses. It is often [120] possible to do almost as well as in the full optimization by restricting the last stages of the iteration to a search in a space spanned by two Gaussian pulses.

These problems can be solved experimentally in the context of optimal control by using experimental data directly, as in the automated feedback control approach to OC [40], to be discussed in detail in Section 13.3. Essentially, in this approach one obtains the optimal laser pulses required to achieve a given task by linking a computer to a pulse-shaping device [149, 150]. By letting the computer vary the laser pulse parameters and performing successive comparisons of the measured outcomes with the preset target, one can get closer and closer (often using "Genetic" search algorithms [156]) to the objective of interest. The strength (and weakness) of this approach is that in adopting it one treats the system as a "black box," avoiding altogether the theoretical treatment of the dynamics.

It is worth mentioning that a different approach to the control problem in finite N-state space was introduced by Harel and Akulin [157]. Rather than search for the optimal $\varepsilon(t)$ field yielding the desired time evolution operator $U(t)$ associated with an optimized $N \times N$ Hamiltonian matrix $\underline{\underline{H}}(t) = \underline{\underline{H}}_0 - \underline{\underline{d}}\varepsilon(t)$, where $\underline{\underline{d}}$ is the $N \times N$ transition-dipole matrix, these authors have shown that it is possible to control the system using just two *fixed-amplitude* instantaneous pulses $\mathcal{E}_A \delta(t - t_n)$ and $\mathcal{E}_B \delta(t - t_{n'})$ applied alternately at N^2 optimally chosen time points t_n and $t_{n'}$. The end result is the representation of the evolution operator at time t as

$$U(t) = e^{-iBt_{N^2}} e^{-iAt_{N^2-1}} \cdots e^{-iBt_2} e^{-iAt_1}, \qquad (4.23)$$

where $A \equiv \underline{\underline{H}}_0 - \underline{\underline{d}}\mathcal{E}_A$ and $B \equiv \underline{\underline{H}}_0 - \underline{\underline{d}}\mathcal{E}_B$. The evolution operator is controlled by the choice of the t_n time points, constrained such that $t = \sum_{n=1}^{N^2} t_n$.

CHAPTER 5

DECOHERENCE AND LOSS OF CONTROL

Thus far we have dealt with the idealized case of isolated molecules that are neither subject to external collisions nor display spontaneous emission. Further, we have assumed that the molecule is initially in a pure state (i.e., described by a wave function) and that the externally imposed electric field is coherent, that is, that the field is described by a well-defined function of time [e.g., Eq. (1.35)]. Under these circumstances the molecule is in a pure state before and after laser excitation and remains so throughout its evolution. However, if the molecule is initially in a mixed state (e.g., due to prior collisional relaxation), or if the incident radiation field is not fully coherent (e.g., due to random fluctuations of the laser phase or of the laser amplitude), or if collisions cause the loss of quantum phase after excitation, then phase information is degraded, interference phenomena are muted, and laser control is jeopardized.

5.1 DECOHERENCE

Loss of quantum information (either of the phase or of the amplitude of a state) due to the interaction of a system with its environment is termed *decoherence*. Examples include the obvious case where a system is actually embedded in an external environment, for example, a molecule in solution, or more subtle cases, for example, where the system is chosen as the center of mass of a body and the environment is the 10^{23} variables associated with the motion of the atoms that comprise the system.

The current view is that certain forms of decoherence can cause the loss of quantum interference in just such a way that the system then obeys classical mechanics [158]. This view does not obviate the possibility that classical mechanics is, in fact, the limit of quantum mechanics when $\hbar \to 0$ (i.e., when the system action becomes very large) [159]. Rather, it proposes an alternate route to classical mechanics for systems in interaction with their environment. Clearly, decoherence effects that change the dynamics from quantum to classical mechanics will destroy quantum phases and hence destroy coherent control. Indeed, most decoherence effects work toward the loss of quantum phase and have deleterious effects on control.

Consider then a system s interacting with an environment. The total Hamiltonian H_{tot} is of the form

$$H_{\text{tot}} = H_s + H_{\text{env}} + H_{\text{int}} \tag{5.1}$$

where H_s is the system Hamiltonian, H_{env} is the Hamiltonian of the environment, and H_{int} is the interaction between them. Ultimately, we intend to focus solely on the properties of the system, that is, the quantities of interest. To do so we have to deal with mixed states and hence we have to invoke the density matrix formulation of quantum mechanics [160]. That is, the system plus environment is described by a density operator $\rho_{\text{tot}}(t)$ whose dynamics is given by the quantum Liouville–von Neumann equation:

$$i\hbar \frac{\partial \rho_{\text{tot}}(t)}{\partial t} = [H_{\text{tot}}, \rho_{\text{tot}}(t)]. \tag{5.2}$$

Equation (5.2) reduces to the Schrödinger equation for the case of a pure state, that is, where $\rho_{\text{tot}}(t) = |\psi(t)\rangle\langle\psi(t)|$. To obtain system properties from known values of $\rho_{\text{tot}}(t)$, say the average value $\bar{A}$ of the operator $A(t)$, requires evaluating $A(t) = \text{Tr}[\rho_{\text{tot}}(t)A(t)]$. Operators that pertain solely to the system, denoted $A_s(t)$, are obtained by averaging only over environmental variables, that is, $A_s(t) = \text{Tr}_{\text{env}}[\rho_{\text{tot}}(t)A(t)]$.

Since it is the dynamics of the system that is of interest, it would be convenient to preaverage over the environment variables and obtain an equation of motion for $\rho_s(t)$, the system component of the density matrix. Formal work of this kind [161, 162] yields the so-called generalized master equation. Deriving the generalized master equation, and extracting the various approximations utilized, goes well astray of the central focus of this book. For this reason we just sketch the models and direct the reader to suitable review articles [161, 162] that provide an appropriate overview.

If the correlation time of the environment is much shorter than the typical time scale for the variation of the system, then the generalized master equation is of the form

$$\frac{\partial \rho_s(t)}{\partial t} = -i\hbar^{-1}[H_s, \rho_s(t)] + F(\rho_s), \tag{5.3}$$

where $F(\rho_s)$ is a functional of ρ_s. Its functional form depends upon the nature of the environment and on the coupling H_{int}, and the resultant equations are quite complex.

As such, a variety of approximations to the generalized master equations have yielded equations that are used to model the effect of the environment on the system dynamics.

5.1.1   Sample Computational Results on Decoherence

To demonstrate the effect of decoherence, we consider computational results [163] using a model, due to Caldeira–Leggett and heavily developed by Zurek [164], that is widely regarded as a paradigm for studies of decoherence. Here the system interacts with a bath, comprising harmonic oscillators in the weak-coupling and high-temperature limit. The harmonic bath thus serves as the source of the decoherence experienced by the system. As a concrete example, we consider the vibrational motion of a model molecule with two degrees of freedom coupled to an harmonic bath.

In doing so it proves convenient to carry out the computations in the Wigner representation. That is, as in quantum mechanics based upon the wave function, it is necessary to deal with a representation of the density operator ρ. The (convenient) Wigner representation ρ^W of ρ is defined, for an N degree of freedom system, by

$$\rho^W \equiv \rho^W(\mathbf{q}, \mathbf{p}) = (\pi\hbar)^N \int d\mathbf{v} \, e^{-2i\mathbf{p}\cdot\mathbf{v}/\hbar} \langle \mathbf{q} - \mathbf{v}|\rho|\mathbf{q} + \mathbf{v}\rangle, \tag{5.4}$$

and the Wigner representation of any operator A, denoted by the superscript W, is given by

$$\mathbf{A}^W = 2^N \int e^{-2i\mathbf{p}\cdot\mathbf{v}/\hbar} \langle \mathbf{q} - \mathbf{v}|A|\mathbf{q} + \mathbf{v}\rangle. \tag{5.5}$$

This Wigner representation of the density $\rho^W(\mathbf{q}, \mathbf{p})$ proves particularly useful since it satisfies a number of properties that are similar to the classical phase-space distribution $\rho_{cl}(\mathbf{q}, \mathbf{p})$. For example, if $\rho = |\psi\rangle\langle\psi|$, that is, if ρ represents a pure state, then $\int d\mathbf{p} \, \rho^W = |\psi(\mathbf{q})|^2$, that is, the integral over $\mathbf{p}$ gives the probability density in coordinate space. Similarly, integrating ρ^W over $\mathbf{q}$ gives the probability density in momentum space. These features are shared by the classical density $\rho_{cl}(\mathbf{p}, \mathbf{q})$ in phase space. Note, however, that ρ^W is not a probability density, as evidenced by the fact that it can be negative, a reflection of quantum features of the dynamics [165].

The Caldeira–Leggett and Zurek models for a two-degree-of-freedom oscillator system with Hamiltonian H_s in contact with a bath is given by [163]

$$\frac{\partial \rho_s^W(t)}{\partial t} = -i\hbar^{-1}[H_s, \rho_s(t)]^W + F_{CL}^W(\rho_s)$$

$$\equiv \left[\{H_s, \rho_s^W\} + \sum_{(l_1+l_2)\text{odd}} \frac{(\hbar/2i)^{(l_1+l_2-1)}}{l_1! l_2!} \frac{\partial^{(l_1+l_2)} V(x, y)}{\partial x^{l_1} \partial y^{l_2}} \frac{\partial^{(l_1+l_2)} \rho_s^W}{\partial p_x^{l_1} \partial p_y^{l_2}} \right] \tag{5.6}$$

$$+ F_{CL}^W(\rho_s).$$

Here

$$F_{CL}^W(\rho_s) = D\left(\frac{\partial^2 \rho_s^W}{\partial p_x^2} + \frac{\partial^2 \rho_s^W}{\partial p_y^2}\right)$$

denotes the Caldeira–Leggett form of $F(\rho_s)$ in the Wigner representation. The term in square brackets in Eq. (5.6) is the Wigner representation $[H_s, \rho_s(t)]^W$ of $[H_s, \rho_s(t)]$. Here (p_x, p_y, x, y) are the system momenta and coordinates, $V(x, y)$ is the potential contribution to the Hamiltonian H, and $\rho_s^W = \rho_s^W(p_x, p_y, x, y; t)$. The first term

$$\{H_s, \rho_s^W\} = \frac{\partial H_s}{\partial x}\frac{\partial \rho_s^W}{\partial p_x} - \frac{\partial H_s}{\partial p_x}\frac{\partial \rho_s^W}{\partial x} + \frac{\partial H_s}{\partial y}\frac{\partial \rho_s^W}{\partial p_y} - \frac{\partial H_s}{\partial p_y}\frac{\partial \rho_s^W}{\partial y}$$

on the right-hand side of Eq. (5.6) is the classical Poisson bracket that generates classical dynamics, the second term is responsible for the difference between quantum and classical mechanics, and the third term induces decoherence.

Numerical calculations on Eq. (5.6) can be compared to classical mechanics by computing the classical dynamics of the phase-space density $\rho_{cl}(x, y, p_x, p_y)$, which is obtained as the solution to the Fokker–Planck equation:

$$\frac{\partial}{\partial t}\rho_{cl}(x, y, p_x, p_y) = \{H_s, \rho_{cl}(x, y, p_x, p_y)\}$$
$$+ D\left[\frac{\partial^2}{\partial p_x^2}\rho_{cl}(x, y, p_x, p_y) + \frac{\partial^2}{\partial p_y^2}\rho_{cl}(x, y, p_x, p_y)\right]. \tag{5.7}$$

Consider, for example, the specific case of the nonlinear oscillator Hamiltonian [166]:

$$H_s = \frac{1}{2}(p_x^2 + p_y^2 + \alpha x^2 y^2) + \frac{\beta}{4}(x^4 + y^4), \tag{5.8}$$

with parameters that can be related to typical molecules: $\beta = 0.01$, $\alpha = 1.0$.

The extent to which quantum effects are diminished in the presence of decoherence is demonstrated in the figures that follow. In particular, Figure 5.1 shows the classical and quantum expectation values, in the absence of decoherence [i.e., $D = 0$ in Eqs. (5.6) and (5.7)] for four moments associated with the y degree of freedom. All figures show qualitatively similar behavior. That is, after an initial period of classical/quantum agreement, the quantum results continue to oscillate while the classical results show smooth relaxation [167]. Note in particular that the quantum results do not always simply oscillate about the classical (e.g., see results for $\langle y^2 \rangle$) and that the quantum fluctuations about the mean are substantial (e.g, 30% in the case of $\langle E_y \rangle$).

Results for the same moments, after introducing decoherence, are shown in Figure 5.2. A comparison of Figures 5.1 and 5.2 shows substantially improved classical–quantum correspondence upon introducing decoherence. Remarkably,

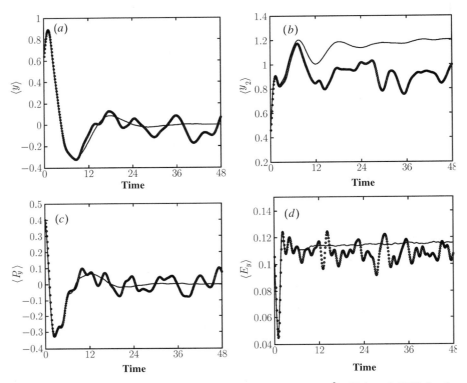

Figure 5.1   Time dependence of four statistical moments ($\langle y \rangle$, $\langle y^2 \rangle$, $\langle P_y \rangle$, and $\langle E_y \rangle$) for the system in the absence of decoherence. Dark dots denote quantum results; thin solid lines are classical results. (From Fig. 1, Ref. [163].)

this is true even for $\langle y^2 \rangle$, where the long-term quantum average in the closed system deviated significantly from the long-term classical average. Qualitatively similar results have been obtained for reactive scattering [168].

These computational results demonstrate the way in which decoherence tends to eliminate quantum effects in the system. As a consequence, quantum control processes must be effectively shielded from decoherence effects in order to survive.

Below we consider a number of approaches to combating decoherence in solutions and other media where collisions are present. An alternative approach, which we do not address, is the method of "decoherence free subspaces" [169]. In this approach one deals with the explicit *design* of systems where a particular subspace is free from decoherence effects. These approaches are of particular interest to the development of subspaces in which to carry out quantum computation, an approach in which the computational machinery follows the laws of quantum mechanics [170]. By contrast, we deal below with the need to curb decoherence effects in traditional preexistent systems.

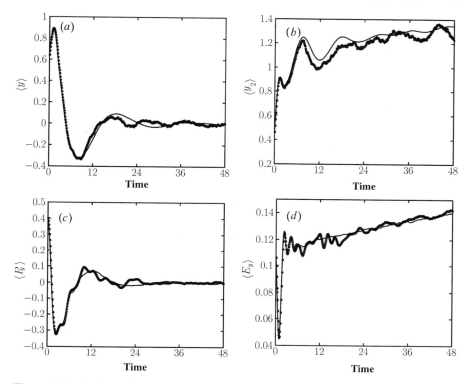

Figure 5.2 As in previous figure but in the presence of decoherence. (From Fig. 2, Ref. [163].)

5.2 COHERENT CONTROL IN EQUILIBRATED CONDENSED PHASES

5.2.1 Optical Bloch Equation

We now consider coherent control as it would apply to liquid-phase chemistry. Here, molecules of species B in solution would be subjected to laser irradiation. Since we are ultimately interested in the fate of the B molecules, B is the system, and the remaining molecules in the solution and the laser are the environment. Decoherence effects can then arise from the collisions of the solvent with the molecule of interest, or from incoherence properties of the laser that cause some loss of quantum phase information. Both of these effects are discussed below, along with proposed methods to reduce the effects of collisional or laser-induced decoherence effects.

The strong role of collisions in decohering a system is readily seen by considering the density matrix ρ_s^T of a system that has reached thermal equilibrium at temperature T through collisional relaxation, that is, $\rho_s^T = Q \exp(-H_s/k_B T)$. Here H_s is the system Hamiltonian, Q is a normalization factor, and k_B is the Boltzmann constant. Considerable insight is obtained if we cast the density matrix in the *energy* representation. Then $\rho_{i,j}^T = \langle E_i | \rho_s^T | E_j \rangle = Q \exp(-E_i/k_B T)\delta_{i,j}$, where $|E_i\rangle$ are energy eigenstates of H_s. That is, the system shows population in each of the energy levels, but no off-diagonal terms. Nonzero off-diagonal terms would represent quantum coherence between the energy eigenstates of the Hamiltonian. To see this, contrast the equilibrium result ρ_s^T with that of the density operator associated with a pure state:

$$|\psi\rangle = \sum_k a_k |E_k\rangle. \tag{5.9}$$

This density operator would be given by

$$\rho_s = |\psi\rangle\langle\psi| = \sum_{k,m} a_k a_m^* |E_k\rangle\langle E_m|, \tag{5.10}$$

which can be confirmed by substituting in Eq. (5.2), but with H and ρ_{tot} replaced by H_s and ρ_s, respectively. In this case,

$$\rho_{m,k} = a_k a_m^*, \tag{5.11}$$

where here and below $\rho_{m,k}$ denotes the m, k element of the density matrix representing ρ_s.

Thus, the fact that there is a well-defined phase relationship between the eigenstates of the Hamiltonian, contained in the wave function, is manifest in the existence of off-diagonal elements in the energy representation. The absence of off-diagonal matrix elements for the thermally equilibrated case makes clear that collisions have destroyed matter coherence manifest as quantum correlations between energy eigenstates.

Recent experimental studies on interference effects in solution, and on collisional vibrational energy transfer between molecules in solution, provide some insight into the molecular time scales of these relaxation events. For example [171], the time scale for transfer of population to the vibrational modes in liquid CH_3OH is on the order of 5 to 15 ps [172]. Further, studies of the preparation of coherent superpositions of states in solution show that phase coherences of molecules exist in solution for time scales greater than 100 fs [173, 174].

Given the significance of collisional effects in solution, we introduce the simplest of models for relaxation and concomitant decoherence. That is, the equation of motion of $\rho_s(t)$ in the energy representation is given the form:

$$\frac{\partial \rho_{i,j}(t)}{\partial t} = -i\hbar^{-1}[H_s, \rho_s(t)]_{i,j} - \frac{1}{T_{i,j}}\rho_{i,j}(t), \tag{5.12}$$

with $T_{i,i} = T_1$ and $T_{i,j} = T_2$ for $i \neq j$. Here T_1 and T_2 are phenomenological relaxation times. In this model, where $F(\rho)_{i,j} = -\rho_{i,j}(t)/T_{i,j}$, the coherent terms $\rho_{i,j}(t)$, $i \neq j$ decay with a rate $1/T_2$ and populations $\rho_{i,i}(t)$ decay with rate $1/T_1$.

If the system is in the presence of a radiation field, then H_s in Eq. (5.12) is augmented by the dipole-electric field interaction H_{MR} [Eq. (2.10)]. The result is the so-called optical Bloch equations. Note that this approach focuses explicitly on decoherence in the energy representation.

The simplest optical Bloch equations result from a system comprised of two eigenstates $|E_1\rangle$, $|E_2\rangle$ of the molecule Hamiltonian H_M that experience the electric field–dipole interaction

$$H_{MR} = -\mathcal{E}\hat{\varepsilon} \cdot \mathbf{d}\cos(\omega t + \phi). \qquad (5.13)$$

The Hamiltonian H_s in Eq. (5.12) is $H_s = H_M + H_{MR}$. Equation (5.12) then becomes [where we suppress the t dependence of $\rho(t)$]:

$$\frac{\partial \rho_{i,j}}{\partial t} = -i\hbar^{-1}\sum_k [H_{i,k}\rho_{k,j} - \rho_{i,k}H_{k,j}] - \frac{1}{T_{i,j}}\rho_{i,j}, \qquad (5.14)$$

where

$$H_{i,k} = E_i\delta_{i,k} - \mathcal{E}\cos(\omega t + \phi)\mathrm{d}_{i,k}(1 - \delta_{i,k}), \qquad (5.15)$$

and $\mathrm{d}_{i,k} = \langle E_i|\hat{\varepsilon} \cdot \mathbf{d}|E_k\rangle$. Noting that $\rho_{i,j} = \rho^*_{j,i}$, we define

$$
\begin{aligned}
R_1 &= 2\,\mathrm{Im}(\rho_{1,2}) = \mathrm{Im}(\rho_{1,2} - \rho_{2,1}), \\
R_2 &= 2\,\mathrm{Re}(\rho_{1,2}) = \mathrm{Re}(\rho_{1,2} + \rho_{2,1}), \\
R_3 &= \rho_{11} - \rho_{22}.
\end{aligned} \qquad (5.16)
$$

Then, with $H_{1,2} = H_{2,1}$ for bound states subjected to Eq. (5.13), Eq. (5.14) becomes

$$
\begin{aligned}
\frac{dR_1}{dt} &= \Delta R_2 - \frac{1}{T_2}R_1 + \frac{2H_{1,2}R_3}{\hbar}, \\
\frac{dR_2}{dt} &= -\Delta R_1 - \frac{1}{T_2}R_2, \\
\frac{dR_3}{dt} &= -\frac{1}{T_1}R_3 - \frac{2H_{1,2}R_1}{\hbar}.
\end{aligned} \qquad (5.17)
$$

Here $\Delta \equiv (E_2 - E_1)/\hbar - \omega \equiv \omega_{2,1} - \omega$, that is, the detuning of ω from the $|E_1\rangle$ to $|E_2\rangle$ transition.

Equation (5.17) constitutes the standard form of the two-level optical Bloch equation.

5.2.2 Countering Collisional Effects

Consider now a scenario that is capable of maintaining coherent control in the presence of collisions for systems that are described by the optical Bloch equations.

In particular, we reconsider the bichromatic control scenario discussed in Section 3.1.1, assuming, however, that the molecules are in solution at temperature T. Further, we irradiate the system so as to saturate the $|E_1\rangle$ to $|E_2\rangle$ transition and simultaneously photodissociate the system.

The initial state, prior to dissociation, is a mixed $|E_1\rangle$, $|E_2\rangle$ state described, in the energy representation, by a 2×2 density matrix with elements $\rho_{i,j}$, $(i,j = 1, 2)$. Photodissociation of this mixed state can be written as a generalization [175] of Eq. (3.12). In Eq. (3.12) we assumed an initial state of the form of Eq. (5.9) (with $k = 1, 2$), so that the corresponding density matrix would be Eq. (5.11). Hence, Eq. (3.12) could be rewritten as

$$P_q(E) = \left(\frac{2\pi}{\hbar}\right)^2 \sum_{i,j=1}^{N} [\rho_{i,j}\bar{\epsilon}(\omega_{E,i})\bar{\epsilon}^*(\omega_{E,j})]\mathrm{d}_q(ji). \tag{5.18}$$

Equation (5.18) is, in fact, the correct generalization to the case where the initial state is mixed and is represented by $\rho_{i,j}$. Below we neglect the z dependence in $\bar{\epsilon}$, replacing it by ϵ [see Eq. (2.9)].

To utilize Eq. (5.18) for the case of interest we determine $\rho_{i,j}$ for the case where two levels $|E_1\rangle$ and $|E_2\rangle$ are continuously subjected to radiation and to collisions, using the optical Bloch approach. We note that if, as in the bichromatic control cases discussed in Chapter 3, $|E_1\rangle$ and $|E_2\rangle$ have the same parity, a one-photon absorption cannot couple these states. We must therefore consider saturating this transition using two-photon absorption through an off-resonant intermediate bound state $|E_0\rangle$ with dipole matrix elements $\mathrm{d}_{0j} = \langle E_0|\mathbf{d} \cdot \hat{\epsilon}|E_j\rangle$, and with $E_2 > E_0 > E_1$. For simplicity we assume that $2\omega = (E_2 - E_1)/\hbar$, so that the transition is two-photon resonant, and we sketch the extension [176] of Eq. (5.17) to two-photon absorption. To carry out this extension, we write the Bloch equations [Eq. (5.14)] for the three level $|E_0\rangle$, $|E_1\rangle$, and $|E_2\rangle$ and adopt the adiabatic approximation for off-resonant transitions to level $|E_0\rangle$. This approximation is equivalent [176] to setting $d\rho_{2,0}/dt = d\rho_{10}/dt = 0$. Substituting the result into the remaining equations gives the set of equations for $d\rho_{i,j}/dt$, with $i,j = 1, 2$. This set can be rewritten, with the help of a modified version of Eq. (5.16), where $\rho_{1,2}$ is replaced by $\rho_{1,2}\exp(-2i\phi)$, as

$$\frac{dR_1}{dt} = -\frac{D_{2,1}R_3}{2} - \frac{R_1}{T_2},$$

$$\frac{dR_2}{dt} = \frac{-R_2}{T_2}, \tag{5.19}$$

$$\frac{dR_3}{dt} = \frac{D_{2,1}R_1}{2} - \frac{R_3 - R_3^e}{T_1},$$

where

$$D_{2,1} = \mathcal{E}^2 \mathrm{d}_{2,0}\mathrm{d}_{0,1}/[2\hbar^2(\omega - \omega_{0,1})]. \tag{5.20}$$

Here we have recognized that the quantity R_3 relaxes to the thermodynamic population difference R_3^e at temperature T, with:

$$R_3^e = [1 - \exp(-\hbar\omega_{2,1}/k_BT)]/[1 + \exp(-\hbar\omega_{2,1}/k_BT)] \tag{5.21}$$

At long times $(t \gg T_2, T_1)$ the system saturates, that is, $dR_3/dt = 0$, and Eq. (5.19) implies that $dR_1/dt = 0$. The equations for R_i can then be readily solved and, in conjunction with Eq. (5.16), gives $\rho_{1,2}$ at saturation:

$$\begin{aligned}
\rho_{1,2} &= R_3^e T_2 D_{2,1} \exp[i(2\phi - \pi/2)]/(4 + D_{2,1}^2 T_1 T_2), \\
\rho_{1,1} &= 0.5[1 + R_3^e/(1 + D_{2,1}^2 T_1 T_2/4)], \\
\rho_{2,2} &= 0.5[1 - R_3^e/(1 + D_{2,1}^2 T_1 T_2/4)].
\end{aligned} \tag{5.22}$$

Consider then excitation of this mixed state with a Gaussian pulse, within the rotating wave approximation. The pulse is of the form

$$\varepsilon(t) = \mathcal{E}e^{-i(\omega_L t + \delta)} e^{-(t-t_0)^2/\tau^2} \tag{5.23}$$

with Fourier transform

$$\begin{aligned}
\epsilon(\omega) &= (\mathcal{E}\tau/\sqrt{\pi})e^{-\tau^2(\omega_L-\omega)^2/4} e^{-i(\omega_L-\omega)t_0} e^{-i\delta} \\
&\equiv \epsilon_\omega e^{-i(\omega_L-\omega)t_0} e^{-i\delta}.
\end{aligned} \tag{5.24}$$

Note that this control arrangement differs from that in Section 3.1.1 insofar as the two frequencies $\omega_1 = (E - E_1)/\hbar$ and $\omega_2 = (E - E_2)/\hbar$ that dissociate the system are components of a single pulse. The temporal width of the pulse is such that $\tau \gg T_1, T_2$.

Inserting the long-time $\rho_{i,j}$ into Eq. (5.18) gives (denoting $\epsilon_{\omega_{qi}}$ by ϵ_i) the probability $P_q(E)$ of forming product in channel q at energy E as

$$P_q(E) = (\pi^2/\hbar^2)[P_{1,1}(E, q) + P_{2,2}(E, q) + P_{1,2}(E, q)], \tag{5.25}$$

where

$$\begin{aligned}
P_{1,1}(E, q) &= \rho_{1,1}\epsilon_1^2 d_q(11) = 0.5[1 + R_3^e/(1 + D_{2,1}^2 T_1 T_2/4)]\epsilon_1^2 d_q(11), \\
P_{2,2}(E, q) &= \rho_{2,2}\epsilon_2^2 d_q(22) = 0.5[1 - R_3^e/(1 + D_{2,1}^2 T_1 T_2/4)]\epsilon_2^2 d_q(22)
\end{aligned}$$

and

$$\begin{aligned}
P_{1,2}(E, q) &= 2|\rho_{1,2}||d_q(12)|\epsilon_1\epsilon_2 \cos[\alpha_q(12) + \omega_{2,1}t_0 + 2\phi - \pi/2] \\
&= 0.5 T_2 D_{2,1} R_3^e/(1 + D_{2,1}^2 T_1 T_2/4)\epsilon_1\epsilon_2|d_q(12)| \cos[\alpha_q(12) \\
&\quad + \omega_{2,1}t_0 + 2\phi - \pi/2],
\end{aligned} \tag{5.26}$$

where $\alpha_q(12)$ is the phase of $d_q(12)$ [see Eq. (3.20)]. From these equations it is evident that coherent control can be achieved in solution by, for example, varying pulse parameters to alter the quantum interference term.

A number of simple qualitative observations are evident. First, $P_q(E)$ depends upon the parameters associated with saturation through the combinations $D_{2,1}T_1$ and $D_{2,1}T_2$ (or their ratio and product T_1/T_2 and $D_{2,1}^2 T_1 T_2$ used below). Control vanishes if $P_{1,2}(E,q) = 0$, which occurs if either the temperature $T \to \infty$ (i.e., $R_3^e \to 0$) or $D_{2,1}T_2 \to 0$. Both these limits correspond to complete loss of coherence. Examination of Eq. (5.26) shows that this is not the case, however, for $D_{2,1}T_1 \to 0$, consistent with the fact that T_1 relates to population, rather than phase, relaxation. Physically [177], however, in collisional environment, $T_1 \gg T_2$ so that the limit $T_1 \to 0$ also implies loss of control. Note also that control vanishes under extremely large pumping rates, $D_{2,1} \to \infty$ for which $\rho_{1,1} = \rho_{2,2}$ and $\rho_{1,2} \to 0$.

Sample computational results are shown in Figures 5.3 to 5.6 for the case of the photodissociation of CH_3I into $CH_3 + I$ vs. $CH_3 + I^*$. In particular, we show control over the ratio $I^*/(I + I^*)$ for the collision-free case in Figure 5.3, and at three successively higher temperatures in the subsequent three figures. The abscissa is $S = \epsilon_1^2/(\epsilon_1^2 + \epsilon_2^2)$ and the ordinate is the angle $\chi_{1,2} = \omega_{2,1}t_0 + 2\phi - \pi/2$. The results clearly show control for the two lower temperatures, including $T = \hbar\omega_{2,1}/k$, and the

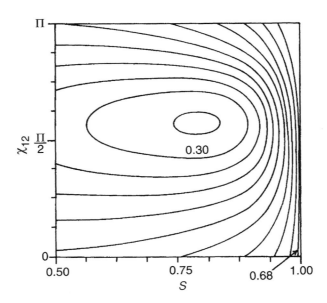

Figure 5.3   Contour plot of I^* yield $[I^*/(I + I^*)]$ for two color photodissociation of a pure CH_3I superposition state composed of bound states with vibrational and rotational quantum numbers $(v, J) = (0, 2)$ and $(1, 2)$ excited with frequencies $\omega_1 = 41,579 \text{ cm}^{-1}$ and $\omega_2 = 41,163 \text{ cm}^{-1}$. Contours increase, in increments of 0.04 from the "center well". (Taken from Fig. 1, Ref. [175].)

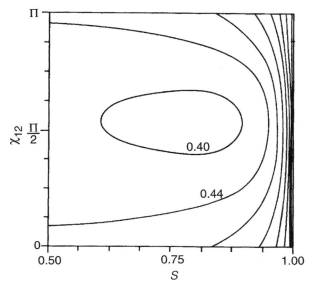

Figure 5.4   As in Figure 5.3 but at $k_B T = 0.2\hbar\omega_{2,1}$. (Taken from Fig. 2, Ref. [175].)

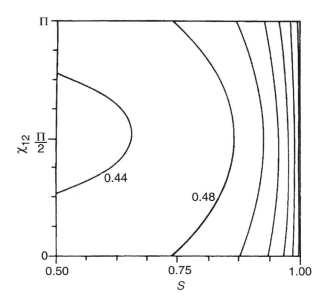

Figure 5.5   As in Figure 5.3 but at $k_B T = \hbar\omega_{2,1}$. (Taken from Fig. 3, Ref. [175].)

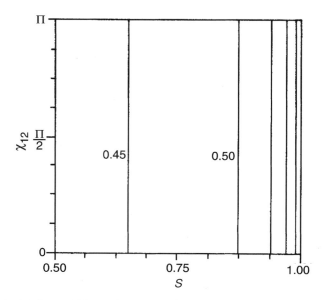

Figure 5.6   As in Figure 5.3 but at $k_B T = 1000 \hbar \omega_{2,1}$. Contours are spaced by 0.05. (Taken from Fig. 4, Ref. [175].)

total loss of control at the highest temperature, as manifest in the lack of dependence of the yield ratio on the phase angle $\chi_{1,2}$.

Thus we see that, although collisional effects do reduce the degree of control relative to the collision-free case, saturation pumping of superposition in the bichromatic control scenario can be used to overcome collisional effects up to some reasonable temperature.

A number of additional theoretical studies, several using optimal or pulse control (see Section 4.1), have been carried out to study control in the presence of solvent and decoherence effects. These include work in the Wilson group on a two-level oscillator model coupled to a background bath [178] and on electronic population transfer in a molecule in solution [179]. This, and more recent related work [171, 180] have generally concluded that some degree of control is indeed possible in solution, depending upon the extent of the coupling between system and solvent, and the degree to which one can manipulate the incident pulses. No quantitative rules have yet emerged on the extent to which control is, in fact, possible.

In addition to model systems, there exists [181] one fully converged computation on bichromatic control in the presence of decoherence. This computation deals with controlled proton transfer between the keto and enol forms of 2-(2′-hydroxyphenyl)-oxazole (see Fig. 5.7). Such computations rely heavily on recent advances in semi-classical mechanics since a full quantum computation is impossible [182, 183]. Here the proton is "the system" and the remaining molecule, comprising 35 coupled degrees of freedom plus 16 out-of-plane vibrational modes, serves as the environment. The results show that despite extensive dephasing, the proton transfer

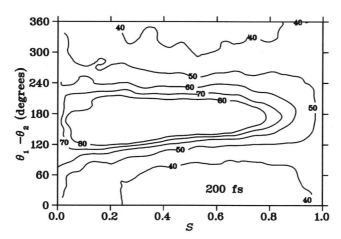

Figure 5.7 Keto and enol forms of 2-(2′-hydroxyphenyl)-oxazole. (From Fig. 1, Ref. [181].)

dynamics is easily controlled using the bichromatic control scenario. For example, consider the case where the initial superposition state involves the ground vibrational state of the oxazole-hydroxyphenyl in-the-plane bending mode, that is, bending motion of the $C_1C_2C_7$ angle, and the first excited state associated with such vibrational mode. Figure 5.8 shows a contour plot of the percentage yield of the reactant at 200 fs after excitation of the system. Here, the degree of yield control is maximum in the $0.2 < S < 0.8$ range, where the amount of the reactant can be reduced from more than 80% to less than 40% by changing the relative phase of the two incident lasers from $120°-180°$ to $0°$.

The extent to which these results are significant is associated with the advent of intrinsic decoherence experienced by the proton during the course of the dynamics. That is, if there is little decoherence, then the system is effectively a small molecule. To this end it is necessary to introduce a quantitative decoherence measure. One such

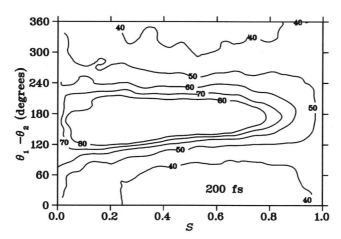

Figure 5.8 Contour plot of reactant for bichromatic coherent control at 200 fs after photo-excitation of 2-(2′-hydroxyphenyl)-oxazole. Here $\theta_i \equiv \phi(\omega_i)$, where $\phi(\omega_i)$ is the phase of the electric field, of frequency ω_i incident on the system [see Eq. (3.16)]. (From Fig. 3, Ref. [181].)

popular measure [184, 185] is $\mathrm{Tr}[\rho_s^2(t)]$ where $\rho_s(t)$ is the system density matrix. In this case, ρ_s would correspond to the density matrix of the proton. If the initial state is chosen to be a pure state, then $\rho_s^2(0) = \rho_s(0)$ and $\mathrm{Tr}[\rho_s^2(0)] = 1$. As the dynamics proceed, and decoherence sets in, $\mathrm{Tr}[\rho_s^2(t)]$ decays. A computation on $\mathrm{Tr}[\rho_s^2(t)]$ on this system shows decay to $\mathrm{Tr}[\rho_s^2(t)] = 0.38$ by 200 fs. Hence decoherence is rapid and effective during the time scale of the control, as shown above.

Finally we note that studies of control in solution [186, 187] indicate that control in the presence of collisional effects is indeed possible. For example, coherent control of the dynamics of I_3^- in ethanol and acetonitrile has been demonstrated. Specifically, I_3^- was excited with a 30-fs ultraviolet (UV) laser pulse to the first excited state. The resultant wave function was comprised of a localized wave function on the ground electronic state and a corresponding depletion of wave function density, that is, a "hole," on the ground electronic state. In this instance the target of the control was the nature of the spectrum associated with the coherences associated with the symmetric stretch. By manipulating various attributes of the exciting pulse (intensity, frequency, and chirp of the excitation pulse), aspects of the spectrum were controlled, despite the decoherence associated with collision effects.

5.3 COUNTERING PARTIALLY COHERENT LASER EFFECTS IN PUMP–DUMP CONTROL

An alternate source of decoherence is in the nature of the laser used to irradiate the system. Specifically, if the laser has random components, then it inputs a degree of randomness into the system, reducing the phase information content and hence decohering the system.

For example, consider Eq. (3.19) corresponding to control of photodissociation of an initial two-level superposition state that is excited to the continuum. If there are sufficiently strong external perturbations, or excessive jitter in the laser phase, then this results in a complete average over $\phi_1 - \phi_2$, and Eq. (3.19) becomes

$$R_{q,q'}(E) \equiv \frac{P_q(E)}{P_{q'}(E)} = \frac{|d_q(1,1)| + x^2 |d_q(2,2)|)}{|d_{q'}(1,1)| + x^2 |d_{q'}(2,2)|}. \tag{5.27}$$

Thus, there is no longer any dependence on any phase characteristics and all phase control would be lost.

Equations (3.19) and (5.27) represent two extremes, where the result is either fully coherent or fully incoherent. It is enlightening to consider [188] the effect of partially coherent laser sources through the example of the pump–dump scenario of Section 3.5 in the case where the laser is not fully coherent.

To characterize partially coherent pulses [189, 190] consider a Gaussian laser pulse with time profile $\varepsilon_x(t)$ and phase $\delta_x(t)$, that is,

$$\varepsilon_x(t) = (\mathcal{E}_{x0}/2)e^{-(t-t_x)^2/\tau_x^2}e^{-i[\omega_x t + \delta_x(t)]}. \tag{5.28}$$

We adopt a modified-phase diffusion model for the partially coherent laser source in which the phase δ_x is allowed to be time dependent and random. Thus, the molecule–laser interaction is modeled by the interaction with an ensemble of lasers, each of different phase. This ensemble is described by a Gaussian correlation for the stochastic phases with a decorrelation time scale τ_{xc}:

$$\langle e^{i\delta_x(t_2)} e^{-i\delta_x(t_1)}\rangle = e^{-(t_2-t_1)^2/2\tau_{xc}^2}. \tag{5.29}$$

Here the angle brackets denote an average over the ensemble of laser phases.

To examine photodissociation given this field requires, as shown below, the frequency–frequency correlation function $\langle \epsilon_x(\omega_2)\epsilon_x^*(\omega_1)\rangle$ where $\epsilon_x(\omega)$ is the Fourier transform of $\varepsilon_x(t)$ for $\delta_x(t)$ equal to a constant δ_x. Given Gaussian pulses [Eqs. (5.28) and (5.29)] we have [189]

$$\langle \epsilon_x(\omega_2)\epsilon_x^*(\omega_1)\rangle = \frac{\mathcal{E}_{x0}^2 \tau_x T_x}{8} e^{i(\omega_2-\omega_1)t_x} e^{-\tau_x^2(\omega_2-\omega_1)^2/8} e^{-T_x^2(\omega_2+\omega_1-2\omega_x)^2/8}, \tag{5.30}$$

where

$$T_x = \tau_x\tau_{xc}/(\tau_x^2 + \tau_{xc}^2)^{1/2}. \tag{5.31}$$

Since the pump–dump control scenario involves two lasers, we adopt a similar description for the dump pulse $\varepsilon_d(t)$, with appropriate change in parameter labels. That is,

$$\varepsilon_d(t) = (\mathcal{E}_{d0}/2)e^{-(t-t_d)^2/\tau_d^2}e^{-i(\omega_d t+\delta_d(t))}, \tag{5.32}$$

$$\langle e^{i\delta_d(t_2)} e^{-i\delta_d(t_1)}\rangle = e^{-(t_2-t_1)^2/2\tau_{dc}^2}, \tag{5.33}$$

$$\langle \epsilon_d(\omega_2)\epsilon_d^*(\omega_1)\rangle = \frac{\mathcal{E}_{d0}^2 \tau_d T_d}{8} e^{i(\omega_2-\omega_1)t_d} e^{-\tau_d^2(\omega_2-\omega_1)^2/8} e^{-T_d^2(\omega_2+\omega_1-2\omega_d)^2/8}, \tag{5.34}$$

where $T_d = \tau_d\tau_{dc}/(\tau_d^2 + \tau_{dc}^2)^{1/2}$. Here τ_{xc} and τ_{dc} define the degree of field coherence, the two limiting cases being a coherent source ($\tau_{ic} \to \infty$, $i = x, d$) and fully incoherent source ($\tau_{ic} \to 0$, $i = x, d$).

The frequency spectrum of the pulse provides a primary means of characterizing the laser–molecule interaction and is given, for the excitation pulse, by

$$I_x(\omega) = \frac{T_x}{\sqrt{2\pi}} e^{-T_x^2(\omega-\omega_x)^2/2}, \tag{5.35}$$

whose full-width at half-maximum (FWHM) is $2\sqrt{2\ln 2}/T_x$. For a fully coherent pulse $T_x = \tau_x$ giving a FWHM of the intensity spectrum equal to $\Delta_x/\sqrt{2}$, where $\Delta_x \equiv 4\sqrt{\ln 2}/\tau_x$ is the FWHM of the frequency profile of the pulse.

Similar definitions apply to characterize the dump pulse. Further, for consistency we define the partial coherence parameter $\Delta_{xc} = 4\sqrt{\ln 2}/\tau_{xc}$ and similar parameters

(Δ_d, Δ_{dc}) for the dump pulse. Note that fitting Eq. (5.35) to the measured pulse frequency spectrum provides a means of obtaining T_x and ω_x from experimental data.

Consider now the pump–dump scenario where, for generality, we assume a pump pulse whose spectral width is sufficiently large to encompass a large number of levels. As in Section 3.5, a molecule with Hamiltonian H_M is subjected to two temporally separated pulses $\varepsilon(t) = \varepsilon_x(t) + \varepsilon_d(t)$. The probability of forming channel q at energy E is now given by the extension of Eq. (3.77) to many level excitations by $\varepsilon_x(t)$, that is,

$$P_q(E) = \left(\frac{2\pi}{\hbar^2}\right) \sum_{\mathbf{m}} | \sum_j b_j \langle E, \mathbf{m}, q^- | \mathbf{d} \cdot \hat{\varepsilon} | E_j \rangle \epsilon_d(\omega_{E,E_j}) |^2, \tag{5.36}$$

where $\omega_{E,E_j} = (E - E_j)/\hbar$ and b_j is given by [Eq. (3.75)]

$$b_j = (2\pi i/\hbar)\langle E_j | \hat{\varepsilon} \cdot \mathbf{d} | E_1 \rangle \epsilon_x(\omega_{j,1}) \equiv c_j \epsilon_x(\omega_{j,1}). \tag{5.37}$$

Equation (5.37) also defines c_j as the field-independent component of b_j.

Expanding the square in Eq. (5.36) allows us to write the probability in a canonical form:

$$
\begin{aligned}
P_q(E) &= \left(\frac{2\pi}{\hbar^2}\right) \sum_{i,j} b_i b_j^* \epsilon_d(\omega_{E,E_i}) \epsilon_d^*(\omega_{E,E_j}) d_q(ij) \\
&\equiv \left(\frac{2\pi}{\hbar^2}\right) \sum_{i,j} c_i c_j^* \epsilon_x(\omega_{ig}) \epsilon_x^*(\omega_{jg}) \epsilon_d(\omega_{E,E_i}) \epsilon_d^*(\omega_{E,E_j}) d_q(ij),
\end{aligned}
\tag{5.38}
$$

where $d_q(ij)$ is defined by Eq. (3.79). To incorporate effects due to a partially coherent source, we independently average Eq. (5.38) over the ensemble of phases of the excitation and dump pulses, giving

$$P_q(E) = \left(\frac{2\pi}{\hbar^2}\right) \sum_{i,j} c_i c_j^* \langle \epsilon_x(\omega_{i,g}) \epsilon_x^*(\omega_{j,g}) \rangle \langle \epsilon_d(\omega_{E,E_i}) \epsilon_d^*(\omega_{E,E_j}) \rangle d_q(ij). \tag{5.39}$$

Note that despite the excitation of multiple levels the only correlation function required is between ϵ at two frequencies. Assuming the phase diffusion model described above, then the frequency–frequency correlation function is given by Eq. (5.30). The probability $P(q)$ of forming product in channel q is then obtained by integrating over the pulse width:

$$
\begin{aligned}
P(q) &= \int dE\, P_q(E) = \left(\frac{2\pi}{\hbar^2}\right) \sum_{i,j} c_i c_j^* \langle \epsilon_x(\omega_{i,g}) \epsilon_x^*(\omega_{j,g}) \rangle \\
&\quad \times \int dE \langle \epsilon_d(\omega_{E,E_i}) \epsilon_d^*(\omega_{E,E_j}) \rangle d_q(ij).
\end{aligned}
\tag{5.40}
$$

It is often the case that the Franck–Condon factors contained in $d_q(ij)$ vary sufficiently slowly over the range of E encompassed by the dump pulse to be regarded as constant [191]. (This assumption, called the *slowly varying continuum approximation* is discussed in detail in Chapter 10). Under these circumstances we can use the following generalized Parseval's equality to show that $P(q)$ is independent of the coherence properties of the dump pulse. Specifically we have

$$\int d\omega \langle \varepsilon^*(\omega - \omega_1)\varepsilon(\omega - \omega_2)\rangle = \frac{1}{2\pi}\int d\omega \iint dt_1\, dt_2 \langle \varepsilon^*(t_1)\varepsilon(t_2)\rangle e^{i\omega(t_2-t_1)}e^{i(\omega_1 t_1 - \omega_2 t_2)}$$

$$= \iint dt_1\, dt_2 \langle \varepsilon^*(t_1)\varepsilon(t_2)\rangle \delta(t_2 - t_1)e^{i(\omega_1 t_1 - \omega_2 t_2)}$$

$$= \int dt \langle \varepsilon^*(t)\varepsilon(t)\rangle e^{i(\omega_1 - \omega_2)t}. \tag{5.41}$$

[The conventional Parseval equality $\int d\omega \langle |\epsilon(\omega)|^2\rangle = \int dt \langle \varepsilon^*(t)\varepsilon(t)\rangle$ is the special case of $\omega_1 = \omega_2$.] Since the right-hand side is independent of the phase of $\varepsilon(t)$, then the frequency-integrated correlation function is independent of the degree of coherence of the dump pulse.

Assuming that $d_q(ij)$ is independent of E over the pulse width allows us to write Eq. (5.40) as

$$P(q) = \left(\frac{2\pi}{\hbar^2}\right) \sum_{i,j} c_i c_j^* \langle \epsilon_x(\omega_{ig})\epsilon_x^*(\omega_{jg})\rangle F(\omega_{j,i})d_q(ij), \tag{5.42}$$

with $F(\omega_{j,i}) = \int dt \langle |\varepsilon_d(t)|^2\rangle e^{i(E_j-E_i)t/\hbar}$. For the Gaussian dump pulse

$$F(\omega) = \left(\frac{\mathcal{E}_{d0}}{2}\right)^2 \sqrt{\frac{\pi}{2}}\exp\left[-\frac{\tau_d^2\omega^2}{8} + i\omega t_d\right]. \tag{5.43}$$

Given Eqs. (5.30), (5.43), and (3.13), Eq. (5.42) assumes the form

$$P(q) = \left(\frac{2\pi}{\hbar^2}\right) \sum_{i,j} c_i c_j^* |\langle \epsilon_x(\omega_{ig})\epsilon_x^*(\omega_{jg})\rangle F(\omega_{j,i})d_q(ij)| \exp[i\omega_{i,j}\Delta_t + i\alpha_q(ij)], \tag{5.44}$$

where $\Delta_t = (t_d - t_x)$. For the simplest case, excitation that encompasses only two levels, $P(q)$ is given by

$$P(q) = (2\pi/\hbar^2)[|c_1|^2 \langle \epsilon_x(\omega_{1g})|^2\rangle d_q(11)F(\omega_{11}) + |c_2|^2 \langle \epsilon_x(\omega_{2g})|^2\rangle d_q(22)F(\omega_{22})$$
$$+ 2|c_1 c_2^* d_q(12)\langle \epsilon_x(\omega_{1g})\epsilon_x^*(\omega_{2g})\rangle F(\omega_{2,1})| \cos(\omega_{2,1}\Delta_t + \alpha_q(12) + \beta)],$$

$$\tag{5.45}$$

with $\langle E_1|\mathbf{d}\cdot\hat{\varepsilon}|E_g\rangle\langle E_g|\mathbf{d}\cdot\hat{\varepsilon}|E_2\rangle \equiv |\langle E_1|\mathbf{d}\cdot\hat{\varepsilon}|E_g\rangle\langle E_g|\mathbf{d}\cdot\hat{\varepsilon}|E_2\rangle|\exp(i\beta)$.

Since partial laser phase coherence affects both the direct terms as well as the cross terms, the extent of control is dependent on the laser properties through the relative magnitudes of $|\langle \epsilon_x(\omega_{1g})\epsilon_x^*(\omega_{2g})\rangle|$ and $\langle |\epsilon_x(\omega_{ig})|^2\rangle$, $i = 1, 2$. To expose the dependence on the coherence of the pump field denote the terms $|c_k c_j^* F(\omega_{jk})d_q(kj)(E)|$ by $a_{k,j}^{(q)}$ and consider the ratio of the $k \neq j$ term in Eq. (5.44) to the associated diagonal terms. That is, consider the contrast ratio:

$$C_{k,j}^{(q)} = \frac{a_{k,j}^{(q)}\langle \epsilon_x(\omega_{kg})\epsilon_x^*(\omega_{jg})\rangle}{a_{k,k}^{(q)}\langle |\epsilon_x(\omega_{kg})|^2\rangle + a_{j,j}^{(q)}\langle |\epsilon_x(\omega_{jg})|^2\rangle}. \tag{5.46}$$

For the Gaussian model of a partially coherent source Eq. (5.46) assumes the form:

$$C_{k,j}^{(q)} = \exp\left[\frac{-\omega_{k,j}^2(\tau_x^2 - T_x^2)}{8}\right]\left[\frac{a_{k,j}^{(q)}\exp(-\omega_{jk}T_x^2\delta_{E_{jk}}/2\hbar)}{a_{k,k}^{(q)} + a_{j,j}^{(q)}\exp(-\omega_{jk}T_x^2\delta_{E_{jk}}/\hbar)}\right]. \tag{5.47}$$

Here $\delta_{E_{jk}} \equiv E_x - E_{av} \equiv \hbar\omega_x - (E_k + E_j)/2$. The second term in Eq. (5.47), in brackets, is a function of $T_x^2\delta_{E_{jk}}$. If two bound state levels dominate the pump excitation, then this term contributes a scaling characteristic to control plots. That is, if we plot contours of constant dissociation probability as a function of Δ_t and $\delta_{E_{jk}}$ then, barring the first term, plots with different T_x will appear similar, with a new range scaled by $\delta_{E'} = (T_x/T_x')^2\delta_{E_{jk}}$.

The first term in Eq. (5.47) can be rewritten as

$$A_{k,j} = \exp\left[\frac{-\omega_{k,j}^2(\tau_x^2 - T_x^2)}{8}\right] = \exp\left[\frac{-\omega_{k,j}^2\tau_x^2}{8}\left(1 - \frac{1}{(\tau_x/\tau_{xc})^2 + 1}\right)\right], \tag{5.48}$$

which affords additional insight into the dependence of control, achieved by varying Δ_t, on coherence characteristics of the pump laser. Note first that Eq. (5.48) implies that for fixed τ_x control comes predominantly from nearby molecular states, that is, those with small $\omega_{k,j}$. Second, for fixed pulse duration τ_x, control is expected to decrease with decreasing τ_{xc}, that is, with decreasing pulse coherence. This is reasonable since decreasing τ_x leads to the preparation of mixed molecular states with increasing degrees of state impurity [192] and hence loss of phase information. Somewhat unexpected, however, is the prediction of improved control with decreasing pulse duration τ_x at fixed (τ_x/τ_{xc}), embodied in the $\exp(-\omega_{k,j}^2\tau_x^2)$ term.

As an example, consider the dependence of the photodissociation the model collinear DH$_2$, discussed in Section 3.5, as a function of Δ_t and of the detuning $\delta_E \equiv E_x - E_{av} \equiv \hbar\omega_x - (E_1 + E_2)/2$, where E_1 and E_2 are two selected neighboring energy levels for variable Δ_{xc}. For small Δ_{xc} only two levels are excited, but as Δ_{xc} increases the excitation encompasses a larger number of levels. Each panel of Figure 5.9 shows a contour plot of the fractional yield of DH [i.e., P(DH)/ (P(DH) + P(H$_2$))] as a function of Δ_t and δ_E, for different values of Δ_{xc}. Panel a, for comparison, shows the case of a coherent pulse ($\Delta_{xc} = 0$); the range of control is

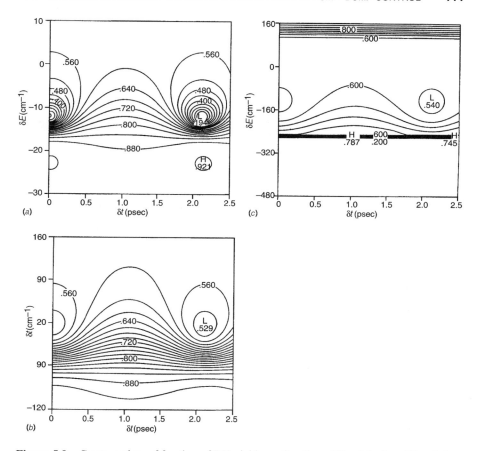

Figure 5.9 Contour plots of fraction of DH yield as a function of the detuning, δE and time delay Δ_t for partially coherent pulsed excitation. Here $\Delta_d = 80 \text{ cm}^{-1}$, $\Delta_x = 20 \text{ cm}^{-1}$; (a) $\Delta_{xc} = 0 \text{ cm}^{-1}$, (b) $\Delta_{xc} = 40 \text{ cm}^{-1}$, and (c) $\Delta_{xc} = 80 \text{ cm}^{-1}$. (Taken from Fig. 1, Ref. [188].)

large with the yield ratio varying from 0.19 to 0.92. Figures 5.9b and 5.9c show control in the same system but with differing amounts of pump laser incoherence. Increasing incoherence is clearly accompanied by a reduction in the range of control, with considerable loss of control by $\Delta_{xc} = 80 \text{ cm}^{-1}$ control (Fig. 5.9c). Note that regions where the yield depends on δ_E but not on Δ_t does not constitute interference-based control. Rather this dependence is a consequence of the (generally uninteresting) predisposition of various bound levels of the excited electronic state to preferentially dissociate to particular products. The pattern shown in Figure 5.9, where the control plot is dominated by a single well-defined peak and valley, is characteristic of the excitation of essentially two levels, that is, two molecular levels under the laser excitation envelope have significant Franck–Condon factors. More complicated behavior is seen upon excitation of additional levels. Figure 5.10 shows results typical of those obtained with the excitation of a large number of levels. The

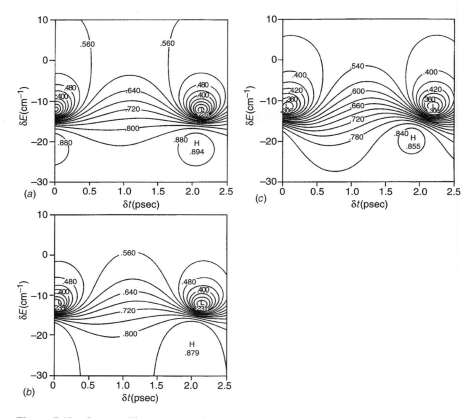

Figure 5.10   Same as Figure 5.9 but for excitation of denser set of bound states [188]. Here $\Delta_x = 200\,\mathrm{cm}^{-1}$ and (a) $\Delta_{xc} = 0\,\mathrm{cm}^{-1}$, (b) $\Delta_{xc} = 200\,\mathrm{cm}^{-1}$, and (c) $\Delta_{xc} = 800\,\mathrm{cm}^{-1}$. (Taken from Fig. 2, Ref. [188].)

resultant dependence of the yield ratio on δ_E and Δ_t is far more complex than the behavior seen in previous figures. However, the reduction of control with increasing Δ_{xc} is clearly evident; control is lost by $\Delta_{xc} = 1000\,\mathrm{cm}^{-1}$ (not shown).

From these and related studies we find that control in the model DH$_2$ system is lost when $(\Delta_{xc}/\Delta_x) = (\tau_x/\tau_{xc}) \approx 4$–5. This precludes the use of typical nanosecond lasers, where $(\Delta_{xc}/\Delta_x) > 10^2$, for pump–dump control.

A more complex analysis of the effect of laser phase diffusion has been applied to the case of one-photon vs. three-photon absorption (i.e., simultaneous absorption of $3\omega_1$ and ω_3 with $3\omega_1 = \omega_3$) (Section 3.3.2) by Camparo and Lambropoulos [193]. They assumed that the ω_3 photon was made by third-harmonic generation from the ω_1 laser. As discussed in Section 3.3.2, current experiments vary the relative phase of two laser beams by passing ω_3 and ω_1 through a gas. If the laser frequency is somewhat unstable, then the relative phase of the two beams will acquire a fluctuating phase that is a source of phase loss in the system. The phase fluctuations of the

ω_1, denoted $\delta\phi$, are assumed proportional to the fluctuations of ω_1, that is, $\delta\phi = \alpha\delta\omega_1$. For typical experimental circumstances it is reasonable to model the $\delta\omega_1$ distribution by the Gaussian probability distribution

$$P(\delta\omega_1) = (2\pi\gamma\beta)^{-1/2} \exp[-(\delta\omega_1)^2/(2\gamma\beta)], \qquad (5.49)$$

where γ and β parameterize the laser frequency fluctuations:

$$\langle\delta\omega_1\delta\omega_1(t - \tau)\rangle = \gamma\beta\exp^{-\beta|t|}. \qquad (5.50)$$

If the one- vs. three-photon scenario is applied to the simple case of He photoionization (i.e., only a single product channel is considered), then in accord with Eq. (3.54) the ionization probability would be given by

$$P_q(E) = F_q(11) - 2x\cos[\phi_3 - 3\phi_1 + \alpha_q(13)]\epsilon_0^2|F_q(13)| + x^2\epsilon_0^4 F_q(33). \qquad (5.51)$$

Assuming that the direct one-photon and direct three-photon parts are unity and the phase assumes a time-dependent fluctuation, then the essential part of the Eq. (5.51) is of the form:

$$P_q(\phi_0, t) = 1 + \cos[\phi_0 + \delta\phi(t)] = 1 \pm \cos[\alpha\delta\omega_1(t)], \qquad (5.52)$$

where $\phi_3 - 3\phi_1 + \alpha_q(13) \equiv \phi_0 + \delta\phi(t)$ and where the plus and minus refer to the cases where $\phi_0 = 0$ and π, respectively. The average probability is then obtained as

$$\langle P_q(E)\rangle = 1 \pm \int P(\delta\omega_1)\cos(\alpha\delta\omega_1)d[\delta\omega_1],$$
$$1 \pm \exp[-\alpha^2\gamma\beta/2]. \qquad (5.53)$$

A useful criteria for the effect of the partially coherent laser on the control is given by the *contrast ratio*, that is, the ratio of the probabilities at its maximum and minimum:

$$\zeta_0 = \log_{10}[1 - \exp[-\alpha^2\gamma\beta/2] - \log_{10}[1 + \exp[-\alpha^2\gamma\beta/2]]. \qquad (5.54)$$

Numerical studies [193] show that Eq. (5.54) provides a zeroth-order approximation to the results of a full computation, which underestimates the degree of possible control in a realistic system. In addition, these results show that even for phase diffusion fields, which have widths on the order of wavenumbers, control is still extensive (e.g., $\zeta_0 \approx 5$). Examination of the experimental results on one-photon vs. three-photon control show, however, contrast ratios on the order of 30% [76, 194]. That is, the main experimental limitation, thus far, is due to experimental issues other than the partial laser coherence.

5.4   COUNTERING CW LASER JITTER

For cw lasers, laser decoherence appears via the jitter and drift of the laser phase ϕ in the field $\mathbf{E}(z, t)$ [e.g., Eq. (3.16)] with a concomitant reduction in control (see Section 5.3). However, suitable design of the control scenario can result in a method that is immune to the effects of laser jitter. In particular, to do so we rely upon the way in which the laser phase enters into control scenarios.

5.4.1   Laser Phase Additivity

The role of the laser phase in controlling molecular dynamics was clear in the examples shown in Chapter 3. For example, in the one- vs. three-photon scenario the relative laser phase $(\phi_3 - 3\phi_1)$ enters directly into the interference term [see, e.g., Eq. (3.53)], as does the relative phase $(\phi_1 - \phi_2)$ in the bichromatic control scenario [Eq. (3.19)]. These results embody two useful general rules about the contribution of the laser phase to coherent control scenarios. The first is that the interference term contains the *difference* between the laser phase imparted to the molecule by one route, and that imparted to the molecule by an alternate route. Second, the phase imparted to the state $|E_m\rangle$ by a light field of the form

$$\hat{\varepsilon} \int_{-\infty}^{\infty} d\omega |\epsilon(\omega)| \exp[i\phi(\omega)] \exp(-i\omega\tau)$$

in an excitation from level $|E_1\rangle$ to $|E_m\rangle$ is $\pm\phi(|\omega_{m,1}|)$. The plus sign applies to light absorption ($E_m > E_1$) and the minus sign to stimulated emission ($E_1 > E_m$). This observation proves very useful in designing schemes that are insensitive to laser phase.

One example is the two-photon plus two-photon scheme discussed in Section 6.1. An alternative, which we sketch here and discuss in further detail in relation to strong-field scenarios (Section 11.2) is called incoherent interference control.

5.4.2   Incoherent Interference Control

Figure 5.11 shows a level scheme where a cw field with frequency ω_1 excites the level $|E_i\rangle$ to the photodissociative continuum. Simultaneously, a stronger cw laser field of frequency ω_2 couples the continuum to the initially empty state $|E_j\rangle$. The phases associated with these two fields are ϕ_1 and ϕ_2, respectively. The effect of the strong field is to cause Rabi cycling of population between $|E_j\rangle$ and the continuum. Thus, in this arrangement, population can be transferred from $|E_i\rangle$ to the continuum by a variety of routes, as shown in Figure 5.12. The method is the multichannel generalization of *laser-induced continuum structure* (LICS) [195–199].

The first panel of Figure 5.12 shows the bichromatic control scenario. The second panel shows the simplest path to the continuum, consisting of one-photon absorption of ω_1. The subsequent panels show the three-photon process to the continuum (absorption of ω_1 followed by stimulated emission and reabsorption of ω_2, etc.),

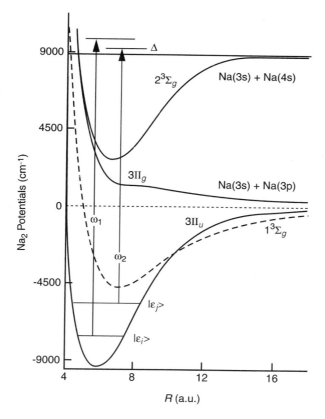

Figure 5.11 Sample scenario for the incoherent interference control of the photodissociation of Na_2. (Taken from Fig. 1, Ref. [200].)

and a five-photon process (absorption of ω_1 followed by stimulated emission and reabsorption of ω_2, twice). This series goes on *ad infinitum*, resulting in an infinite number of interfering pathways.

In accord with Section 5.4.1 the phase imparted to the continuum state by the first route in Figure 5.12 is ϕ_1, and by the second is $\phi_1 - \phi_2 + \phi_2$. The $-\phi_2$ contribution to the latter phase is due to the stimulated emission step, and the following $+\phi_2$ is due to the absorption. Hence both routes impart the overall phase ϕ_1 to the continuum state. It is clear that this is also the case for all additional routes to the continuum since they must contain an equal number of stimulated emission and absorption steps.

Examination of all previous described scenarios makes clear, however, that it is the relative phase imparted to the routes that affects control. In the case described

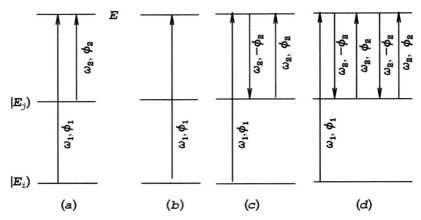

Figure 5.12   Interfering pathways from $|E_i\rangle$ to the continuum associated with the scenario in Figure 5.11. The frequency and phase of the lasers are ω_i and ϕ_i. (a) Bichromatic control. (b) One-photon absorption. (c) Three-photon process in which initially unpopulated state $|E_j\rangle$ is coupled to the continuum at energy E and interferes with one-photon absorption from state $|E_i\rangle$. (d) Same as in (c) but for a five-photon process. Notice that in processes depicted in (c) and (d) the phase ϕ_2 gets canceled at the completion of each stimulated emission followed by absorption cycle.

here, the relative phase of the routes is $\phi_1 - \phi_1 = 0$, so that control is independent of the laser phase. As a consequence, even lasers with extreme laser jitter and drift can be used in this scenario. Note, however, the additional consequence that, in this scenario, control is achieved by varying the frequencies ω_1 and ω_2.

An experimental realization with the pulsed laser version of this approach [201] is discussed in detail in Section 11.2 where control over both relative populations into different channels, as well as over the photodissociation yield into both channels is computed and demonstrated.

CHAPTER 6

CASE STUDIES IN COHERENT CONTROL

Chapter 3 provided an introduction to the principles of coherent control and included a number of examples of scenarios that embody that principle. In this chapter we consider several other scenarios that both shed further light on these principles and that suggest a number of useful experimental scenarios.

6.1 TWO-PHOTON VS. TWO-PHOTON CONTROL

The M vs. N photon scenarios, where both routes are nonresonant, were discussed in Section 3.3. Here we consider resonantly enhanced routes and show, in particular, that a resonantly enhanced two-photon vs. two-photon excitation (see Fig. 6.1) provides a means of maintaining control in a molecular system in thermal equilibrium. The resonant character of the excitations ensure that only a particular initial state, out of the thermal distribution of molecular levels, participates in the photodissociation. Hence coherence is established by the excitation, and maintained throughout the process. The design of the proposed control scenario also relies upon the phase additivity arguments in Section 5.4.1 to provide a method for overcoming the reduction in the interference term due to phase jitter in the laser source. In addition, we show that this approach allows one to reduce the contributions from uncontrolled satellite routes. Thus, this scenario is capable of overcoming three of the major decoherence mechanisms discussed in Chapter 5, thereby enabling the execution of coherent control in "natural" thermal environments.

The two-photon control scenario that we describe is completely general, but we focus here on the photodissociation of diatomic molecules. Consider first photodissociation along *one* of the paths shown in Figure 6.1a, that is, two-photon dissociation of a molecule where the first laser is resonant with an intermediate bound level.

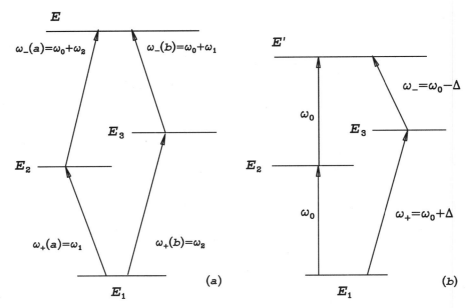

Figure 6.1 Resonantly enhanced two-photon vs. two-photon control scenarios: (*a*) using four frequencies and (*b*) using three frequencies.

The molecule, initially in a state $|E_i, J_i, M_i\rangle$, is subjected to two cw fields. Once again J_i, M_i denote the angular momentum and its projection along the z axis. The matter–radiation interaction term is of the form

$$H_{\mathrm{MR}}(t) = -2\mathbf{d} \cdot \mathrm{Re}[\hat{\varepsilon}_2 \bar{\epsilon}_2 \exp(-i\omega_2 t) + \hat{\varepsilon}_1 \bar{\epsilon}_1 \exp(-i\omega_1 t)]. \tag{6.1}$$

As a result of irradiation by this field the molecule photodissociates, yielding a number of different product channels labeled by q. Absorption of the first photon of frequency ω_1 lifts the system to an energy close to the energy E_m of an intermediate bound state $|E_m, J_m, M_m\rangle$. A second photon of frequency ω_2 carries the system to the dissociating states $|E, \hat{\mathbf{k}}, q^-\rangle$.

Extending Eq. (3.35) to the case of two different frequencies, ω_1 and ω_2, it is possible to show that [202] the probability-amplitude $D_{\hat{\mathbf{k}},q,i}(E, E_i J_i M_i, \omega_2, \omega_1)$ for resonantly enhanced two-photon dissociation is

$$D_{\hat{\mathbf{k}},q,i}(E, E_i J_i M_i, \omega_2, \omega_1)$$
$$= \sum_{E_m, J_m, e'} \frac{\langle E, \hat{\mathbf{k}}, q^- |d_{e,e'} \bar{\epsilon}_2 |E_m, J_m, M_i\rangle \langle E_m, J_m, M_i |d_{e',g} \bar{\epsilon}_1 |E_i J_i M_i\rangle}{\hbar\omega_1 - (E_m + \Delta_m - E_i) + i\Gamma_m/2}. \tag{6.2}$$

Here $E = E_i + (\omega_1 + \omega_2)\hbar$, Δ_m and Γ_m are, respectively, the radiative shift and the full width at half maximum (FWHM) of the intermediate state. (The derivation of these quantities is discussed in detail in Section 6.3.1.) We assume that the lasers are linearly polarized and that their electric-field vectors are parallel to one another.

The term $[\hbar\omega_1 - (E_m + \Delta_m - E_i) + i\Gamma_m]$ in the denominator of Eq. (6.2) allows us to tune ω_1 so that only a select few of the thermally populated levels $|E_i, J_i, M_i\rangle$ are excited. That is, making this denominator small (i.e. achieving this resonance condition) establishes coherence, despite the thermal environment.

The probability of producing the fragments in channel q is obtained by integrating the square of Eq. (6.2) over the scattering angles $\hat{\mathbf{k}}$:

$$P_q(E, E_iJ_iM_i, \omega_2, \omega_1) = \int d\hat{\mathbf{k}} |D_{\hat{\mathbf{k}},q,i}(E, E_iJ_iM_i, \omega_2, \omega_1)|^2. \qquad (6.3)$$

We consider now the scenarios, shown in Figure 6.1, of simultaneously exciting a molecule by two resonantly enhanced two-photon routes. For example (Fig. 6.1b), a molecule is irradiated with three interrelated frequencies, $\omega_0, \omega_+, \omega_-$ with $\omega_\pm = \omega_0 \pm \Delta$. The interference occurs between the two 2-photon dissociation routes leading to identical final energies $E = E_i + 2\hbar\omega_0 = E_i + \hbar(\omega_+ + \omega_-)$, where ω_0 and ω_+ are chosen resonant with intermediate bound state levels. The associated field amplitudes are $\bar{\epsilon}_0, \bar{\epsilon}_+$, and $\bar{\epsilon}_-$, whose phases are denoted ϕ_0, ϕ_+, ϕ_-. These fields generate two independent routes to the continuum at energy E. Thus, the probability of photodissociation at energy E into arrangement channel q is given by the square of the sum of the D matrix elements from pathway a (absorption of $\omega_0 + \omega_0$) and pathway b (absorption of $\omega_+ + \omega_-$). That is, the probability into channel q is

$$\begin{aligned} P_q(E, E_iJ_iM_i; \omega_0, \omega_+, \omega_-) &= \int d\hat{\mathbf{k}} |D_{\hat{\mathbf{k}},q,i}(E, E_iJ_iM_i, \omega_0, \omega_0) \\ &\quad + D_{\hat{\mathbf{k}},q,i}(E, E_iJ_iM_i, \omega_+, \omega_-)|^2 \\ &\equiv P_q(a) + P_q(b) + P_q(ab). \end{aligned} \qquad (6.4)$$

Here $P_q(a)$ and $P_q(b)$ are the independent photodissociation probabilities associated with routes a and b, respectively, and $P_q(ab)$ is the interference term between them, discussed below.

The interference term $P_q(ab)$ can be written as

$$P_q(ab) = 2|F_q(ab)| \cos[\alpha_q(ab) + \delta\phi], \qquad (6.5)$$

where

$$\delta\phi = 2\phi_0 - \phi_+ - \phi_-, \qquad (6.6)$$

and the amplitude $|F_q(ab)|$ and the "molecular phase" $\alpha_q(ab)$ are defined via the cross term in Eq. (6.4). [See also Section 3.1, Eq. (3.20)].

Consider now the quantity of interest, the channel branching ratio $R_{qq'}$. Noting that in the weak-field case $P_q(a)$ is proportional to $|\bar{\epsilon}_0|^4$, $P_q(b)$ to $\bar{\epsilon}_+\bar{\epsilon}_-|^2$, and $P_q(ab)$ to $|\bar{\epsilon}_0^2\bar{\epsilon}_+\bar{\epsilon}_-|$, we can write

$$R_{qq'} = \frac{d_q(a) + x^2 d_q(b) + 2x|d_q(ab)|\cos[\alpha_q(ab) + \delta\phi] + (B_q/|\bar{\epsilon}_0|^4)}{d_{q'}(a) + x^2 d_{q'}(b) + 2x|d_{q'}(ab)|\cos[\alpha_{q'}(ab) + \delta\phi] + (B_{q'}/|\bar{\epsilon}_0|^4)}, \tag{6.7}$$

where $d_q(a) = P_q(a)/|\bar{\epsilon}_0|^4$, $d_q(b) = P_q(b)/(|\bar{\epsilon}_+\bar{\epsilon}_-|^2)$, and $|d_q(ab)| = |F_q(ab)|/(|\bar{\epsilon}_0^2\bar{\epsilon}_+\bar{\epsilon}_-|)$, and $x = |\bar{\epsilon}_+\bar{\epsilon}_-/\bar{\epsilon}_0^2|$. The terms with $B_q, B_{q'}$, described below, correspond to resonantly enhanced photodissociation routes to energies other than $E = E_i + 2\hbar\omega_0$. That is, they are satellite terms that do not coherently interfere with the a and b pathways. These uncontrollable terms should be minimized, and we discuss how this can be done below. Here we just note that the product ratio in Eq. (6.7) depends upon both the laser intensities and relative laser phase, which are therefore control parameters in this scenario.

This scenario, embodied in Eq. (6.7), also provides a means by which control can be improved by eliminating effects due to laser jitter. Specifically, the $\delta\phi$ term of Eq. (6.6) can be subject to the phase fluctuations arising from laser instabilities, substantially reducing control. One can, however, design an experimental implementation of the two-photon plus two-photon scenario that readily compensates for this problem. For example, consider generating the two frequencies $\omega_\pm = \omega_0 \pm \Delta$ as the "signal" and the "idler" beams in a nonlinear down-conversion process that occurs when a beam of frequency $2\omega_0$ passes through an optical parametric oscillator (OPO) [203]. The $2\omega_0$ beam is assumed generated by second-harmonic generation from the laser ω_0 with the phase ϕ_0. Because ω_+ and ω_- are generated in this way, the phase difference $\delta\phi$ between the $(\omega_0 + \omega_0)$ and $(\omega_+ + \omega_-)$ routes is a constant [204]. That is, fluctuations in ϕ_0 cancel and have no effect on $\delta\phi$ of Eq. (6.6), nor on the interference term $P_q(ab)$ of Eq. (6.5).

The above control scheme also allows for the systematic reduction of the satellite contributions B_q and $B_{q'}$ [Eq. (6.7)]. These contributions, explicitly given in Ref. [202], include terms from the resonantly enhanced photodissociation processes due to absorption of $(\omega_0 + \omega_-)$, $(\omega_0 + \omega_+)$, $(\omega_+ + \omega_0)$, or $(\omega_+ + \omega_+)$ that lead to photodissociation at energies $E_- = E_i + (\omega_0 + \omega_-)$, $E_+ = E_i + \hbar(\omega_0 + \omega_+)$, $E_{++} = E_i + \hbar(\omega_+ + \omega_+)$, respectively. Other nonresonant pathways are possible but are negligible by comparison. Controllable reduction of this background term is indeed possible because B_q and $B_{q'}$ are functions of $|\bar{\epsilon}_+|$ or $|\bar{\epsilon}_-|$, while the photodissociation products resulting from paths a and b depend on the product $|\bar{\epsilon}_+\bar{\epsilon}_-|$. Thus, changing $|\bar{\epsilon}_+|$ (or $|\bar{\epsilon}_-|$) while keeping $|\bar{\epsilon}_+\bar{\epsilon}_-|$ fixed will not affect the yield from the controllable paths a and b, but will affect B_q. To this end we introduce the parameters $\bar{\epsilon}_b^2 = \bar{\epsilon}_+\bar{\epsilon}_-$ with $|\bar{\epsilon}_-|^2 = \eta|\bar{\epsilon}_b|^2$, $|\bar{\epsilon}_+|^2 = |\bar{\epsilon}_b^2|/\eta$. The terms B_q and $B_{q'}$ are the only terms dependent upon η and can be reduced by appropriate choice of this parameter. Numerical examples are provided below.

To examine the range of control afforded by this scheme consider the photodissociation of Na$_2$ (Fig. 6.2). For simplicity we focus on the regime below the

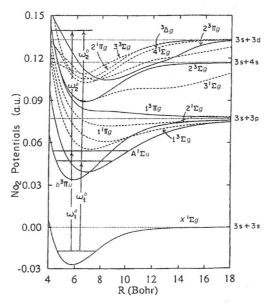

Figure 6.2 Na_2 potential energy surfaces included in the two-photon vs. two-photon control scenario. Arrows indicate the resonantly enhanced two-photon vs. two-photon pathways included in the computation discussed below. (From Fig. 2, Ref. [205].)

Na(3d) + Na(3s) threshold, where dissociation is to the two-product channels Na(3s) + Na(3p) and Na(3s) + Na(4s). Results above this energy, where the channel Na(3s) + Na(3d) is open, are described in the literature [200, 336].

Resonantly enhanced two-photon dissociation of Na_2 from a bound state of the ground electronic state occurs [202] by initial excitation to an excited intermediate bound state $|E_m, J_m, M_m\rangle$. The latter is a superposition of states of the $A^1\Sigma_u^+$ and $b^3\Pi_u$ electronic curves, a consequence of spin-orbit coupling. The continuum states reached in the two-photon excitation can have either a singlet or a triplet character, but, despite the multitude of electronic states involved in the computation reported below, the predominant contributions to the products Na(3s) + Na(3p) and Na(3s) + Na(4s) are found to come from the $^3\Pi_g$ and $^3\Sigma_g^+$ electronic states, respectively. The resonant character of the two-photon excitation allows the selection of a single initial state from a thermal ensemble; here results for $v_i = J_i = 0$, where v_i, J_i denote the vibrational and rotational quantum numbers of the initial state, are discussed.

The ratio $R_{qq'}$ depends on a number of laboratory control parameters including the relative laser intensities x, relative laser phase, and the ratio of $|\bar{\epsilon}_+|$ and $|\bar{\epsilon}_-|$ via η. In addition, the relative cross sections can be altered by modifying the detuning. Typical control results are shown in Figures 6.3 and 6.4, which provide contour plots of the Na(3s) + Na(3p) yield as a function of the ratio of the laser amplitudes x, and of the relative laser phase $\delta\phi$ of Eq. (6.6). Consider first Figure 6.3 resulting from

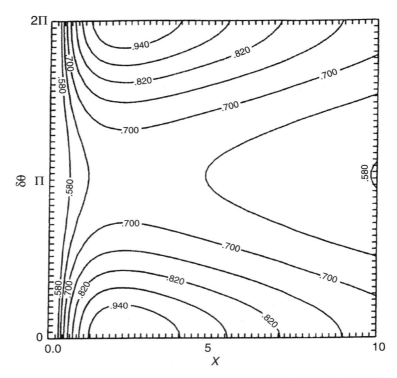

Figure 6.3   Contours of equal Na(3p) yield. Ordinate is the relative laser phase and abscissa is the field intensity ratio x. Here for $\lambda_0 = 623.367$ nm, $\lambda_+ = 603.491$ nm, $\lambda_- = 644.596$ nm, and $\eta = 1$. (From Fig. 1, Ref. [206].)

excitation with $\lambda_0 = 2\pi c/\omega_0 = 623.367$ nm and $\lambda_+ = 2\pi c/\omega_+ = 603.491$ nm, which are in close resonance with the intermediate states $v = 13$ and 18, $J_m = 1$ of $^1\Sigma_u^+$, respectively. The corresponding $\lambda_- = 2\pi c/\omega_-$ is 644.596 nm. The yield of Na(3p) is seen to vary from 58 to 99%, with the Na(3p) atom predominant in the products. Although this range is large, variation of the η parameter should allow for improved control by minimizing the background contributions. This improvement is, in fact, not significant in this case since reducing η decreases B_q from the $(\omega_0 + \omega_-)$ route but increases the contribution from the $(\omega_0 + \omega_+)$ route. By contrast, the background can be effectively reduced for the frequencies shown in Figure 6.4. Here $\lambda_0 = 631.899$ nm, $\lambda_+ = 562.833$ nm, and $\lambda_- = 720.284$ nm. In this example, with $\eta = 5$, a larger range of control is achieved, from 30% Na(3p) to 90% as $\delta\phi$ and x are varied.

This control scenario is not limited to the specific frequency scheme discussed above. Essentially all that is required is that two or more resonantly enhanced photodissociation routes interfere and that the cumulative laser phases of the two routes be independent of laser jitter. As one sample extension, consider the case

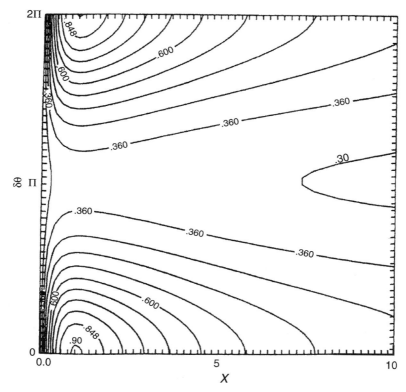

Figure 6.4   As in Figure 6.3 but for $\lambda_0 = 631.899$ nm, $\lambda_+ = 562.833$ nm, $\lambda_- = 720.284$ nm, and $\eta = 5$. (From Fig. 2, Ref. [206].)

(Fig. 6.1a) where paths a and b are composed of totally different photons, $\omega_+(a)$ and $\omega_-(a)$, and $\omega_+(b)$ and $\omega_-(b)$, with $\omega_+(a) + \omega_-(a) = \omega_+(b) + \omega_-(b)$. Both these sets of frequencies can be generated, for example, by passing $2\omega_0$ light through two nonlinear crystals, hence yielding two pathways whose relative phase is independent of laser jitter in the initial $2\omega_0$ source. A sample control result is shown in Figure 6.5 where $\lambda_+(a) = 599.728$ nm, $\lambda_-(a) = 652.956$ nm, $\lambda_+(b) = 562.833$ nm and $\lambda_-(b) = 703.140$ nm, and $\eta_a = \frac{1}{2}$, $\eta_b = 10$, where η_a and η_b are the analog of η for the a and b paths, respectively. Here the range of the control is from 14 to 95%, that is, a substantial fraction of total product control and an improvement over the three-frequency approach.

This method has been implemented experimentally [207] in an attempt to control the branching $Na_2 \rightarrow Na + Na(3p)$, $Na + Na(3d)$ reaction. In this case one starts with two dye lasers of frequencies $\omega_+(a) = \omega_1$ and $\omega_+(b) = \omega_2$ with two, totally uncorrelated phases ϕ_1 and ϕ_2. By mixing these two beams (using two nonlinear crystals) with a third frequency ω_0 (in the actual experiment this was the fundamental frequency of a Nd–Yag laser operating at $\lambda_0 = 1.06 \, \mu m$), one generates two

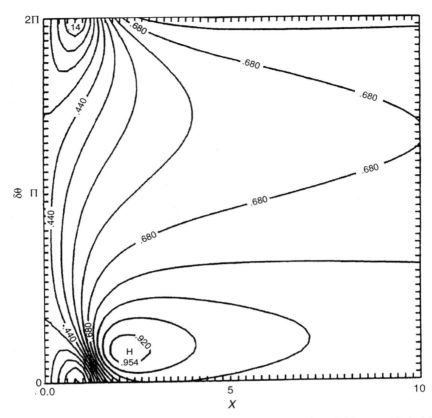

Figure 6.5 As in Figure 6.3 but for the Na(4s) product in the four-field case with $\lambda_+(a) =$ 599.728 nm, $\lambda_-(a) = 652.956$ nm, $\lambda_+(b) = 562.833$ nm, $\lambda_-(b) = 703.140$ nm, $\eta_a = \frac{1}{2}$ and $\eta_b = 10$. (From Fig. 3, Ref. [206].)

additional frequencies $\omega_-(b) = \omega_1 + \omega_0$ and $\omega_-(a) = \omega_2 + \omega_0$. As depicted in Figure 6.1a, one interferes a two-photon route composed of absorption of ω_1 and $\omega_2 + \omega_0$ with the ω_2 and $\omega_1 + \omega_0$ two-photon route. Using the phase additivity, which is realized quite well in the nonlinear mixing process, the phase of the first route is $\phi_1 + (\phi_2 + \phi_0)$ and the phase of the second route is $\phi_2 + (\phi_1 + \phi_0)$. That is, the overall phase associated with each route is the same. This is the case, even though the four phases that make up the two 2-photon routes are totally uncorrelated. By adding a delay line to either of the two-photon pathways, one can introduce any desired *relative phase* $\delta\phi$ between the two routes in a controlled way, which then serves as a laboratory knob in the control experiment.

A simpler, though much more limited implementation of two-photon vs. two-photon control, entails the use of just two frequencies, ω_1 and ω_2. It is a special case of the four-frequency scenario depicted in Figure 6.1a, where $\omega_0 = 0$ (i.e., no mixing with ω_0 is performed). Because any phase change in either ω_1 or ω_2 will automatically affect the two routes, it is not possible in this two-frequency scenario to introduce an externally controlled relative phase change between the two routes. It is, however, possible to alter the $\alpha_q(ab)$ "molecular phase" by detuning either ω_1 or ω_2 off their respective resonances. Due to the presence of the $i\Gamma_m$ term in the denominator of Eq. (6.2), the detuning of just one frequency from resonance results in a phase change (in addition to the much more noticeable amplitude change) of one two-photon matrix element relative to another.

This type of two-photon vs. two-photon phase control has been implemented experimentally in molecules by Pratt [208], who studied the photoionization of the $A^2\Sigma^+, v' = 1$ state of NO, and in atoms by Elliott's group [209], who demonstrated control over the branching ratio for photoionization of Ba into the $6s_{1/2}, 5d_{3/2}$ and $5d_{5/2}$ states of Ba^+. Quite unexpectedly, the interference between the two-photon routes turned out to be destructive when both lasers were resonant with their respective transitions. A theoretical treatment of effect was provided by Luc-Koenig et al. [210].

An experimental implementation of a three-color two-photon vs. two-photon control scheme of Figure 6.1b was performed by Georgiades et al. [211]. Their experimental setup is shown in Figure 6.6 where the ω_0 beam is generated by a Ti: Sapphire laser, and the $\omega_\pm$ beams are generated from an optical parametric oscillator (OPO), which is pumped by a frequency-doubled Ti: Sapphire laser beam at $2\omega_0$. Thus, the conditions set out in Figure 6.1b, that $\omega_+ + \omega_- = 2\omega_0$, are satisfied. Using a piezoelectric transducer, it is possible to change the optical path of the ω_0 beam relative to the ω_+ beams. The three beams serve to excite the $6D_{5/2}, F'' = 6$ state of Cs whose population is measured by monitoring the $6D_{5/2}, F'' = 6 \rightarrow 6P_{3/2}, F' = 5$ fluorescence. As shown in Figure 6.7, modulation

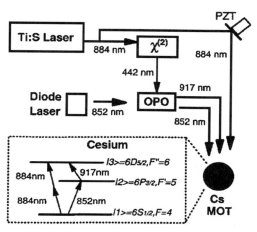

Figure 6.6 Experimental arrangement for $(\omega_+ + \omega_i)$ vs. $2\omega_0$ two-photon vs. two-photon control of population of Cs in the $6D_{5/2}, F'' = 6$ state. PZT is the piezoelectric transducer, OPO is optical parametric oscillator, and MOT is the magneto-optical trap of Cs atoms. The inset shows the Cs levels of interest. (From Fig. 1, Ref. [211].)

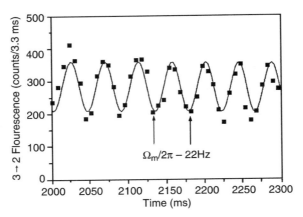

Figure 6.7   Fluorescence due to the $|3\rangle \rightarrow |2\rangle$ transition, where $|3\rangle = 6D_{5/2}, F'' = 6\rangle$ and $|2\rangle = |6P_{3/2}, F' = 5\rangle$ as the PZT is scanned in time, thereby inducing a proportional phase change. The solid curve is a fit of a constant plus a sinusoidal function. (From Fig. 2, Ref. [211].)

of the fluorescence signal as a function of the relative phase between the $\omega_+ + \omega_-$ and the $2\omega_0$ routes is obtained, indicating a modulation in the population of the $6D_{5/2}, F'' = 6$ state.

6.2   CONTROL OVER THE REFRACTIVE INDEX

Photodissociation is but one of many processes that are amenable to control. A host of other processes that have been studied are discussed later in this book, such as asymmetric synthesis, control of bimolecular reactions, strong-field effects, and so forth. Also of interest is control of nonlinear optical properties of materials [203], particularly for device applications. In this section we describe an application of the bichromatic control scenario discussed in Section 3.1.1) to the control of refractive indices.

The real and imaginary parts of the refractive index n quantify the scattering and absorption (or amplification) properties of a material. The refractive index is best derived from the susceptibility tensor $\underline{\underline{\chi}}$ of the material, defined below, which describes the response of a macroscopic system to incident radiation [212]. Specifically, an incident electric field $\mathbf{E}(\mathbf{r}, t)$, where $\mathbf{r}$ denotes the location in the medium, tends to displace charges, thereby polarizing the medium. The change in $\mathbf{d}^{\text{ind}}(\mathbf{r}, t)$, the induced dipole moment, from point $\mathbf{r}$ to point $\mathbf{r} + d\mathbf{r}$ is given in terms of the polarization vector $\mathbf{P}(\mathbf{r}, t)$, defined as

$$d\mathbf{d}^{\text{ind}}(\mathbf{r}, t) = \mathbf{P}(\mathbf{r}, t) \, d\mathbf{r}. \qquad (6.8)$$

It is customary to relate the polarization to the external field by defining a susceptibility $\underset{=}{\chi}(t)$ tensor via the relation

$$\mathbf{P}(\mathbf{r}, t) = \epsilon_0 \int_0^\infty d\tau \, \underset{=}{\chi}(\mathbf{r}, \tau) \cdot \mathbf{E}(\mathbf{r}, t - \tau). \tag{6.9}$$

The term $\underset{=}{\chi}$ is a complex tensor whose real part relates to light scattering and whose imaginary part describes absorption or amplification of light. In the weak-field (linear) domain it is independent of the field $\mathbf{E}(\mathbf{r}, t)$. As the field gets stronger, $\underset{=}{\chi}$ may become dependent on the field, in which case we say that $\underset{=}{\chi}$ has nonlinear contributions. Below, for convenience, we suppress the spatial dependence of $\underset{=}{\chi}$.

For cw light, $\mathbf{E}(z, t) = \mathcal{E}/2\hat{\epsilon} \exp(-i\omega t + i\mathbf{k} \cdot \mathbf{r}) + \text{c.c.}$, where c.c. denotes the complex conjugate of the preceding term, and Eq. (6.9) becomes

$$\mathbf{P}(\mathbf{r}, t) = \frac{\epsilon_0 \mathcal{E}}{2} [\underset{=}{\chi}(\omega) \exp(-i\omega t + i\mathbf{k} \cdot \mathbf{r}) + \underset{=}{\chi}(-\omega) \exp(i\omega t - i\mathbf{k} \cdot \mathbf{r})] \cdot \hat{\epsilon}, \tag{6.10}$$

where $\underset{=}{\chi}(\omega)$ is the Fourier transform of $\underset{=}{\chi}(t)$.

To see the relationship to the refractive index, we focus on the case where the tensor $\underset{=}{\chi}$ reduces to a scalar. This is the case, for example, if the electronic response of the medium is isotropic, in which case $\underset{=}{\chi}(\omega)$ reduces to a scalar $\chi(\omega)$, or where the field is, for example, along the laboratory z axis and only the single $\chi_{zz}(\omega)$ component is of interest. In the former case, the complex refractive index $n(\omega)$ is given by

$$n^2(\omega) = 1 + \chi(\omega). \tag{6.11}$$

Analogous expressions hold for radiation incident on *individual* molecules, but here the polarizability $\underset{=}{\alpha}(t)$ replaces the susceptibility $\underset{=}{\chi}(t)$. That is, the induced molecular dipole $\mathbf{d}^{\text{ind}}(t)$ is given as

$$\mathbf{d}^{\text{ind}}(t) = \int d\tau \, \underset{=}{\alpha}(\tau) \cdot \mathbf{E}(\mathbf{r}, t - \tau). \tag{6.12}$$

Therefore, given the system and field $\mathbf{E}(t)$, to obtain either the susceptibility or the polarizability requires that we compute the induced polarization, or induced dipole, respectively.

The susceptibility and polarizability differ in that the latter deals with single molecules, and the former with an entire medium. Further, the polarizability can be defined for the system in a particular quantum state, whereas the susceptibility generally refers to the bulk system, often in thermodynamic equilibrium. Typically, then [212] the susceptibility includes the number of particles per unit volume ρ and an average over populated system energy levels. Below we compute the polarizability and susceptibility for neither of these cases. Rather, we extend the standard definition of the polarizability to include a molecule initially in a superposition state.

In this case the polarizability and susceptibility are virtually identical, being related as

$$\epsilon_0 \underline{\underline{\chi}}(\omega) = \rho \underline{\underline{\alpha}}(\omega), \tag{6.13}$$

where $\underline{\underline{\alpha}}(\omega)$ is the Fourier transform of $\underline{\underline{\alpha}}(t)$.

6.2.1 Bichromatic Control

Here we show that an application of bichromatic control (Section 3.1.1) allows us to control both the real and imaginary parts of the refractive index. In doing so we consider isolated molecules [213, 214], or molecules in a very dilute gas, where collisional effects can be ignored and time scales over which radiative decay occurs can be ignored.

Consider then the case of bichromatic control where a system prepared in a superposition of bound Hamiltonian eigenstates $|E_m\rangle$,

$$|\Phi(t)\rangle = c_1 |E_1\rangle \exp(-iE_1 t/\hbar) + c_2 |E_2\rangle \exp(-iE_2 t/\hbar), \tag{6.14}$$

is subjected to two cw fields,

$$\mathbf{E}(t) = \sum_{i=1}^{2} 2\hat{\mathbf{e}} \, \mathrm{Re}[\bar{\epsilon}(\omega_i) \exp(-i\omega_i t)]. \tag{6.15}$$

In accord with Section 3.1.1, $\omega_{2,1} \equiv (\omega_2 - \omega_1) = (E_1 - E_2)/\hbar$, so that excitation of $|E_1\rangle$ by ω_1 and of $|E_2\rangle$ by ω_2 lead to the same energy $E = E_1 + \hbar\omega_1 = E_2 + \hbar\omega_2$. Here, however, there is no dissociation at E. We focus attention on obtaining $\underline{\underline{\chi}}(\omega_1)$ and $\underline{\underline{\chi}}(\omega_2)$.

Given the interaction Hamiltonian of Eq. (3.17),

$$H_{\mathrm{MR}}(t) = -\sum_{i=1}^{2} 2\mathbf{d} \cdot \hat{\mathbf{e}} \, \mathrm{Re}[\bar{\epsilon}(\omega_i) \exp(-i\omega_i t)], \tag{6.16}$$

we have, according to perturbation theory, that the wave function $|\psi(t)\rangle$ resulting from the interaction of the superposition state of Eq. (6.14) with the field is

$$|\psi(t)\rangle = c_1 \exp\left(\frac{-iE_1 t}{\hbar}\right) |E_1\rangle + c_2 \exp\left(\frac{-iE_2 t}{\hbar}\right) |E_2\rangle + \sum_{m>2} c_m^{(1)}(t) \exp\left(\frac{-iE_m t}{\hbar}\right) |E_m\rangle, \tag{6.17}$$

with expansion coefficients given by

$$
c_m^{(1)}(t) = \frac{1}{\hbar} c_1 \sum_{k=1}^{2} \mathbf{d}_{m,1}^{\varepsilon_k} \left[\bar{\epsilon}(\omega_k) \frac{e^{i(\omega_{m,1}-\omega_k)t}}{\omega_{m,1} - \omega_k - i\gamma} + \bar{\epsilon}(\omega_k)^* \frac{e^{i(\omega_{m,1}+\omega_k)t}}{\omega_{m,1} + \omega_k - i\gamma} \right]
$$
$$
+ \frac{1}{\hbar} c_2 \sum_{k=1}^{2} \mathbf{d}_{m,2}^{\varepsilon_k} \left[\bar{\epsilon}(\omega_k) \frac{e^{i(\omega_{m,2}-\omega_k)t}}{\omega_{m,2} - \omega_k - i\gamma} + \bar{\epsilon}(\omega_k)^* \frac{e^{i(\omega_{m,2}+\omega_k)t}}{\omega_{m,2} + \omega_k - i\gamma} \right].
$$

$$(6.18)$$

Here γ is the average radiative line half-width at half maximum (HWHM) of bound levels (which has been introduced phenomenologically), $\omega_{m,n} = (E_m - E_n)/\hbar$, and $\mathbf{d}_{j,m}^{\varepsilon_k} = \langle E_j | \mathbf{d} \cdot \hat{\varepsilon}_k | E_m \rangle$. In obtaining this result, we have assumed that (a) the cw fields are turned on at $t \to -\infty$, at which time the system is in its initial superposition state [Eq. (6.14)], (b) the medium has no permanent dipole moment, and (c) $\langle E_1 | \mathbf{d} | E_2 \rangle = 0$ due to the fact that $|E_1\rangle$ and $|E_2\rangle$ are assumed to have the same parity.

The expectation value of the induced dipole is given according to perturbation theory by

$$
\langle \mathbf{d}^{\mathrm{ind}}(t) \rangle = \sum_m c_m^{(1)}(t)[c_1^* \mathbf{d}_{1,m} \exp(-i\omega_{m,1}t) + c_2^* \mathbf{d}_{2,m} \exp(-i\omega_{m,2}t)] + \text{c.c.}, \quad (6.19)
$$

where $\mathbf{d}_{j,m} = \langle E_j | \mathbf{d} | E_m \rangle$. The sum above is over all $|E_m\rangle$ states, including $m = 1, 2$. Inserting Eq. (6.18) in Eq. (6.19) gives 32 terms contributing to $\langle \mathbf{d}^{\mathrm{ind}}(t) \rangle$. Half of these terms are proportional to $|c_1|^2$ or $|c_2|^2$ and hence correspond to the independent effects of ω_1 and ω_2. The other half are proportional to $c_i c_j^*(i \neq j)$ and are interference terms resulting from the irradiation of the initial coherent superposition of two $|E_i\rangle$ states. Of these 16 interference terms, 8 do not oscillate with frequency ω_1 or ω_2 and hence do not contribute to the susceptibility at these frequencies. Identifying the terms that contribute at ω_1 and ω_2 gives the susceptibilities, for ω_i, $i = 1, 2$, as

$$
\underline{\underline{\chi}}(\omega_i) = \underline{\underline{\chi}}^n(\omega_i) + \underline{\underline{\chi}}^{in}(\omega_i) \tag{6.20}
$$

with

$$
\frac{\epsilon_0 \underline{\underline{\chi}}^n(\omega_i)}{\rho} \equiv \underline{\underline{\alpha}}^n(\omega_i) = \frac{|c_1|^2}{\hbar} \sum_m \mathbf{d}_{1,m} \otimes \mathbf{d}_{m,1} \left(\frac{1}{\omega_{m,1} - \omega_i - i\gamma} + \frac{1}{\omega_{m,1} + \omega_i - i\gamma} \right)
$$
$$
+ \frac{|c_2|^2}{\hbar} \sum_m \mathbf{d}_{2,m} \otimes \mathbf{d}_{m,2} \left(\frac{1}{\omega_{m,2} - \omega_i - i\gamma} + \frac{1}{\omega_{m,2} + \omega_i - i\gamma} \right),
$$
$$
\frac{\epsilon_0 \underline{\underline{\chi}}^{in}(\omega_1)}{\rho} \equiv \underline{\underline{\alpha}}^{in}(\omega_1) = \frac{c_1^* c_2 \bar{\epsilon}(\omega_2)}{\hbar \bar{\epsilon}(\omega_1)} \sum_m \left(\frac{\mathbf{d}_{1,m} \otimes \mathbf{d}_{m,2}}{\omega_{m,1} - \omega_1 - i\gamma} + \frac{\mathbf{d}_{m,2} \otimes \mathbf{d}_{1,m}}{\omega_{m,2} + \omega_1 - i\gamma} \right),
$$
$$
\frac{\epsilon_0 \underline{\underline{\chi}}^{in}(\omega_2)}{\rho} \equiv \underline{\underline{\alpha}}^{in}(\omega_2) = \frac{c_1 c_2^* \bar{\epsilon}(\omega_1)}{\hbar \bar{\epsilon}(\omega_2)} \sum_m \left(\frac{\mathbf{d}_{2,m} \otimes \mathbf{d}_{m,1}}{\omega_{m,2} - \omega_2 - i\gamma} + \frac{\mathbf{d}_{m,1} \otimes \mathbf{d}_{2,m}}{\omega_{m,1} + \omega_2 - i\gamma} \right).
$$

$$(6.21)$$

The term χ^n is the noninterfering component of χ and χ^{in} is the interfering component of χ. Here the $\otimes$ symbol denotes the *outer product* of two vectors. For example, $\mathbf{a} \otimes \mathbf{b}$, where $\mathbf{a}$ and $\mathbf{b}$ each have x, y, and z components, is a 3×3 matrix whose elements are $a_x b_x$, $a_x b_y$, and so on.

Below we assume that all of the incident light is linearly polarized along the z axis and denote the laboratory zz component of χ as χ_{zz}. Hence, as above [Eq. (6.11)], the desired index of refraction n is obtained from the susceptibility as $n(\omega_i) = \sqrt{1 + \chi_{zz}(\omega_i)}$.

Examination of Eq. (6.21) shows that $\chi(\omega)$ is comprised of two terms that are proportional to $|c_i|^2$ and that are associated with the traditional contribution to the susceptibility from state $|E_1\rangle$ and $|E_2\rangle$ independently, plus two field-dependent terms, proportional to $a_{i,j} = c_i^* c_j \bar{\epsilon}(\omega_j)/\bar{\epsilon}(\omega_i)$, which results from the coherent excitation of both $|E_1\rangle$ and $|E_2\rangle$ to the same total energy $E = E_1 + \hbar\omega_1 = E_2 + \hbar\omega_2$. As a consequence, changing $a_{i,j}$ alters the interference between excitation routes and allows for coherent control over the susceptibility. As in all bichromatic control scenarios, this control is achieved by altering the parameters in the state preparation in order to affect c_1, c_2 and/or by varying the relative intensities of the two laser fields. Note that control over $\chi(\omega_i)$ is expected to be substantial if $\bar{\epsilon}(\omega_j)/\bar{\epsilon}(\omega_i)$ is large. However, under these circumstances control over $\chi(\omega_j)$ is minimal since the corresponding interference term is proportional to $\bar{\epsilon}(\omega_i)/\bar{\epsilon}(\omega_j)$. Hence, effective control over the refractive index is possible only at one of ω_1 or ω_2.

Sample control results for $n(\omega)$, both off-resonance and near-resonance, for gaseous N_2 are shown below. Control is shown as a function of the relative laser phase $\delta\phi = \theta_{1,2} + \phi_2 - \phi_1$ where $\theta_{1,2}$ is the initial phase of $c_1^* c_2$ and ϕ_i ($i = 1, 2$) is the phase of $\bar{\epsilon}(\omega_i)$.

Figures 6.8 and 6.9 show the dependence of the real and imaginary parts of $n(\omega_1) = n'(\omega_1) + in''(\omega_1)$ on $|F_2/F_1| \equiv |\bar{\epsilon}(\omega_2)/\bar{\epsilon}(\omega_1)|$ for various different values of $\delta\phi$. Results are shown for the N_2 molecule in the $v_1 = 0$ ground vibrational level, and a superposition of rotational states, $|\Phi\rangle = \sqrt{0.8}|J_1 = 0, M_1 = 0\rangle + \sqrt{0.2}|J_2 = 2, M_2 = 0\rangle$, using $\omega_1 = 3 \times 10^{15}$ Hz, and $\omega_2 = 2.99775 \times 10^{15}$ Hz. The quantities (v_i, J_i, M_i) denote quantum numbers for vibration, rotation, and for the projection of the angular momentum along the z axis.

Consider first the case of $\delta\phi = -\pi/2$. Here $n''(\omega_1) = 0$ (corresponding to no absorption of the field) and $n'(\omega_1)$ is seen to grow linearly on the log–log plot for $|F_2/F_1| > 10$, that is, once the interference term in Eq. (6.21) dominates. Extensive control over n' is evident; for example, n' has changed by well over 10% by $|F_2/F_1| \sim 10^4$. This is in sharp contrast with the tiny refractive index changes associated, for example, with the optical Kerr effect, or self-focusing [215] (which change the index of refraction of N_2 by as little as 10^{-6} and which require laser intensities of $> 10^{12}$ W/cm^2).

Figures 6.8 and 6.9 display a broad range of behavior of n' and n''. For example, for the case of $\delta\phi = 0$ and π, the n' increases for $|F_2/F_1| > 1100$. For $\delta\phi = 0$ this increase is accompanied by positive n'', and hence by the absorption of the field by the molecules. By contrast, the case of $\delta\phi = \pi$ shows negative n'', that is, the field is amplified. Also of interest is the case of $\delta\phi = \pi/2$, which shows rapidly decreasing

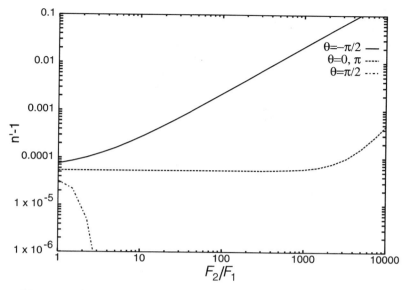

Figure 6.8 Dependence of the real part of $n(\omega)$ on F_2/F_1 in N_2 (in the superposition state described in the text) for different values of relative laser phase $\delta\phi$. Here $\delta\phi = -\pi/2$ (solid), $\delta\phi = 0$ and π (dashed), and $\delta\phi = \pi/2$ (dot–dash). (From Fig. 1, Ref. [213], where $\delta\phi$ was denoted θ.)

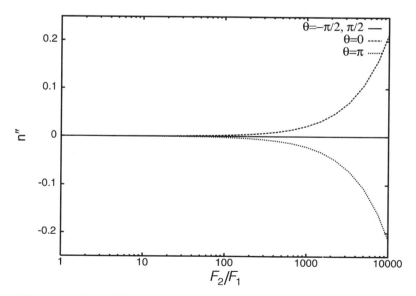

Figure 6.9 Dependence of imaginary part of $n(\omega)$ on F_2/F_1 in N_2 (in the superposition state described in the text) for different values of relative laser phase $\delta\phi$. Here $\delta\phi = -\pi/2$ and $\pi/2$ (solid), $\delta\phi = 0$ (dashed) and $\delta\phi = \pi$ (dotted). (From Fig. 1, Ref. [213] where $\delta\phi$ was denoted θ.)

n' with increasing $|F_2/F_1|$, accompanied by zero n''. Qualitatively similar results are attained for thermally distributed initial populations.

Consider now near-resonant excitation of the superposition state used in Figure 6.8. Excitation with $\omega_1 = 1.900884 \times 10^{16}$ Hz, and $\omega_2 = 1.900659 \times 10^{16}$ Hz excites the system, on resonance, to the $|v = 0, J = 1, M = 0\rangle$ bound state, of energy E_b, of the $b^1\Pi_u$ electronic state of N_2. Figures 6.10 and 6.11 show n' and n'' as a function of the detuning Δ (i.e., $E = E_1 + \hbar\omega_1 = E_2 + \hbar\omega_2 = E_b - \hbar\Delta$) for $\delta\phi = -\pi/2$, $F_2/F_1 = 1000$. Here large values of the index of refraction are seen to be associated with negligible absorption at $\Delta > 20$ GHz. Further, this significant change in the index of refraction leads to a substantial change in the speed with which light travels through the medium. Specifically, we can calculate the group velocity of light [217] as $v_g = c/(n' + \omega_1 dn'/d\omega_1)$, which, in this regime, can be estimated to be 150 m/s. More dramatic examples of slow light have been demonstrated by Hau and co-workers [217] using the electromagnetically induced transparency (EIT) effect (discussed in Section 9.1), in a Bose–Einstein condensate where they initially obtained a group velocity of 17 m/s.

The sensitivity to the control parameters is evident by changing $\delta\phi$ to $\pi/2 + 10^{-6}$, shown in Figure 6.11. Here the resultant n' is negative, corresponding to amplification of the beam. One should note then that a rich range of behavior is possible in near-resonance cases as the control variables $a_{i,j}$ are altered. Numerous examples are provided in Ref. [214].

An interference-based scheme of increasing the refractive index characterized by low absorption near resonances, was originally proposed by Scully and co-workers

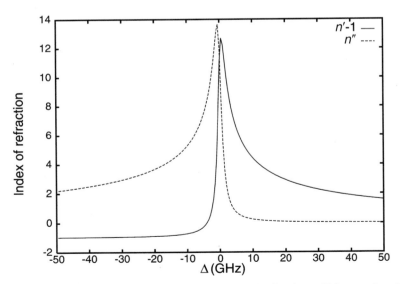

Figure 6.10   Real and imaginary parts of the index of refraction of N_2 as a function of detuning Δ for the initial superposition state described in the text with $F_2/F_1 = 1000$, for $\delta\phi = -\pi/2$. (From Fig. 3, Ref. [213].)

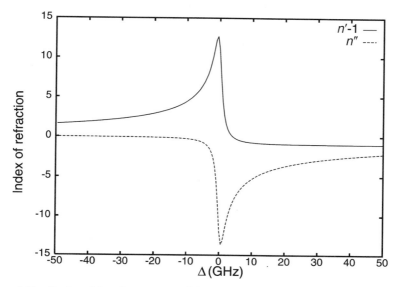

Figure 6.11 Real and imaginary parts of the index of refraction of N_2 as a function of detuning Δ for the initial superposition state described in the text with $F_2/F_1 = 1000$, for $\delta\phi = \pi/2 + 10^{-6}$. (From Fig. 3, Ref. [213].)

[218]. The scheme is similar to the one above, but relies upon excitation of two very closely spaced levels $|E_1\rangle$ and $|E_2\rangle$ using a pulsed laser. Once again, control is extensive.

6.3 MOLECULAR PHASE IN PRESENCE OF RESONANCES

Spectroscopy has long had, as its central goal, the use of light to extract molecular information, such as potential surfaces, system properties, and the like. Gordon and Seideman [219–221] have noted that because coherent control depends upon interference effects it introduces a new spectroscopic tool to extract previously unattainable information regarding the molecular continuum. In particular, they theoretically analyzed the physical origin of, and information contained in, the molecular phase term $\alpha_q(13)$ [Eq. (3.52)]. Recall that $\alpha_q(13)$ is defined as the phase of cumulative molecular matrix elements and appears prominently in the one-photon vs. three-photon interference term [Eq. (3.53)]. Further, Gordon and Seideman have focused experimentally and theoretically upon the information contained in the "phase lag" $\delta(q, q') = \alpha_q(13) - \alpha_{q'}(13)$ between two product channels q and q'. This would correspond, for example, to the phase difference between the two curves shown in Figure 3.12.

A formal analysis of the physical origins of the molecular phase and phase lag in the one- vs. three-photon scenario is provided below. To appreciate these results

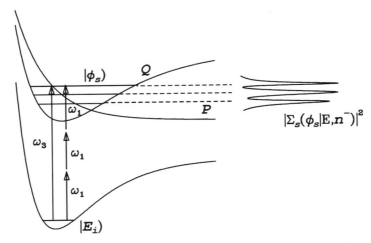

Figure 6.12   Schematic illustration of formation of a resonance from a bound state $|\phi_s\rangle$ and the way it is probed by a competing three-photon vs. one-photon transition from initial state $|E_i\rangle$.

requires that we recall the concept of a resonance [8], that is, a state that results from a bound state $|\phi_s\rangle$ coupled to a continuum. From the time-dependent viewpoint, such states decay into the continuum over some lifetime. The complementary picture in energy space (depicted in Fig. 6.12) is that the resonance is characterized by a rapid change, for example, a peak, in the cross section for the process as a function of energy. The energy range over which this change occurs defines the width of the resonance. If the resonances are sufficiently far apart in energy from one another so that their energy regions of influence do not overlap, they are called *isolated* resonances. Otherwise, a set of closely spaced resonances that affect the same energy region are called *overlapping* resonances.

The results of the analysis of Seideman and Gordon [221] show that a nonzero molecular phase $\alpha_q(13)$ can arise from a number of circumstances, discussed below. These include (1) multichannel scattering that displays coupling between the continua associated with different product channels, (2) the presence of a resonance connecting to a product channel, or (3) the presence of a resonance at an energy lower than the continuum that contributes to the phase of one of the interfering pathways. Thus, experimental evidence for a nonzero molecular phase provides information on the nature of the continuum. Further, the functional form of the energy dependence of the molecular phase provides insight into the character of the dynamics.

In several other simpler cases, discussed below, the molecular phase vanishes. We note in passing that, in accord with Eq. (3.53), the vanishing of the molecular phase does not imply that control is lost. However, a significant phase lag, from the viewpoint of control, is advantageous.

To see the origin of the molecular phase lag in the one- vs. three-photon control scenario, we reconsider the formalism discussed in Section 3.3.2. However, for notational simplicity, we denote the set of scattering eigenstates of the full Hamiltonian at energy E and fragment quantum numbers $\mathbf{n}$ in channel q as $|E, \mathbf{n}^-\rangle$, that is, we subsume the q within the labels $\mathbf{n}$.

6.3.1 Theory of Scattering Resonances

To understand how resonances affects the molecular phase, we briefly outline the basics of the theory of scattering resonances. We consider bound states $|\phi_s\rangle$ interacting with a set of continuum states denoted $|E, \mathbf{n}^-; 1\rangle$, where, as for the full scattering states $|E, \mathbf{n}^-\rangle$ [see Eq. (2.66)], the states $|E, \mathbf{n}^-; 1\rangle$ approach the free asymptotic solutions at infinite time:

$$\lim_{t\to\infty} e^{-i(E-i\epsilon)t/\hbar}|E, \mathbf{n}^-; 1\rangle = e^{-i(E-i\epsilon)t/\hbar}|E, \mathbf{n}^-; 0\rangle. \tag{6.22}$$

The interaction between the bound and continuum parts is depicted schematically in Figure 6.12. The emergence of resonances is derived most naturally via the use of *partitioning* technique [7], which focuses attention on either the bound or continuum subspace. In this approach one defines two projection operators Q and P, satisfying the equalities

$$QQ = Q, \qquad PP = P, \qquad PQ = QP = 0, \qquad P + Q = I, \tag{6.23}$$

where I is the identity operator. The Q and P operators are chosen to project out the subspaces spanned by the bound states and the continuum states, respectively. Further, as Eq. (6.23) indicates, they are orthogonal, for example, they may project onto two different electronic states, or any two spaces previously known to be orthogonal.

The full scattering incoming states $|E, \mathbf{n}^-\rangle$, introduced in Chapter 2, are eigenstates of the Schrödinger equation $[E - i\epsilon - H]|E, \mathbf{n}^-\rangle = 0$, where the $-i\epsilon$ serves to remind us of the incoming boundary conditions [Eq. (2.66)]. This equation can be rewritten as

$$[E - i\epsilon - H][P + Q]|E, \mathbf{n}^-\rangle = 0. \tag{6.24}$$

Multiplying this equation once by P and once by Q, we obtain, using Eq. (6.23), two coupled equations:

$$[E - i\epsilon - PHP]P|E, \mathbf{n}^-\rangle = PHQ|E, \mathbf{n}^-\rangle, \tag{6.25}$$

$$[E - i\epsilon - QHQ]Q|E, \mathbf{n}^-\rangle = QHP|E, \mathbf{n}^-\rangle. \tag{6.26}$$

We define two basis sets, $|E, \mathbf{n}^-; 1\rangle$ and $|\phi_s\rangle$, which are the solutions of the *homogeneous* (decoupled) parts of Eqs. (6.25) and (6.26). That is

$$[E - i\epsilon - PHP]|E, \mathbf{n}^-; 1\rangle = 0, \tag{6.27}$$

$$[E_s - QHQ]|\phi_s\rangle = 0. \tag{6.28}$$

Implicit in Eqs. (6.27) and (6.28) is that $|E, \mathbf{n}^-; 1\rangle \in P$ and $|\phi_s\rangle \in Q$, and as such they are orthogonal to one another. We, in fact, assume that each basis set spans the entire subspace to which it belongs, hence we can write an explicit representation of Q and P as

$$Q = \sum_s |\phi_s\rangle\langle\phi_s|, \tag{6.29}$$

$$P = \sum_{\mathbf{n}} \int dE |E, \mathbf{n}^-; 1\rangle\langle E, \mathbf{n}^-; 1|. \tag{6.30}$$

Using Eqs. (6.29) and (6.30) we can therefore write $|E, \mathbf{n}^-\rangle = [P + Q]|E, \mathbf{n}^-\rangle$ in terms of Q and P as

$$|E, \mathbf{n}^-\rangle = \sum_s |\phi_s\rangle\langle\phi_s|E, \mathbf{n}^-\rangle + \sum_{\mathbf{n'}} \int dE' |E', \mathbf{n'}^-; 1\rangle\langle E', \mathbf{n'}^-; 1|E, \mathbf{n}^-\rangle. \tag{6.31}$$

For this expansion to be useful it is necessary that we have explicit expressions for $\langle\phi_s|E, \mathbf{n}^-\rangle$ and for $\langle E', \mathbf{n'}^-; 1|E, \mathbf{n}^-\rangle$. These are obtained below.

We first solve for $P|E, \mathbf{n}^-\rangle$ by writing it as a sum of the homogeneous solution of Eq. (6.27) and a particular solution of Eq. (6.25) obtained by inverting $[E - i\epsilon - PHP]$,

$$P|E, \mathbf{n}^-\rangle = P|E, \mathbf{n}^-; 1\rangle + [E - i\epsilon - PHP]^{-1}PHQ|E, \mathbf{n}^-\rangle. \tag{6.32}$$

Substituting this solution into Eq. (6.26) we obtain that

$$[E - i\epsilon - QHQ]Q|E, \mathbf{n}^-\rangle = QHP|E, \mathbf{n}^-; 1\rangle + QHP[E - i\epsilon - PHP]^{-1}PHQ|E, \mathbf{n}^-\rangle. \tag{6.33}$$

Reordering terms in this equation gives

$$[E - i\epsilon - Q\mathcal{H}Q]Q|E, \mathbf{n}^-\rangle = QHP|E, \mathbf{n}^-; 1\rangle \tag{6.34}$$

where

$$Q\mathcal{H}Q \equiv QHQ + QHP[E - i\epsilon - PHP]^{-1}PHQ. \tag{6.35}$$

Equation (6.34) can be solved to yield

$$Q|E, \mathbf{n}^-\rangle = [E - i\epsilon - Q\mathcal{H}Q]^{-1}QHP|E, \mathbf{n}^-; 1\rangle. \qquad (6.36)$$

An explicit representation of Eq. (6.36) is obtained by using the well-known identity (obtained by a similar contour integration to that depicted in Fig. 2.1),

$$[E - i\epsilon - PHP]^{-1} = \mathbf{P}_v[E - PHP]^{-1} + i\pi\delta(E - PHP), \qquad (6.37)$$

with $\mathbf{P}_v$ denoting a Cauchy principal value integral:

$$\mathbf{P}_v\int_a^b dE' \frac{f(E')}{E - E'} \equiv \lim_{\epsilon\to 0}\int_a^{E-\epsilon} dE' \frac{f(E')}{E - E'} + \int_{E+\epsilon}^b dE' \frac{f(E')}{E - E'}. \qquad (6.38)$$

The $\mathbf{P}_v[E - PHP]^{-1}$ operator above is given, using the spectral resolution of an operator, as

$$\mathbf{P}_v[E - PHP]^{-1} \equiv \sum_{\mathbf{n}} \mathbf{P}_v \int \frac{dE'}{E - E'} |E', \mathbf{n}^-; 1\rangle\langle E', \mathbf{n}^-; 1|. \qquad (6.39)$$

Using Eqs. (6.37) and (6.39) we can write $Q\mathcal{H}Q$ of Eq. (6.35) as

$$Q\mathcal{H}Q = QHQ + QHP\mathbf{P}_v[E - PHP]^{-1}PHQ + i\pi QHP\delta(E - PHP)PHQ. \qquad (6.40)$$

Assuming for simplicity the case of "noninteracting" overlapping resonances in which (by definition) $Q\mathcal{H}Q$ is diagonal (the case of overlapping and interacting resonances is dealt with in Section 9.1), we can use Eqs. (6.39) and (6.40) to write the representation of $[E - i\epsilon - Q\mathcal{H}Q]$ in the $\{|\phi_s\rangle\}$ basis as

$$\langle\phi_s|[E - i\epsilon - Q\mathcal{H}Q]|\phi_s\rangle = [E - E_s - \Delta_s(E) - i\Gamma_s(E)/2], \qquad (6.41)$$

where

$$V(s|E, \mathbf{n}) \equiv \langle\phi_s|QHP|E, \mathbf{n}^-; 1\rangle \equiv \langle\phi_s|H|E, \mathbf{n}^-; 1\rangle, \qquad (6.42)$$

$$\Gamma_s(E) \equiv \sum_{\mathbf{n}} 2\pi|V(s|E, \mathbf{n})|^2, \qquad (6.43)$$

$$\Delta_s(E) \equiv \mathbf{P}_v \sum_{\mathbf{n}} \int \frac{dE'}{E - E'} |V(s|E', \mathbf{n})|^2. \qquad (6.44)$$

It follows from Eqs. (6.41), (6.43), and (6.44) that the $\langle\phi_s|E, \mathbf{n}^-\rangle$ overlap integrals are given as

$$\langle\phi_s|E, \mathbf{n}^-\rangle = \frac{V(s|E, \mathbf{n})}{E - E_s - \Delta_s(E) - i\Gamma_s(E)/2}. \qquad (6.45)$$

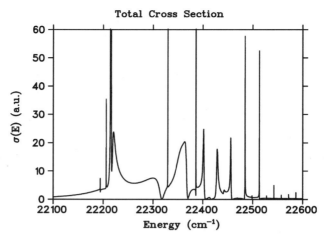

Figure 6.13 Absorption spectrum resulting from the sum of overlapping resonances.

Using Eqs. (6.32) and (6.37) we obtain the expression

$$\langle E', \mathbf{m}^-; 1 | E, \mathbf{n}^- \rangle = \delta(E - E')\delta_{\mathbf{n,m}} + \sum_s V(E', \mathbf{m}|s)$$

$$\times \left[\mathbf{P}_v \frac{1}{E - E'} + i\pi\delta(E - E') \right] \langle \phi_s | E, \mathbf{n}^- \rangle. \qquad (6.46)$$

Given Eqs. (6.45) and (6.46) we can express, via Eq. (6.31), the full scattering wave function $|E, \mathbf{n}^-\rangle$ in terms of $|\phi_s\rangle$ and $|E, \mathbf{n}^-; 1\rangle$.

The form given in Eq. (6.45) gives rise to a Lorentzian shape that is depicted schematically in Figure 6.12. A spectrum resulting from the excitation to a collection of overlapping resonances is shown in Figure 6.13.

6.3.2 Three-Photon vs. One-Photon Coherent Control in Presence of Resonances

We now apply the above partitioning method to the three-photon vs. one-photon coherent control treated in detail in Section 3.3.2 to consider control in the presence of resonances. Consider a molecule initially in a bound energy eigenstate $|E_i\rangle$ subjected to two co-propagating pulses,

$$\mathbf{E}(t) = 2 \int d\omega_3 \, \mathrm{Re}[\hat{\varepsilon}_3 \bar{\varepsilon}_3(\omega_3) \exp(-i\omega_3 t) + \hat{\varepsilon}_1 \bar{\varepsilon}_1(\omega_1) \exp(-i\omega_1 t)], \qquad (6.47)$$

where $3\omega_1 = \omega_3$. After the pulse is over, some of the molecules have absorbed either a single photon or three photons from the field, and the excited wavepacket is given by

$$|\Psi(t)\rangle = \sum_{\mathbf{n}} \int dE |E, \mathbf{n}^-\rangle A_{\mathbf{n}}(E) \exp\left(\frac{-iEt}{\hbar}\right), \qquad (6.48)$$

where $E = E_i + \hbar\omega_3 = E_i + 3\hbar\omega_1$. Assuming electric-dipole interaction, $H_{MR}(t) = -\mathbf{d} \cdot \mathbf{E}(t)$, and that $\langle E_i|E, \mathbf{n}^-\rangle = 0$, $A_{\mathbf{n}}(E)$, the continuum preparation coefficients are given by

$$A_{\mathbf{n}}(E) = (2\pi i/\hbar)\langle E, \mathbf{n}^-|[\bar{\epsilon}_3(\omega_{E,i})\mathbf{d}_{e,g} + \bar{\epsilon}_1^3(\omega_{E,i}/3)T_{e,g}]|E_i\rangle. \qquad (6.49)$$

Here, $T_{e,g}$ is the three-photon transition operator, given in Eq. (3.44) as

$$T_{e,g} = \sum_{e'e''} \mathbf{d}_{e,e'}(E_i - H_{e'} + 2\hbar\omega_1 - i\gamma_2)^{-1}\mathbf{d}_{e',e''}(E_i - H_{e''} + \hbar\omega_1 - i\gamma_1)^{-1}\mathbf{d}_{e'',g}. \qquad (6.50)$$

It follows from Eqs. (6.49) and (6.31) that $A_{\mathbf{n}}(E)$ can be written as

$$A_{\mathbf{n}}(E) = \left(\frac{2\pi i}{\hbar}\right)\left\{\sum_s \langle E, \mathbf{n}^-|\phi_s\rangle \mathcal{T}_{s,i}(\bar{\epsilon}_3, \bar{\epsilon}_1)\right.$$
$$\left. + \sum_{\mathbf{m}} \int dE' \, \langle E, \mathbf{n}^-|E', \mathbf{m}^-; 1\rangle \mathcal{T}_{\mathbf{m},i}(E', \bar{\epsilon}_3, \bar{\epsilon}_1)\right\}, \qquad (6.51)$$

where

$$\mathcal{T}_{s,i}(\bar{\epsilon}_3, \bar{\epsilon}_1) \equiv \langle\phi_s|[\bar{\epsilon}_3(\omega_{E,i})\mathbf{d}_{e,g} + \bar{\epsilon}_1^3(\omega_{E,i}/3)T_{e,g}]|E_i\rangle,$$
$$\mathcal{T}_{\mathbf{m},i}(E', \bar{\epsilon}_3, \bar{\epsilon}_1) \equiv \langle E', \mathbf{m}^-; 1|[\bar{\epsilon}_3(\omega_{E,i})\mathbf{d}_{e,g} + \bar{\epsilon}_1^3(\omega_{E,i}/3)T_{e,g}]|E_i\rangle.$$

Substituting Eq. (6.46) into Eq. (6.51) gives the amplitude as

$$A_{\mathbf{n}}(E) = \frac{2\pi i}{\hbar} \mathcal{T}_{n,i}(E, \bar{\epsilon}_3, \bar{\epsilon}_1) + \frac{2\pi i}{\hbar}\sum_s \langle E, \mathbf{n}^-|\phi_s\rangle$$
$$\times \left[\mathcal{T}_{s,i}(\bar{\epsilon}_3, \bar{\epsilon}_1) + \sum_{\mathbf{m}} \int \frac{dE'}{E + i\epsilon - E'} V^*(s|E', \mathbf{m})\mathcal{T}_{\mathbf{m},i}(E', \bar{\epsilon}_3, \bar{\epsilon}_1)\right]. \qquad (6.52)$$

Inserting Eqs. (6.45) and (6.46) into Eq. (6.52) we obtain that

$$
A_{\mathbf{n}}(E) = \left(\frac{2\pi i}{\hbar}\right)\langle E, \mathbf{n}^-; 1|\left[\bar{\epsilon}_3(\omega_{E,i})\mathrm{d}_{e,g} + \bar{\epsilon}_1^3\left(\frac{\omega_{E,i}}{3}\right)T_{e,g}\right]|E_i\rangle
$$

$$
+ \left(\frac{2\pi i}{\hbar}\right)\sum_s \frac{V^*(E, \mathbf{n}|s)}{E - E_s - \Delta_s(E) + i\Gamma_s(E)/2}
$$

$$
\times \left\{\langle\phi_s|\left[\bar{\epsilon}_3(\omega_{E,i})\mathrm{d}_{e,g} + \bar{\epsilon}_1^3\left(\frac{\omega_{E,i}}{3}\right)T_{e,g}\right]|E_i\rangle\right.
$$

$$
+ \sum_{\mathbf{m}} -i\pi V^*(s|E, \mathbf{m})\langle E, \mathbf{m}^-; 1|\left[\bar{\epsilon}_3(\omega_{E,i})\mathrm{d}_{e,g} + \bar{\epsilon}_1^3\left(\frac{\omega_{E,i}}{3}\right)T_{e,g}\right]|E_i\rangle
$$

$$
\left. + \mathbf{P}_v\int\frac{dE'}{E - E'}V^*(s|E', \mathbf{m})\langle E', \mathbf{m}^-; 1|\left[\bar{\epsilon}_3(\omega_{E,i})\mathrm{d}_{e,g} + \bar{\epsilon}_1^3\left(\frac{\omega_{E,i}}{3}\right)T_{e,g}\right]|E_i\rangle\right\}.
$$

$$(6.53)$$

The probability $P_{\mathbf{n}}(E)$ of observing the final state $\mathbf{n}$ at energy E is given as $P_{\mathbf{n}}(E) = |A_{\mathbf{n}}(E)|^2$. In accord with Eq. (3.48) we identify the components associated with one-photon excitation $P_{\mathbf{n}}^{(1)}(E)$, three-photon excitation $P_{\mathbf{n}}^{(3)}(E)$, and one-photon/three-photon interference $P_{\mathbf{n}}^{(13)}(E)$ so that

$$
P_{\mathbf{n}}(E) = P_{\mathbf{n}}^{(1)}(E) + P_{\mathbf{n}}^{(3)}(E) + P_{\mathbf{n}}^{(13)}(E). \tag{6.54}
$$

From Eq. (6.53) we have that the interference term is

$$
P_{\mathbf{n}}^{(13)}(E) = 2\bar{\epsilon}_3(\omega_{E,i})\bar{\epsilon}_1^3\left(\frac{\omega_{E,i}}{3}\right)\left(\frac{2\pi}{\hbar}\right)^2
$$

$$
\times \mathrm{Re}\left\{T(i|E, \mathbf{n}) + \sum_{s'}\frac{V(s'|E, \mathbf{n})[T(i, s') + i\Gamma_{s'}^T(i|E)/2 + \Delta_{s'}^T(i|E)]}{E - E_{s'} - \Delta_{s'}(E) - i\Gamma_{s'}(E)/2}\right\}
$$

$$
\times \left\{\mathrm{d}(E, \mathbf{n}|i) + \sum_s \frac{V^*(E, \mathbf{n}|s)[\mathrm{d}(s|i) - i\Gamma_s^d(E|i)/2 + \Delta_s^d(E|i)]}{E - E_s - \Delta_s(E) + i\Gamma_s(E)/2}\right\}, \tag{6.55}
$$

where

$$
\mathrm{d}(E, \mathbf{n}|i) \equiv \langle E, \mathbf{n}^-; 1|\mathrm{d}_{e,g}|E_i\rangle, \qquad \mathrm{d}(s|i) \equiv \langle\phi_s|\mathrm{d}_{e,g}|E_i\rangle,
$$

$$
T(i|E, \mathbf{n}) \equiv \langle E_i|T_{e,g}^*|E, \mathbf{n}^-; 1\rangle, \qquad T(i|s') \equiv \langle E_i|T_{e,g}^*|\phi_{s'}\rangle,
$$

$$
\Delta_s^{\mathrm{d}}(E|i) \equiv \sum_{\mathbf{m}}\mathbf{P}_v\int\frac{dE'}{E - E'}V^*(s|E', \mathbf{m})\mathrm{d}(E', \mathbf{m}|i),
$$

$$
\Gamma_s^{\mathrm{d}}(E|i) \equiv 2\pi\sum_{\mathbf{m}}V^*(s|E, \mathbf{m})\mathrm{d}(E, \mathbf{m}|i)
$$

$$
\Delta_s^T(i|E) \equiv \sum_{\mathbf{m}}\mathbf{P}_v\int\frac{dE'}{E - E'}T(i|E', \mathbf{m})V(E', \mathbf{m}|s),
$$

$$
\Gamma_s^T(i|E) \equiv 2\pi\sum_{\mathbf{m}}T(i|E, \mathbf{m})V(E, \mathbf{m}|s). \tag{6.56}
$$

These equations are completely general. They allow us to look at a variety of limiting cases, discussed below.

Case A: Indirect Transition to Isolated Resonance   Here the photodissociation occurs by excitation to a single resonance, followed by a transition from the resonance to the continuum. In this case the sum over s reduces to a single term, and the direct optical transitions to the continuum are suppressed. That is,

$$d(E, \mathbf{n}|i) = T(i|E, \mathbf{n}) = 0, \tag{6.57}$$

and therefore

$$\Gamma_s^d(E|i) = \Delta_s^d(E|i) = \Delta_s^T(i|E) = \Gamma_s^T(i|E) = 0. \tag{6.58}$$

Equation (6.55) therefore becomes

$$P_{\mathbf{n}}^{(13)}(E) = 2\bar{\epsilon}_3(\omega_{E,i})\bar{\epsilon}_1^3\left(\frac{\omega_{E,i}}{3}\right)\left(\frac{2\pi}{\hbar}\right)^2 \frac{|V(s|E, \mathbf{n})|^2 \, \text{Re}[T(i|s)d(s|i)]}{[E - E_s - \Delta_s(E)]^2 + \Gamma_s^2(E)/4}. \tag{6.59}$$

$P_{\mathbf{n}}^{(13)}(E)$ is seen to be real since both $|\phi_s\rangle$ and $|E_i\rangle$ are bound wavefunctions and can be chosen as real functions. Hence the molecular phase, that is, the phase of this term, is zero.

Case B: Purely Direct Transition to Continuum   In this case the resonances are not optically coupled to the initial state $|E_i\rangle$, that is, $d(s|i) \equiv \langle\phi_s|d_{e,g}|E_i\rangle = T(i|s') \equiv \langle E_i|T^*_{e,g}|\phi_{s'}\rangle = 0$, and only direct transitions to the continuum survive. We obtain that

$$P_{\mathbf{n}}^{(13)}(E) = 2\bar{\epsilon}_3(\omega_{E,i})\bar{\epsilon}_1^3(\omega_{E,i}/3)(2\pi/\hbar)^2 \, \text{Re}[T(i|E, \mathbf{n})d(E, \mathbf{n}|i)]. \tag{6.60}$$

Here there are two different possibilities. If the physics is such that $|E, \mathbf{n}^-; 1\rangle$ is a solution of a single-channel problem, then its coordinate space representation can always be written [8] as

$$\langle r|E, \mathbf{n}^-; 1\rangle = e^{i\delta}\langle r|E, \mathbf{n}^-; 1\rangle_R, \tag{6.61}$$

where $\langle r|E, \mathbf{n}^-; 1\rangle_R$ is a real function and δ is a phase that is independent of r. Under these circumstances the phase of the $T(i|E, \mathbf{n})$ exactly cancels the phase of $d(E, \mathbf{n}|i)$ and the phase of the $T(i|E, \mathbf{n})d(E, \mathbf{n}|i)$ products vanishes. As a result, the molecular phase is zero. If, on the other hand, the scattering involves coupling between many channels, then $|E, \mathbf{n}^-; 1\rangle$ is a solution of a multichannel problem, the factorization in Eq. (6.61) no longer holds, and the molecular phase is both nonzero and a function of $\mathbf{n}$.

Case C: Indirect Transition to Set of Overlapping Resonances   Here the dynamics occurs by excitation to a set of overlapping resonances, with subsequent decay into the continuum. In this case there is a sum over the resonances in Eq. (6.55), but there is no direct transition to the continuum. That is, Eq. (6.57) still holds and Eq. (6.55) becomes

$$P_{\mathbf{n}}^{(13)}(E) = 2\bar{\epsilon}_3(\omega_{E,i})\bar{\epsilon}_1^3(\omega_{E,i}/3)(2\pi/\hbar)^2 \, \text{Re}[\mathcal{D}(E, \mathbf{n})T^*(E, \mathbf{n})], \qquad (6.62)$$

where

$$
\begin{aligned}
\mathcal{D}(E, \mathbf{n}) &\equiv \sum_s \frac{V(s|E, \mathbf{n})d(s|i)}{E - E_s - \Delta_s(E) - i\Gamma_s(E)/2} \\[2mm]
T^*(E, \mathbf{n}) &\equiv \sum_s \frac{V^*(E, \mathbf{n}|s)T(i|s)}{E - E_s - \Delta_s(E) + i\Gamma_s(E)/2}.
\end{aligned}
\qquad (6.63)
$$

Here no factorization of the $V(s|E, \mathbf{n})$ terms out of the sum is possible, the molecular phase $\alpha_{\mathbf{n}}(E)$ is now a function of $\mathbf{n}$. Note that the energy dependence of Eq. (6.62) is distinctly different than that in Eq. (6.60) providing insight into the nature of the continuum.

Case D: Sum of Direct and Indirect Transition to Isolated Resonance

Here the product is reached either via a single resonance or directly via the continuum. Hence the sum over s in Eq. (6.55) reduces to a single term, but the direct optical transitions to the continuum are *not* suppressed. This gives the case of a Fano-type interference [222] to an isolated resonance, where the two pathways to the continuum interfere with one another. In this case Eq. (6.55) becomes

$$
\begin{aligned}
&P_{\mathbf{n}}^{(13)}(E) \\[2mm]
&= 2\bar{\epsilon}_3(\omega_{E,i})\bar{\epsilon}_1^3\left(\frac{\omega_{E,i}}{3}\right)\left(\frac{2\pi}{\hbar}\right)^2 \text{Re}\Bigg\{ d(E, \mathbf{n}|i)T(i|E, \mathbf{n}) \\[2mm]
&\quad + \frac{d(E, \mathbf{n}|i)V(s|E, \mathbf{n})[T(i, s) + i\Gamma_s^T(i|E)/2 + \Delta_s^T(i|E)]}{E - E_s - \Delta_s(E) - i\Gamma_s(E)/2} \\[2mm]
&\quad + \frac{T(i|E, \mathbf{n})V^*(E, \mathbf{n}|s)[d(s|i) - i\Gamma_s^d(E|i)/2 + \Delta_s^d(E|i)]}{E - E_s - \Delta_s(E) + i\Gamma_s(E)/2} \\[2mm]
&\quad + \frac{|V(s|E, \mathbf{n})|^2[d(s|i) - i\Gamma_s^d(E|i)/2 + \Delta_s^d(E|i)][T(i, s) + i\Gamma_s^T(i|E)/2 + \Delta_s^T(i|E)]}{[E - E_s - \Delta_s(E)]^2 + \Gamma_s^2(E)/4} \Bigg\}.
\end{aligned}
\qquad (6.64)
$$

Once again, the molecular phase $\alpha_{\mathbf{n}}(E)$ is nonzero and is a function of $\mathbf{n}$.

Thus, we see, in accord with extensive work by Gordon and Seideman, that the presence of a nonzero molecular phase provides insight into features of the continuum [221]. Further, the detailed nature of the energy dependence of the molecular phase assists in distinguishing between the various cases discussed above.

As an example consider, once again, the one- vs. three-photon excitation of HI, which undergoes two competitive processes:

$$HI \rightarrow HI^+ + e^-,$$
$$HI \rightarrow H + I. \tag{6.65}$$

As noted in Section 3.3.2, Gordon and co-workers have measured (see Fig. 3.12) the modulation of the HI^+ and I^+ (from the H + I channel) as the relative laser phase of ω_1 and ω_3 is varied. Experiments of this kind provide the phase lag $\delta(q, q')$ plotted as a function of energy in Figure 6.14. In particular, this figure shows the phase lag data for the above two channels in the region of the overlapping $5d\pi$ and $5d\delta$ Rydberg resonances of HI at 356 nm. Also shown is the phase lag for the mixture of HI^+ and H_2S^+, where the phase lag of the latter species is known to vanish. Both phase lag curves show a maximum at 356.1 nm, in the region of the resonance, as well as another maximum at 355.2 nm. The peaks are atop an almost zero background, indicating that in this region the continua are elastic. Thus, this figure demonstrates both the resonance and nonresonant contributions to the phase lag.

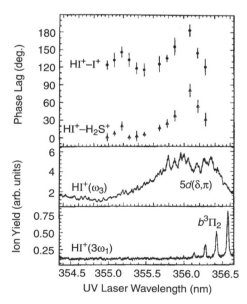

Figure 6.14 Phase lag spectrum (top) for the photodissociation and photoionization of HI (circles) and for photoionization of a mixture of HI and H_2S (triangles). Bottom two panels are the one- and three-photon ionization spectra of HI. (From Fig. 7, Ref. [221].)

6.4 CONTROL OF CHAOTIC DYNAMICS

Studies in classical nonlinear mechanics over the past few decades have shown that systems can be categorized as lying between two limits: that of integrable dynamics and that of chaotic dynamics [2, 223–225]. In the integrable case the dynamics of a system of N degrees of freedom possesses N conserved integrals of motion and is stable with respect to small external perturbations. In the chaotic case the system dynamics usually possesses only symmetry-based integrals of motion, such as the total energy and angular momentum, and the dynamics is extremely sensitive to initial conditions and external perturbations. This categorization extends to quantum mechanics in the sense that a system is said to be quantum mechanically chaotic if its classical counterpart is classically chaotic. Numerous computational studies [226] have shown that quantum systems do display characteristics of classical chaos if they are sufficiently close to the classical limit, a manifestation of the correspondence principle [159]. It is expected that the vast majority of realistic systems are sufficiently complex so as to display some degree of chaotic behavior.

Considering the sensitivity of classical chaotic systems to external perturbations, and the ubiquitous nature of chaotic dynamics in larger systems, it is important to establish that quantum mechanics allows for control in chaotic systems as well.

One simple molecular system that displays quantum chaos is the rotational excitation of a diatomic molecule using pulsed microwave radiation [227]. Under the conditions adopted below, this system is a molecular analog of the "delta-kicked rotor," that is, a rotor that is periodically kicked by a delta function potential, which is a paradigm for chaotic dynamics [228, 229]. The observed energy absorption of such systems is called *quantum chaotic diffusion*.

If the orientation of a diatomic molecule is described by two angles θ and ϕ [230], then the corresponding Hamiltonian is

$$H = \frac{\hat{J}^2}{2I} + \mathbf{d} \cdot \mathbf{E}_0 \cos\theta \sum_n \Delta\left(\frac{t}{T} - n\right), \tag{6.66}$$

where $\hat{J}$ is the angular momentum operator in three dimensions:

$$\hat{J}^2 = -\hbar^2 \left[\frac{1}{\sin\theta} \frac{\partial}{\partial\theta}\left(\sin\theta \frac{\partial}{\partial\theta}\right) + \frac{1}{\sin^2\theta} \frac{\partial^2}{\partial\phi^2} \right]. \tag{6.67}$$

Here $\mathbf{d}$ is the molecular electric dipole moment, $\mathbf{E}_0$ is the amplitude of the driving field whose polarization direction defines the z direction, I is the moment of inertia of the molecule about an axis perpendicular to the symmetry axis, and $\Delta(t/T - n)$ is the pulse shape function of the form

$$\Delta\left(\frac{t}{T} - n\right) = 1 + 2\sum_{m=1}^{m=7} \cos\left[2m\pi\left(\frac{t}{T} - n - \frac{1}{2}\right)\right]. \tag{6.68}$$

Eigenstates of the Hamiltonian H are $|n_J, m_J\rangle$, where n_J is the angular momentum quantum number with projection m_J along the z axis.

As shown by Blümel and co-workers [227], the kicked CsI molecule is particularly appropriate candidate for this study since it has a large dipole moment, which increases the molecule–field coupling strength, and the rotation–vibration coupling is small at low excitation energies so that one may consider solely rotational excitation. We consider then the dynamics of CsI in the indicated pulsed field, in a parameter range known to display classical chaos [231].

To demonstrate control of chaotic dynamics, we assume that an initial superposition of Hamiltonian eigenstates of the form [231]

$$|\psi(0)\rangle = \cos\alpha\,|j_1, 0\rangle + \sin\alpha\,\exp(-i\beta)|j_2, 0\rangle \tag{6.69}$$

has been previously prepared. This system is now subjected to pulsed microwave irradiation, and the rotational energy absorption is measured. In particular, we define the dimensionless rotational energy $\tilde{E} \equiv \sum_j P_j j(j+1)\tau^2/2$, $\tau = \hbar T/I$, where P_j is the occupation probability of the $|j, 0\rangle$ state, as a measure of the absorbed energy.

To anticipate the result of pulsed excitation of a superposition state, note from Eqs. (6.66) and (6.68) that the Hamiltonian is strictly periodic in time. We denote the time evolution operator associated with one period T as $\hat{F}$. Although it is not possible to give an explicit form of $\hat{F}$ in the kicked molecule case, the existence of this formal solution yields a stroboscopic description of the dynamics,

$$|\psi(nT)\rangle = \hat{F}^{n-1}|\psi[(n-1)T]\rangle = \hat{F}^n|\psi(0)\rangle, \tag{6.70}$$

where n is an integer.

The operator $\hat{F}$ can be formally diagonalized by a unitary transformation $\underline{\underline{U}}$ so that,

$$\langle j_a, 0|\hat{F}|j_b, 0\rangle = \sum_{j_c} \exp(-i\phi_{j_c}) U^*_{j_c, j_a} U_{j_c, j_b}, \tag{6.71}$$

where $U_{j_c, j_a} \equiv \langle j_c, 0|\hat{U}|j_a, 0\rangle$ ($j_a = 0, 1, 2, \ldots$) is the eigenvector with eigenphase ϕ_{j_c}. Moreover, since the basis states $|j, 0\rangle$ are time-reversal invariant, one can prove that the matrix elements U_{j_c, j_b} can be chosen as real numbers [232], that is,

$$U^*_{j_c, j_a} = U_{j_c, j_a}, \qquad j_a, j_c = 0, 1, 2, \ldots. \tag{6.72}$$

Further, evaluating $\tilde{E}$ at $t = NT$ with Eqs. (6.69), (6.71), and (6.72) gives

$$
\begin{aligned}
\frac{2\tilde{E}}{\tau^2} &= \langle \psi(0) | \hat{F}^{-N} \frac{\hat{J}^2}{\hbar^2} \hat{F}^N | \psi(0) \rangle \\
&= \cos^2 \alpha \sum_{\substack{jj_a j_b}} j(j+1) U_{j_a j_1} U_{j_b j} U_{j_a j} U_{j_b j_1} e^{iN(\phi_{j_a} - \phi_{j_b})} \\
&\quad + \sin^2 \alpha \sum_{\substack{jj_a j_b}} j(j+1) U_{j_a j_2} U_{j_b j} U_{j_a j} U_{j_b j_2} e^{iN(\phi_{j_a} - \phi_{j_b})} \\
&\quad + \frac{1}{2} \sin(2\alpha) \left(e^{-i\beta} \sum_{\substack{jj_a j_b}} j(j+1) U_{j_a j_1} U_{j_b j_2} U_{j_a j} U_{j_b j} e^{iN(\phi_{j_a} - \phi_{j_b})} + \text{c.c.} \right).
\end{aligned}
\tag{6.73}
$$

Evidently, the first two terms are incoherent since they do not depend on the value of the phase β in Eq. (6.69). They represent quantum dynamics associated with each of the states $|j_1, 0\rangle$ and $|j_2, 0\rangle$ independently. The last two terms represent interference effects due to initial-state coherence between $|j_1, 0\rangle$ and $|j_2, 0\rangle$. Hence, the absorption of rotational energy in this system, that is, quantum chaotic diffusion, can be controlled by manipulating the quantum phase β in the initial state [Eq. (6.69)], which corresponds to manipulating the interference term in Eq. (6.73).

In Figure 6.15 we present a representative example of phase control in this system. In the chosen parameter region the underlying classical dynamics of rotational excitation is strongly chaotic [231] and the excitation is far off-resonance, with many levels excited. We choose $j_1 = 1$ and $j_2 = 2$ to create the initial superposition state $(|1, 0\rangle \pm 2, 0\rangle)/\sqrt{2}$, that is, $\alpha = \pi/4$ and $\beta = 0, \pi$ in Eq. (6.69). (Such states can be prepared experimentally by, for example, STIRAP, a technique discussed in detail in Section 9.1.) The results, shown in Figure 6.15, display striking phase

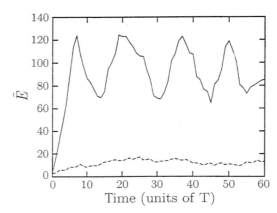

Figure 6.15 Dimensionless rotational energy of the kicked diatomic molecule $\tilde{E} = \sum_j P_j j(j+1)\tau^2/2$ versus time (in units of T). Solid line and dashed lines are for initial states $(|1, 0\rangle + |2, 0\rangle)/2^{1/2}$ and $(|1, 0\rangle - |2, 0\rangle)/2^{1/2}$, respectively, for $\tau = 1.2$, $k = 4.8$. (From Fig. 2, Ref. [231].)

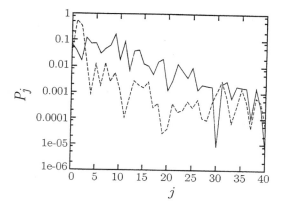

Figure 6.16 Occupation probability P_j versus the rotational quantum number j at $t = 60T$. Solid line and dashed lines are for initial states $(|1, 0\rangle + |2, 0\rangle)/2^{1/2}$ and $(|1, 0\rangle - |2, 0\rangle)/2^{1/2}$, respectively, for $\tau = 1.2$, $k = 4.8$. (From Fig. 3, Ref. [231].)

control. That is, $(|1, 0\rangle - 2, 0\rangle)/\sqrt{2}$ shows almost no energy absorption at all, whereas the $(|1, 0\rangle + 2, 0\rangle)/\sqrt{2}$ case shows extraordinarily fast energy absorption [231] before it essentially stops at $t \approx 10T$. Note (1) that this huge difference is achieved solely by changing the initial relative phase between the two participating states $|1, 0\rangle$ and $|2, 0\rangle$ in the initial superposition state, and (2) that by contrast, each of $|1, 0\rangle$ or $|2, 0\rangle$ individually would give very similar diffusion behavior lying between the solid and dashed lines in Figure 6.15. This shows that the two participating states $|1, 0\rangle$ and $|2, 0\rangle$ can either constructively or destructively interfere with one another, even though the underlying classical dynamics is strongly chaotic. A

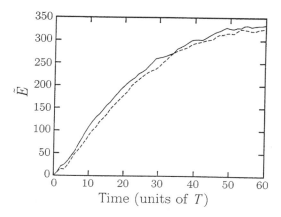

Figure 6.17 Occupation probability P_j versus the rotational quantum number j at $t = 60T$. Solid line and dashed lines are for the initial states $(|1, 0\rangle + |2, 0\rangle)/2^{1/2}$ and $(|1, 0\rangle - |2, 0\rangle)/2^{1/2}$, respectively, for $\tau = 1.2$, $k = 4.8$. (From Fig. 5, Ref. [231].)

detail of the respective wave functions at $t = 60T$ is shown in Figure 6.16 in terms of the occupation probability P_j vs. j. One sees vividly that changing β from 0 to π alters the occupation probability of many states by almost an order of magnitude.

The quantum dynamics of the kicked molecule depends on two parameters, $\tau \equiv \hbar T / I$ and $k \equiv \mathbf{d} \cdot \mathbf{E}_0 / \hbar$. However, as shown elsewhere [231] the classical dynamics depends solely on the product $k\tau$. Thus, by decreasing the magnitude of $\tau \equiv \hbar T / I$ while keeping $k\tau$ fixed, we can approach the classical limit while keeping the underlying classical dynamics unaffected. This is a useful tool to show that the demonstrated phase control is indeed quantal in nature. Specifically we show, in Figure 6.17, the CsI quantum dynamics after reducing the effective Planck constant τ by 50 times, while keeping $k\tau$ constant. Here, with $\tau = 0.024$ and $k = 240$, the energy diffusion only shows slight dependence on β. That is, the phase control disappears, clearly demonstrating the quantum nature of the control in the figures above.

CHAPTER 7

COHERENT CONTROL OF BIMOLECULAR PROCESSES

The results described in the preceding chapters deal with control of unimolecular processes, that is, processes beginning with a single molecule that subsequently undergo excitation and dynamics. However, the vast majority of chemical reactions occur via bimolecular processes, for example,

$$A + B \rightarrow C + D, \tag{7.1}$$

where A, B, C, D are, in general, molecules of mass M_A, M_B, M_C, and M_D. Here C and D can be identical to A and B (nonreactive scattering) or different from A and B (reactive scattering). We label A + B as arrangement q and C + D as arrangement q'. Below we describe coherent control of bimolecular collisions, demonstrating that coherent control is possible in bimolecular scattering [233–240].

In accord with Eq. (2.65), the amplitude for scattering between the asymptotic states $|E, q, \mathbf{m}; 0\rangle$ of A + B (labeled q) and $|E, q', \mathbf{n}; 0\rangle$ of C + D (labeled q') is given by the matrix element $\langle E, q', \mathbf{n}; 0|S|E, q, \mathbf{m}; 0\rangle$. The probability of making this transition is therefore

$$P_E(\mathbf{n}, q'; \mathbf{m}, q) = |\langle E, q', \mathbf{n}; 0|S|E, q, \mathbf{m}; 0\rangle|^2. \tag{7.2}$$

Alternatively, using Eq. (2.65), we can write the probability in terms of the potential. Specifically, we define the cross section $\sigma_E(\mathbf{n}, q'; \mathbf{m}, q)$ for forming $|E, q', \mathbf{n}; 0\rangle$ having initiated the scattering in $|E, q, \mathbf{m}; 0\rangle$, as

$$\sigma_E(\mathbf{n}, q'; \mathbf{m}, q) = |\langle E, q', \mathbf{n}^-|V_q|E, q, \mathbf{m}; 0\rangle|^2. \tag{7.3}$$

Here $|E, q', \mathbf{n}^-\rangle$ denotes the incoming scattering solutions associated with the product in state $|E, q', \mathbf{n}; 0\rangle$, and V_q is the component of the total potential that vanishes as the A to B distance becomes arbitrarily large. The cross section for scattering into arrangement q', independent of the product internal state $\mathbf{n}$, is then

$$\sigma_E(q'; \mathbf{m}, q) = \sum_{\mathbf{n}} |\langle E, q', \mathbf{n}^- | V_q | E, q, \mathbf{m}; 0\rangle|^2. \tag{7.4}$$

Assorted other cross sections may be defined, depending upon which of the elements of $\mathbf{n}$ are summed over. For example, by not including the scattering angles θ, ϕ in the sum, we obtain $\sigma_E(q', \theta, \phi; \mathbf{m}, q)$, corresponding to scattering into the q' product channel and into scattering angles (θ, ϕ). Similarly, $\sigma_E(q', \theta; \mathbf{m}, q)$ is the traditional differential cross section $\sigma_E(q', \theta; \mathbf{m}, q) = \int_0^{2\pi} d\phi \, \sigma_E(q', \theta, \phi; \mathbf{m}, q)$ for scattering into angle θ.

Note that Eqs. (7.2) to (7.4) describe motion in the center-of-mass coordinate system, that is, they arise in scattering theory after separating out the motion of the center of mass of A–B, a feature discussed in greater detail in Section 7.1.

Control of bimolecular collisions is achieved by constructing an initial state $|E, q, \{a_\mathbf{m}\}\rangle$ composed of a superposition of N energetically degenerate asymptotic states $|E, q, \mathbf{m}; 0\rangle$:

$$|E, q, \{a_\mathbf{m}\}\rangle = \sum_\mathbf{m} a_\mathbf{m} |E, q, \mathbf{m}; 0\rangle. \tag{7.5}$$

The cross section associated with using Eq. (7.5) as the initial state, obtained by replacing $|E, q, \mathbf{m}; 0\rangle$ in Eq. (7.3) by Eq. (7.5), is

$$\begin{aligned}
\sigma_E(\mathbf{n}, q'; \{a_\mathbf{m}\}, q) &= \left| \left\langle E, q', \mathbf{n}^- \left| V_q \sum_\mathbf{m} a_\mathbf{m} \right| E, q, \mathbf{m}; 0 \right\rangle \right|^2 \\
&= \sum_\mathbf{m} |a_\mathbf{m}|^2 |\langle E, q', \mathbf{n}^- | V_q | E, q, \mathbf{m}; 0\rangle|^2 \\
&\quad + \sum_{\mathbf{m}'} \sum_{\mathbf{m} \neq \mathbf{m}'} a_\mathbf{m} a_{\mathbf{m}'}^* \langle E, q, \mathbf{m}'; 0 | V_q | E, q', \mathbf{n}^- \rangle \\
&\quad \times \langle E, q', \mathbf{n}^- | V_q | E, q, \mathbf{m}; 0\rangle \\
&\equiv \sum_\mathbf{m} |a_\mathbf{m}|^2 \sigma(\mathbf{n}, q'; \mathbf{m}, q) + \sum_{\mathbf{m}'} \sum_{\mathbf{m} \neq \mathbf{m}'} a_\mathbf{m} a_{\mathbf{m}'}^* \sigma(\mathbf{n}, q'; \mathbf{m}', \mathbf{m}, q),
\end{aligned} \tag{7.6}$$

where $\sigma(\mathbf{n}, q'; \mathbf{m}', \mathbf{m}, q)$ is defined via Eq. (7.6). The total cross section into arrangement q' is then given by

$$\sigma_E(q'; \{a_\mathbf{m}\}, q) = \sum_\mathbf{n} \sigma_E(\mathbf{n}, q'; \{a_\mathbf{m}\}, q). \tag{7.7}$$

Note that Eq. (7.6), and hence Eq. (7.7), are of a standard coherent control form, that is, direct contributions from each individual member of the superposition, proportional to $|a_\mathbf{m}|^2$, plus interference terms that are proportional to $a_\mathbf{m} a_{\mathbf{m}'}^*$. It is clear that if we control the $a_\mathbf{m}$, through assorted preparation methods, then we can control the interference term, and hence the scattering cross section.

7.1 ISSUES IN PREPARATION OF THE SCATTERING SUPERPOSITION

To describe how the required superposition state [Eq. (7.5)] can be constructed in the laboratory requires some introductory remarks. Note first that Eqs. (7.2) to (7.7) and the $|E, q, \mathbf{m}; 0\rangle$ states are understood to be in the center-of-mass coordinate system and describe the relative translational motion as well as the internal state of A and B. In typical A–B scattering, separating out the center-of-mass motion comes about in a straightforward way. That is, let $\mathbf{r}_A$ and $\mathbf{r}_B$ denote the laboratory position of A and B and $\hbar\mathbf{k}^A$, $\hbar\mathbf{k}^B$ denote their laboratory momenta. The relative momentum $\mathbf{k}$, relative coordinate $\mathbf{r}$, center-of-mass momentum $\mathbf{K}$ and position $\mathbf{R}_{cm}$ are defined as

$$\mathbf{K} = \mathbf{k}^A + \mathbf{k}^B; \qquad \mathbf{R}_{cm} = (M_A\mathbf{r}_A + M_B\mathbf{r}_B)/(M_A + M_B),$$
$$\mathbf{k} = (M_B\mathbf{k}^A - M_A\mathbf{k}^B)/(M_A + M_B); \qquad \mathbf{r} = \mathbf{r}_A - \mathbf{r}_B. \tag{7.8}$$

In the case where A and B are initially in internal states $|\phi_A(i)\rangle$ and $|\phi_B(j)\rangle$, of energies $e_A(i)$ and $e_B(j)$, and the initial A and B translational motion are described by plane waves of momenta $\mathbf{k}_i^A$ and $\mathbf{k}_j^B$ then the incident wave function ψ_{in} is the product

$$\psi_{in} = |\phi_A(i)\rangle|\phi_B(j)\rangle \exp(i\mathbf{k}_i^A \cdot \mathbf{r}_A)| \exp(i\mathbf{k}_j^B| \cdot \mathbf{r}_B)$$
$$= |\phi_A(i)\rangle|\phi_B(j)\rangle \exp(i\mathbf{k} \cdot \mathbf{r}) \exp(i\mathbf{K} \cdot \mathbf{R}_{cm}). \tag{7.9}$$

The second equality follows from Eqs. (7.8). Since the interaction potential V_q between A and B depends only upon the relative coordinates of A–B, the center-of-mass momentum is conserved in the collision, allowing us to separate out the center-of-mass motion and to describe the dynamics in the center-of-mass coordinate system, that is, in terms of $|\phi_A(i)\rangle|\phi_B(j)\rangle \exp(i\mathbf{k} \cdot \mathbf{r})$. This state is, in fact, $\langle\mathbf{r}|E, q, \mathbf{m}; 0\rangle$, where the relative motion is in the coordinate representation.

Note that the scattering in Eq. (7.9) occurs at a fixed value of the center-of-mass momentum $\mathbf{K}$. Scattering may also occur from a state comprised of different $\mathbf{K}$ values. For example, the incident wave function may be of the form

$$|\psi_{in}\rangle = \sum_{l\mathbf{m}} d_{l\mathbf{m}}|E, q, \mathbf{m}; 0\rangle|\mathbf{K}_l\rangle, \qquad (\mathbf{K}_{l'} \neq \mathbf{K}_l). \tag{7.10}$$

Since the center-of-mass momentum is conserved and can be measured, components of the wave function with different values of $|\mathbf{K}_l\rangle$ contribute independently to the reaction cross section and cannot interfere with one another. That is, the cross section for scattering into $|E, q', \mathbf{n}; 0\rangle$ in this case is given by

$$\sigma_E(\mathbf{n}, q'; \{d_{l\mathbf{m}}\}, q) = \sum_l \left| \langle E, q', n^- | V_q \sum_{\mathbf{m}} d_{l\mathbf{m}} | E, q, \mathbf{m}; 0 \rangle \right|^2. \qquad (7.11)$$

Consider now preparation of the generalized superposition states [Eq. (7.5)] where for simplicity we limit consideration to a superposition of two states. To do so we examine the scattering of A and B, each previously prepared in the laboratory in a superposition state. The wave functions of A and B in the laboratory frame, ψ_A and ψ_B, are of the general form:

$$|\psi_A\rangle = a_1|\phi_A(1)\rangle \exp(i\mathbf{k}_1^A \cdot \mathbf{r}_A) + a_2|\phi_A(2)\rangle \exp(i\mathbf{k}_2^A \cdot \mathbf{r}_A), \qquad (7.12)$$

$$|\psi_B\rangle = b_1|\phi_B(1)\rangle \exp(i\mathbf{k}_1^B \cdot \mathbf{r}_B) + b_2|\phi_B(2)\rangle \exp(i\mathbf{k}_2^B \cdot \mathbf{r}_B). \qquad (7.13)$$

The incident wave function is then the product

$$\begin{aligned}
|\psi_{in}\rangle = |\psi_A\rangle|\psi_B\rangle &= [a_1|\phi_A(1)\rangle \exp(i\mathbf{k}_1^A \cdot \mathbf{r}_A) + a_2|\phi_A(2)\rangle \exp(i\mathbf{k}_2^A \cdot \mathbf{r}_A)] \\
&\times [b_1|\phi_B(1)\rangle \exp(i\mathbf{k}_1^B \cdot \mathbf{r}_B) + b_2|\phi_B(2)\rangle \exp(i\mathbf{k}_2^B \cdot \mathbf{r}_B)] \qquad (7.14) \\
&= \sum_{i,j=1}^{2} A_{ij} \exp(i\mathbf{k}_{ij} \cdot \mathbf{r}) \exp(i\mathbf{K}_{ij} \cdot \mathbf{R}_{cm}),
\end{aligned}$$

where $A_{ij} = a_i b_j |\phi_A(i)\rangle|\phi_B(j)\rangle$, $\mathbf{k}_{ij} = (M_B \mathbf{k}_i^A - M_A \mathbf{k}_j^B)/(M_A + M_B)$, and $\mathbf{K}_{ij} = \mathbf{k}_i^A + \mathbf{k}_j^B$.

As constructed, Eq. (7.14) is composed of four independent noninterfering incident states since each has a different center-of-mass wave vector $\mathbf{K}_{ij}$. However, we can set conditions so that interference, and hence control, is allowed. That is, we can require the equality of the center-of-mass motion of two components, plus energy degeneracy:

$$\begin{aligned}
\mathbf{K}_{12} &= \mathbf{K}_{21} \\
\hbar^2 k_{12}^2/2\mu + e_A(1) + e_B(2) &= \hbar^2 k_{21}^2/2\mu + e_A(2) + e_B(1),
\end{aligned} \qquad (7.15)$$

with $\mu = M_A M_B/(M_A + M_B)$. Equation (7.14) then becomes

$$\begin{aligned}
\psi_{in} = &[A_{12} \exp(i\mathbf{k}_{12} \cdot \mathbf{r}) + A_{21} \exp(i\mathbf{k}_{21} \cdot \mathbf{r})] \exp(i\mathbf{K}_{12} \cdot \mathbf{R}_{cm}) \\
&+ A_{11} \exp(i\mathbf{k}_{11} \cdot \mathbf{r}) \exp(i\mathbf{K}_{11} \cdot \mathbf{R}_{cm}) + A_{22} \exp(i\mathbf{k}_{22} \cdot \mathbf{r}) \exp(\mathbf{K}_{22} \cdot \mathbf{R}_{cm}),
\end{aligned} \qquad (7.16)$$

where the term in the first bracket, due to Eq. (7.15), is a linear superposition of two *degenerate* states. We therefore expect that the scattering cross section will be composed of noninterfering contributions from three components with differing

$\mathbf{K}_{ij}$, but where the first term allows for control via the interference between the A_{12} and A_{21} terms. The two remaining terms, proportional to A_{11} and A_{22}, are uncontrolled satellite contributions.

For example, if we design the experiment so that $\mathbf{k}_1^A = -\mathbf{k}_2^B$ and $\mathbf{k}_2^A = -\mathbf{k}_1^B$, then $\mathbf{K}_{12} = \mathbf{K}_{21} = 0$, and $\mathbf{k}_{12} = \mathbf{k}_1^A$, $\mathbf{k}_{21} = -\mathbf{k}_1^B$, so that the degeneracy requirement [Eq. (7.15)] becomes

$$\hbar^2 (k_1^A)^2/2\mu + e_A(1) + e_B(2) = \hbar^2 (k_1^B)^2/2\mu + e_A(2) + e_B(1). \tag{7.17}$$

Note also that we can implement Eq. (7.17) for the case of atom–diatomic-molecule scattering by setting $|\phi_A(1)\rangle = |\phi_A(2)\rangle = |\phi_A(g)\rangle$, where $|\phi_A(g)\rangle$ is the, for example, ground electronic state of atom A. In this case the degeneracy condition [Eq. (7.17)] is

$$\frac{\hbar^2}{2\mu}[(k_1^A)^2 - (k_2^A)^2] = [e_B(1) - e_B(2)]. \tag{7.18}$$

In general, these conditions demand a method of preparing $|\psi_A\rangle$ and $|\psi_B\rangle$ that correlate the internal states $|\phi_A(i)\rangle$ and $|\phi_B(i)\rangle$ with their associated momenta $\mathbf{k}_i^A$, $\mathbf{k}_i^B$ so as to obtain Eq. (7.17). Since the overall phase of the wave function is irrelevant to the state of the system, the dynamics is not sensitive to the overall phase of $|\psi_A\rangle|\psi_B\rangle$. However, the phases of the interference term must be well defined or the control will average to zero.

Specifically, coherent control of bimolecular processes requires the production of states [Eqs. (7.12) and (7.13)] where the translational and internal states are "entangled," that is, composed of components in which two or more translational and internal states are correlated. Such states do result from photodissociation processes [see, e.g., Eq. (2.73), which is a sum over $|E, \mathbf{n}^-\rangle$ states, each going over in the long-time limit to a translational-internal product state $|E, \mathbf{n}; 0\rangle$. Their sum goes over to a state in which the translational and the internal motions are entangled]. However, they are not necessarily suitable for our purposes. Below we introduce three tentative suggestions for constructing such states whose realization would require an extension of current laboratory techniques.

For example, entangled states of internal and translational degrees of freedom that might be useful for bimolecular control have been prepared in atoms in a relatively straightforward way [241]. Consider, for example, a system with two levels $|E_1\rangle$, $|E_2\rangle$, initially in the lower state $|E_1\rangle$ and moving with kinetic energy $E_t(1)$. Passing the system through a spatially dependent field with off-resonant frequency $\omega = (E_2 - E_1)/\hbar - \delta$ results in excitation to the state $|E_2\rangle$ with kinetic energy $E_t(2)$. Conservation of energy requires, however, that

$$E_1 + E_t(1) + \hbar\omega = E_2 + E_t(2), \tag{7.19}$$

or $E_t(2) = E_t(1) - \hbar\delta$. That is, the created superposition state has two internal states correlated with two different translational energy states, precisely as required for bimolecular coherent control. Tuning δ above or below the resonance results in an increase, or decrease, of kinetic energy upon excitation. The extension of this technique to most cases of interest to us will, however, be difficult since the $\hbar\delta$ required here is far larger than that in the atomic case.

Similarly, the momentum transfer associated with a collision of photons with atoms is used regularly to cool atoms [242], that is, to alter the translational energy of an atom. Indeed, the momentum of large numbers of photons (over 140-photon momenta) have been successfully transferred coherently to atoms [243]. This suggests the possibility of preparing an initial superposition of internal states of a molecule, followed by the state-specific absorption of photon momenta of one of the internal states in order to form the required entangled superposition of the translational and internal states.

Finally, we note that a number of experiments have shown that it is possible to accelerate or decelerate molecules using time-varying electric fields [244]. In this case the molecule is passed through an array of synchronously pulsed electric field stages that interact with the molecular dipole. Since the dipole is a function of the state of the system, it may be possible to prepare a superposition of internal states and then selectively accelerate one of the two internal states to produce the desired superposition.

7.2 IDENTICAL PARTICLE COLLISIONS

It is important to note that the situation simplifies enormously for the case of identical particle collisions, that is, when $B = A$ [239]. Specifically, consider

$$A + A' \rightarrow C + D \tag{7.20}$$

with $\mathbf{k}_i^A = \mathbf{k}_i^{A'}$. Here we have used A' to denote the molecule A, but in a superposition state that is not necessarily the same as A. If we prepare each of the two initial A and A' superposition states from the same molecular bound states, for example, $|\phi_A(1)\rangle = |\phi_{A'}(1)\rangle$ and $|\phi_A(2)\rangle = |\phi_{A'}(2)\rangle$ then the requirement for conservation of energy in the center of mass [Eq. (7.15)] becomes

$$k_{12}^2 = k_{21}^2. \tag{7.21}$$

For the case of $A + A'$ collisions, this condition is always satisfied.

This scenario opens up a wide range of possible experimental studies of control in bimolecular collisions. Specifically, we need only prepare A and A' in a controlled superposition of two states [e.g., by resonant laser excitation of $|\phi_A(1)\rangle$] to produce a superposition with $|\phi_A(2)\rangle$, direct them antiparallel in the laboratory, and vary the coefficients in the superposition to affect the reaction probabilities. Control originates in quantum interference between two degenerate states associated with the contributions of $|\phi_A(1)\rangle|\phi_{A'}(2)\rangle$ and $|\phi_A(2)\rangle|\phi_{A'}(1)\rangle$. This is accompanied by two uncontrolled scattering contributions corresponding to the contributions of

$|\phi_A(1)\rangle|\phi_{A'}(1)\rangle$ and $|\phi_A(2)\rangle|\phi_{A'}(2)\rangle$. Control is achieved by varying the four coefficients $a_i, b_i, i = 1, 2$. Stimulated rapid adiabatic passage (STIRAP) [245, 246] to be discussed in detail in Section 9.1, provides one choice for such state preparation.

The control approach described above can be generalized to a superposition of N levels in each of the two A and A' reactants. Specifically, choosing all $\mathbf{k}_i^A = \mathbf{k}^A$ and with $\mathbf{k}_i^{A'} = -\mathbf{k}^A$ we have

$$|\psi_A\rangle = \exp(i\mathbf{k}^A \cdot \mathbf{r}_A)\left[\sum_{i=1}^{N} a_i|\phi_A(i)\rangle\right],$$
$$|\psi_{A'}\rangle = \exp(-i\mathbf{k}^A \cdot \mathbf{r}_{A'})\left[\sum_{j=1}^{N} b_j|\phi_{A'}(j)\rangle\right]. \tag{7.22}$$

The scattering wave function is then

$$|\psi_{in}\rangle = |\psi_A\rangle|\psi_{A'}\rangle = \exp(i\mathbf{k}\cdot\mathbf{r})\left[\sum_{i=1}^{N} a_i|\phi_A(i)\rangle\right]\left[\sum_{j=1}^{N} b_j|\phi_{A'}(j)\rangle\right]. \tag{7.23}$$

Since $M_A = M_{A'}$, $\mathbf{k} = (\mathbf{k}^A - \mathbf{k}^{A'})/2 = \mathbf{k}^A$. The kinetic energy $k^2/2\mu$ is the same for each term in Eq. (7.23) so that degenerate states in the center-of-mass frame correspond to states $|\phi_A(i)\rangle|\phi_{A'}(j)\rangle$ in Eq. (7.23), which are of equal internal energy $e_A(i) + e_{A'}(j)$. Expanding the product in Eq. (7.23) gives N^2 terms, N terms of which are of differing energy $2e_A(i)$, $i = 1, \ldots, N$ and $(N^2 - N)$ states of energy $e_A(i) + e_{A'}(j)$, $i \neq j$. Of the latter terms, each is accompanied by another term of equal energy [i.e., $e_A(i) + e_{A'}(j) = e_A(j) + e_{A'}(i)$]. Hence the N^2 terms are comprised of N direct terms plus $(N^2 - N)/2$ degenerate pairs, which are a source of interference, and hence control. Here control is achieved by altering the $2N$ coefficients a_i, b_i in the initially prepared state [Eq. (7.23)], for example, by shaped pulsed laser excitation of A and A'.

Computational examples of this approach have been restricted to control over rotational excitation in $H_2 + H_2$, a consequence of limitations on the ability to perform quantum computations on AB + AB scattering. A careful analysis of the scattering [247] requires consideration both of the interference effects as well as the nature of the identical particle scattering. Typical results are shown in Figure 7.1 for various low-energy scattering cases. Specifically, we show the differential cross section into scattering angles θ and final states $j_1' = j_2' = 2$, arising from scattering of para H_2 + para H_2, where each H_2 is in an initial superposition, with either a plus or minus sign, of $j_1 = 4$ and $j_2 = 0$. The cross term contributing to the scattering in these cases is

$$|\psi_{j_1 j_2}^{\pm}\rangle = \frac{1}{\sqrt{2}}[|j_1\rangle|j_2\rangle \pm |j_2\rangle|j_1\rangle]|m_1 = 0, v_1 = 0\rangle|m_2 = 0, v_2 = 0\rangle, \tag{7.24}$$

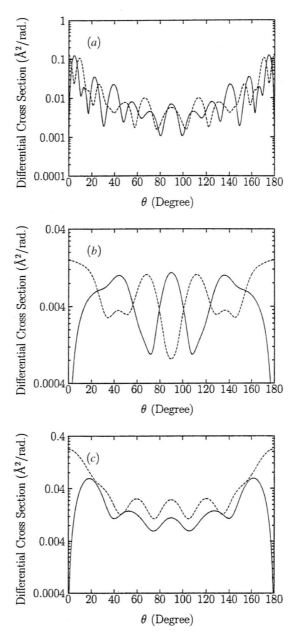

Figure 7.1 Inelastic differential cross section for para H_2 + para H_2, where the collision energy is (a) $400 \, cm^{-1}$, (b) $40 \, cm^{-1}$, and (c) $4 \, cm^{-1}$. Dashed and solid lines are for incoming free entangled states $|\psi_{j_1 j_2}^+\rangle$ and $|\psi_{j_1 j_2}^-\rangle$. Here $j_1 = 4$, $j_2 = 0$, $j_1' = j_2' = 2$. (From Fig. 2, Ref. [247].)

where v_i, m_i denote the vibrational state and angular momentum projection along the z axis. Note that results for $m_i = m_2$ are essentially independent of the value of m_i.

The results of this computation (Fig. 7.1—note the logarithmic ordinate scale) clearly show that the phase of the j_1, j_2 superposition has a significant effect on the differential cross section. This translates into considerable control over the total inelastic cross section, that is, the integral of the differential cross section over θ. For example, for the case in Figure 7.1c, the total inelastic cross section is 0.057 for $|\psi_{j_1 j_2}^+\rangle$ and 0.032 for $|\psi_{j_1 j_2}^-\rangle$.

Results on more complex scattering, for example, reactive scattering, await further computational (or experimental) developments.

7.3 *m* SUPERPOSITIONS

For the case of nonidentical particle scattering, it is clear that the easiest way to implement bimolecular control in the laboratory is to start with a superposition of degenerate states of the fragments. In atom–diatomic-molecule scattering, to which we now restrict our attention, one possibility is to use the $(2j + 1)$ diatomic-molecule rotational states $|j, m\rangle$ associated with fixed j, where m is the projection of diatomic-molecule angular momentum j along a space-fixed axis. In this case [238] we show below that control over the differential cross section is possible, but control over the total cross section is not. The argument is related to that given in our discussion of polarization control of photodissociation processes (Section 3.4).

Consider first superimposing two m states of a diatomic molecule in atom + diatomic-molecule scattering. For this case the initial state [Eq. (7.5)] assumes the form:

$$|E, q, \{a_{\mathbf{m}}\}\rangle = \sum_{i=1,2} a_i |qvjm_i\rangle |E_q^{kin}\rangle, \tag{7.25}$$

where $|qvjm_i\rangle$ is an eigenstate of the diatomic-molecule internal Hamiltonian in the q channel, of energy $e_q(vj)$, with v denoting the diatomic-molecule vibrational quantum number. The state $|E_q^{kin}\rangle$ is a plane wave of energy $E_q^{kin} = E - e_q(vj)$, describing the free motion of the atom relative to the diatomic molecule in the q arrangement.

Consider scattering into a final state $\langle E, q', \mathbf{n}; 0|$ with $\mathbf{n}$ defined by v', j', λ' and scattering angles θ, φ from the initial state $|E, q, \{a_{\mathbf{m}}\}\rangle$ in Eq. (7.25). Here λ' is the helicity, that is, the projection of the product diatomic-molecule angular momentum onto the final relative translational velocity vector. Then the resultant differential cross section can be written as [10]

$$\sigma(q'v'j'\lambda' \leftarrow q, v, j, m_1, m_2 | \theta, \varphi) = \left| \sum_{i=1,2} a_i f_{q'v'j'\lambda' \leftarrow qvjm_i}(\theta, \varphi) \right|^2, \tag{7.26}$$

where f (the so-called scattering amplitude) is given by

$$f_{q'v'j'\lambda' \leftarrow qvjm_i}(\theta, \varphi) = \frac{i^{j-j'+1}e^{im_i\varphi}}{2k_q(vj)} \sum_j (2J+1)d^J_{\lambda'm_i}(\pi - \theta)$$

$$\times [S^J_{q'v'j'\lambda',qvjm_i} - \delta_{q'q}\delta_{v'v}\delta_{j'j}\delta_{\lambda'm_i}] \tag{7.27}$$

with $d^J_{\lambda,m}(\theta)$ being the reduced Wigner rotation matrices [248]. The $S^J_{q'v'j'\lambda',qvjm_i}$ of the above are the elements of the scattering S matrix in the so-called helicity representation [249, 250], that is, where helicity is one of the final quantum numbers. The quantity $k_q(vj) = \sqrt{2\mu_q[E - e_q(vj)]}/\hbar$, with μ_q being the atom–diatomic-molecule reduced mass in the q channel.

Expanding the square in Eq. (7.26) gives the reactive differential scattering cross section as (where we drop the initial state labels for convenience)

$$\sigma^R(v'j'\lambda'|\theta, \varphi) = |a_1|^2\sigma^R_{11}(v'j'\lambda'|\theta) + |a_2|^2\sigma^R_{22}(v'j'\lambda'|\theta) + 2\operatorname{Re}\{a_1^*a_2\sigma^R_{12}(v'j'\lambda'|\theta, \varphi)\}, \tag{7.28}$$

where

$$\sigma^R_{ii}(v'j'\lambda'|\theta) = |f_{q'v'j'\lambda' \leftarrow qvjm_i}(\theta, \varphi)|^2$$

$$= [2k_q(vj)]^{-2} \sum_{J,J'}(2J+1)(2J'+1)d^J_{\lambda'm_i}(\pi - \theta)d^{J'}_{\lambda'm_i}(\pi - \theta) \tag{7.29}$$

$$\times S^J_{q'v'j'\lambda',qvjm_i}[S^{J'}_{q'v'j'\lambda',qvjm_i}]^*, \qquad q \neq q', \ i = 1, 2,$$

and

$$\sigma^R_{12}(v'j'\lambda'|\theta, \varphi) = f_{q'v'j'\lambda' \leftarrow qvjm_1}(\theta, \varphi)f^*_{q'v'j'\lambda' \leftarrow qvjm_2}(\theta, \varphi)$$

$$= \frac{e^{i(m_1-m_2)\varphi}}{4k_q^2(vj)} \sum_{J,J'}(2J+1)(2J'+1)$$

$$\times d^J_{\lambda'm_1}(\pi - \theta)d^{J'}_{\lambda'm_2}(\pi - \theta)S^J_{q'v'j'\lambda',qvjm_1}[S^{J'}_{q'v'j'\lambda',qvjm_2}]^*, \qquad q \neq q'. \tag{7.30}$$

Here, the superscript R denotes reactive scattering into a specific final arrangement channel $q' \neq q$. The total differential cross section, $\sigma^R(\theta, \varphi)$, for reaction out of a state in Eq. (7.25) is given by the sum over final states at energy E as

$$\sigma^R(\theta, \varphi) = \sum_{v', j', \lambda'} \sigma^R(v'j'\lambda'|\theta, \varphi). \tag{7.31}$$

Note that the φ dependence of the measurable cross sections is due solely to the interference term. Thus, traditional (uncontrolled) scattering is φ independent.

Integration of Eq. (7.28) or (7.31) over angles $\theta \in [0, \pi]$ and $\varphi \in [0, 2\pi]$ gives the state-resolved integral reactive cross section $\sigma^R(v'j'\lambda')$. However, the integral of $\sigma^R_{12}(v'j'\lambda'|\theta, \varphi)$ over φ is zero so that the interference term, and hence control over the integral cross section, disappears. Indeed, the integral over $0 < \varphi < \pi$ exactly cancels the integral over $\pi < \varphi < 2\pi$. For this reason we consider control over scattering into hemisphere $0 < \varphi < \pi$, giving the state-resolved integral cross section, which we denote as $\sigma^R(v'j'\lambda'; \varphi \leq \pi)$. This can also be written as three terms, as in Eq. (7.28), but with the σ^R_{ij} replaced by $\sigma^R_{ij}(v'j'\lambda'; \varphi \leq \pi)$, $(i, j = 1, 2)$, where

$$\sigma^R_{ik}(v'j'\lambda'; \varphi \leq \pi) = \int_0^\pi \sin\theta \, d\theta \int_0^\pi d\varphi \, \sigma^R_{ik}(v', j', \lambda'|\theta, \varphi). \tag{7.32}$$

Summing over the final v', j', λ' at energy E, gives the total integral cross section for total scattering into the hemisphere as

$$\sigma^R(\varphi \leq \pi) = \sum_{v', j', \lambda'; \varphi \leq \pi} \sigma^R(v', j', \lambda'; \varphi \leq \pi). \tag{7.33}$$

It is important to stress that state-to-state cross sections $\sigma^R_{ii}(v'j'\lambda'|\theta)$ and $\sigma^R_{ii}(v'j'\lambda'; \varphi \leq \pi)$ in Eqs. (7.29) and (7.32), as well as the corresponding total cross sections $\sigma^R_{ii}(\theta)$ and σ^R_{ii}, appear in standard scattering theory (see, e.g., [9]), while $\sigma^R_{12}(v'j'\lambda'|\theta, \varphi)$ and $\sigma^R_{12}(v'j'\lambda'; \varphi \leq \pi)$ are new types of interference terms that allow for control, through the a_i, over the atom–diatomic-molecule collision process. As in the case of photodissociation, significant control requires substantial σ^R_{12}, which follows from the Schwartz inequality $[|\sigma^R_{12}| \leq \sqrt{\sigma^R_{11}\sigma^R_{22}}]$. That is, large σ^R_{12} requires also large σ^R_{11} and σ^R_{22}.

To examine the extent of control over the reaction we rewrite the reactive cross section in the form (where we refer to scattering into a hemisphere, but drop the notation "$\varphi \leq \pi$" for convenience),

$$\sigma^R = [\sigma^R_{11} + x^2\sigma^R_{22} + 2x|\sigma^R_{12}|\cos(\delta^R_{12} + \phi_{12})]/(1 + x^2), \tag{7.34}$$

where $x = |a_2/a_1|$, $\phi_{12} = \arg(a_2/a_1)$, and $\delta^R_{12} = \arg(\sigma^R_{12})$ with the branching ratio between the reactive and nonreactive total cross sections given by

$$\frac{\sigma^R}{\sigma^{NR}} = \frac{\sigma^R_{11} + x^2\sigma^R_{22} + 2x|\sigma^R_{12}|\cos(\delta^R_{12} + \phi_{12})}{\sigma^{NR}_{11} + x^2\sigma^{NR}_{22} + 2x|\sigma^{NR}_{12}|\cos(\delta^{NR}_{12} + \phi_{12})}. \tag{7.35}$$

Here NR refers to nonreactive scattering; definitions of the nonreactive cross sections are analogous to their reactive counterparts.

It follows from Eqs. (7.34) and (7.35) that by varying the a_i coefficients (i.e., varying either the relative magnitude, x, or the relative phase, ϕ_{12}) in Eq. (7.25) through the initial preparation step, we can directly alter the interference term σ^R_{12}

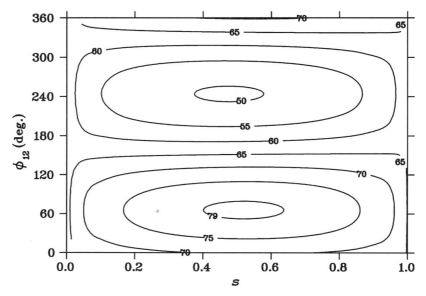

Figure 7.2   Contour plot of the ratio $\sigma^R(\varphi \leq \pi)/\sigma^{NR}(\varphi \leq \pi)$ ($\times 10^3$) for D + H$_2$ at $E = 0.93$ eV as a function of ϕ_{12} and s. Initial state is a superposition of $v = 0$, $j = 2$, $m_1 = 1$, $m_2 = 0$. (From Fig. 2, Ref. [238].)

(and/or σ_{12}^{NR}) and hence control the scattering cross sections. Such a preparation might be carried out, for example, by a suitably devised molecular beam experiment where the diatomic molecule is excited with elliptically polarized light to a collection of well-defined m states.

As an example of this approach consider control over the reaction D + H$_2$ (for other isotopes see Ref. [238]). Typical results corresponding to D + H$_2$ → H + HD at $E = 0.93$ eV, with scattering from an initial superposition of ($v = 0$, $j = 2$, $m_1 = 1$, $m_2 = 0$) are shown in Figure 7.2.

Results are reported as contour plots of the cross-section ratios vs. the phase ϕ_{12} and the parameter $s = x^2/(1 + x^2)$. The value $s = 0$ corresponds to scattering from the state with $m_1 = 1$ and $s = 1$ corresponds to scattering from the state with $m_2 = 0$. The ratio of cross sections is seen to vary from 0.05 to 0.079, showing maximum and minimum that are well outside (up to factors of 1.22 and 1.26, respectively) of the range of results for scattering from a single m state. Thus, the ratio of cross sections can be increased or decreased by approximately 20% through coherent control effects.

Thus, superposing two m levels provides some degree of control over the differential cross sections. Nonetheless, control is far from extensive. The origin of this behavior is evident from an examination of the σ_{12} compared to σ_{11} and σ_{22}. In particular, the cross term σ_{12} is found to be 4 to 10 times smaller than the σ_{ii} for the reactive case and far smaller for the nonreactive case. As a consequence, the extent of control is rather limited. Similar control, in this case over the total cross section, was found for superpositions of rovibrational states of the diatomic [251].

Two alternatives for improved control suggest themselves. The first is to seek alternate linear superpositions, or possibly different chemical reactions, with larger σ_{12}. The second is to examine the extent of control resulting from the inclusion of more than two degenerate reactant states in the initial superposition, as discussed below.

7.3.1   Optimal Control of Bimolecular Scattering

We can readily extend bimolecular control to superpositions composed of more than two states. Indeed, we can introduce a straightforward method to *optimize* the reactive cross section as a function of $a_{\mathbf{m}}$ for any number of states [252]. Doing so is an example of optimal control theory, a general approach to altering control parameters to optimize the probability of achieving a desired goal, introduced in Chapter 4.

Consider scattering from incident state $|E, q, i; 0\rangle$ to final state $|E, q', f; 0\rangle$. The label f includes the angles θ, ϕ into which the products scatter. Hence summations below over f imply integrations over these angles. In accord with Eq. (7.2), the probability $P(f, q'; i, q)$ of producing product in final state $|E, q', f; 0\rangle$ having started in the initial state $|E, q, i; 0\rangle$ is

$$P(f, q'; i, q) = |S_{fi}|^2, \tag{7.36}$$

where $S_{fi} = \langle E, q', f; 0|S|E, q, i; 0\rangle$ and where S is the scattering matrix for the process. In the general case we consider Eq. (7.36) separately for each total angular momentum contributing to the scattering. The total probability $P(q'; i, q)$ of scattering into arrangement channel q', assuming n open product states, is then given by

$$P(q'; i, q) = \sum_{f=1}^{n} |S_{fi}|^2. \tag{7.37}$$

To simplify the notation we have not carried an E label in the probabilities: Fixed energy E is understood. [Note that Eqs. (7.36) and (7.37) are quite general and can be applied to a host of processes, other than just scattering, since one may, in general, write the probability of transitions between states in terms of a generalized S matrix.]

If we now consider scattering from an initial state $|E, q, \{a_i\}; 0\rangle$ comprised of a linear superposition of N states, then the probability of forming $|E, q', f; 0\rangle$ from this initial state is

$$P(f, q'; \mathbf{a}, q) = \left| \sum_{i=1}^{N} a_i S_{fi} \right|^2, \tag{7.38}$$

where $\mathbf{a} \equiv \{a_i\}$. The total reactive scattering probability into channel q', $P(q'; \mathbf{a}, q)$, is therefore

$$P(q'; \mathbf{a}, q) = \sum_{f=1}^{n} \left| \sum_{i=1}^{N} a_i S_{fi} \right|^2. \tag{7.39}$$

To simplify the notation we introduce the matrix $\boldsymbol{\sigma} = \mathbf{S}_{q'}^{\dagger} \mathbf{S}_{q'}$ with elements $\boldsymbol{\sigma}_{ij} = \sum_{f=1}^{n} S_{fj}^* S_{fi}$, which allows us to rewrite Eq. (7.39) as

$$P(q'; \mathbf{a}, q) = \mathbf{a}^{\dagger} \boldsymbol{\sigma} \mathbf{a}. \tag{7.40}$$

Here † denotes the Hermitian conjugate, and the q' subscript on the $\mathbf{S}$ indicates that we are dealing with the submatrix of the S matrix associated with scattering into the product manifold defined by the q' quantum number.

One can optimize scattering into arrangement channel q', with the normalization constraint $\sum_{i=1}^{N} |a_i|^2 = 1$, by requiring

$$\frac{\partial}{\partial a_k^*} \left[P(q'; \mathbf{a}, q) - \lambda \sum_{i=1}^{N} |a_i|^2 \right] = \frac{\partial}{\partial a_k^*} [\mathbf{a}^{\dagger} \boldsymbol{\sigma} \mathbf{a} - \lambda \mathbf{a}^{\dagger} \mathbf{a}] = 0, \qquad k = 1, \ldots, N,$$
$$\tag{7.41}$$

where λ is a Lagrange multiplier. Explicitly taking the derivative gives the result that the vector of optimal coefficients $\mathbf{a}_{\lambda}$ satisfies the eigenvalue equation

$$\boldsymbol{\sigma} \mathbf{a}_{\lambda} = \lambda \mathbf{a}_{\lambda}. \tag{7.42}$$

Additional labels may be necessary to account for degeneracies of the eigenvectors $\mathbf{a}_{\lambda}$. Optimization is now equivalent to solving Eq. (7.42).

This approach has been used to obtain optimal $\mathbf{a}_{\lambda}$ for various isotopic variants of the $H + H_2$ reaction [238]. In general, the range of control was found to increase with j, that is, with the increasing number of available m states. Thus, for example, for scattering into $H' + HD$, where H and H' are assumed distinguishable, σ^R / σ^{NR} could be varied between 1.15×10^{-2} and 2.62×10^{-2} for initial $j = 2$, and between 1.85×10^{-5} and 6.29×10^{-3} for initial $j = 10$.

The optimization procedure yields a set of coefficients a_i. Of considerable interest is the question of whether these coefficients merely define a new vector that is simply a vector in a rotated coordinate system. If so, this would indicate that the optimum solution corresponds to a simple classical reorientation of the diatomic-molecule angular momentum vector. Examination of the optimal results [238] indicate that this is not the case. That is, control is the result of quantum interference effects.

Optimized Bimolecular Scattering: Total Suppression of Reactive Event

In Section 7.3.1 we considered optimizing reactive scattering by varying the coefficients a_i of a superposition of states. In this section we show that when the number of initial open states in the reactant space exceeds the number of open states in the product space it is possible to find a particular set of a_i coefficients such that one can totally *suppress* reactive scattering. This result is proven below and applied to display the total suppression of tunneling [252]. Note that in general each total angular momentum contributing to the scattering must be treated separately.

To see this result consider Eq. (7.42). Note that if $\lambda = 0$ is an eigenvalue of this equation with eigenfunctions $\mathbf{a}_0$ then by inserting Eq. (7.42) into Eq. (7.40) we have that $P(q'; \mathbf{a}_0, q) = 0$. That is, if $\lambda = 0$ is a solution to Eq. (7.42), then the coefficients $\mathbf{a}_0$ completely suppress reaction into arrangement channel q'.

Clearly, $\lambda = 0$ is a solution if

$$\det(\boldsymbol{\sigma}) = \det(\mathbf{S}_{q'}^{\dagger}\mathbf{S}_{q'}) = 0, \tag{7.43}$$

which is the case if the number of initial states N participating in the initial superposition is greater than the number M of open-product states. To see this result note that, under these circumstances, $\boldsymbol{\sigma}$ is a matrix of order $N \times N$ and $\mathbf{S}_{q'}$ is of order $N \times M$. If $N > M$, we can construct an $N \times N$ matrix $\mathbf{A}_{q'}'$ by adding a submatrix of $(N - M)$ rows of zeroes to the lower part of $\mathbf{S}_{q'}$. This procedure does not change the $(\mathbf{S}_{q'}^{\dagger}\mathbf{S}_{q'})$ product; hence

$$\det(\boldsymbol{\sigma}) = \det(\mathbf{S}_{q'}^{\dagger}\mathbf{S}_{q'}) = \det(\mathbf{A}_{q'}^{\dagger}\mathbf{A}_{q'}) = \det(\mathbf{A}_{q'}^{\dagger})\det(\mathbf{A}_{q'}) = 0. \tag{7.44}$$

The last equality holds because the determinants of $\mathbf{A}_{q'}$ and $\mathbf{A}_{q'}^{\dagger}$ are zero.

As an example of the kind of results that are possible, consider optimizing a barrier penetration problem modeled by a set of multichannel Schrödinger equations of the type:

$$\boldsymbol{\Psi}''(r) = \frac{2\mu}{\hbar^2}(\hat{\underline{E}} - \underline{V})\boldsymbol{\Psi}(r), \tag{7.45}$$

where μ is the relevant mass, $\underline{V}$ is a potential matrix and $\hat{\underline{E}}$ is a diagonal matrix with elements $E - e_i$. Sample results for four open channels (that is, channels satisfying the $E - e_i > 0$ condition) were obtained for a model for which $\underline{V}$ is a matrix of Eckart potentials given by

$$V_{ij}(r) = -\frac{a_{ij}\xi}{1 - \xi} - \frac{b_{ij}\xi}{(1 - \xi)^2} + c_{ij}, \qquad (i, j = 1, \ldots, 4), \tag{7.46}$$

where $\xi = -\exp(2\pi r/t)$, with t a distance potential parameter. These potentials are shown in Figure 7.3. Scattering results, obtained from numerically integrating Eq. (7.45) are shown in Figure 7.4. Here reactivity is shown as a function of energy for the case where the number of populated initial states N_{pop} is less than N: Here $N_{pop} = 3$. The curves labeled P_i correspond to the standard $P(q'; i, q)$, that is, total reaction probability from each of the individual initial (i) states. The quantities P_1 and P_3, which are open asymptotically at all energies, show a gradual rise with increasing energy, whereas P_2, which is closed on the product side until $E_{th}(3) = 0.008$ a.u., stays rather small until $E = E_{th}(3)$, where it displays a very rapid rise to near unity. Total reaction probability reaches unity above $E_{th}(4) = 0.010$ a.u., the threshold for the opening of the fourth channel.

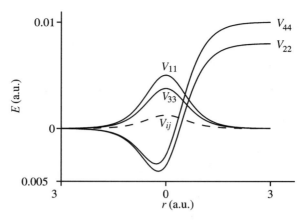

Figure 7.3   Elements of the potential matrix for the model scattering problem. V_{ij} denotes the three off-diagonal matrix elements $i \neq j$, which are all chosen to be equal. The nonzero a_{ij}, b_{ij}, c_{ij} elements [Eq. (7.46)] are given (in a.u.) by $b_{11} = 0.02$, $a_{22} = 0.008$, $b_{22} = -0.03$, $b_{33} = 0.015$, $a_{44} = 0.01$, $b_{44} = -0.03$, $b_{ij} = 0.005$, $i \neq j$ with $t = 2.5$ and $\mu = 0.6666$ a.m.u. (From Fig. 1, Ref. [252].)

Of particular relevance here are the solid curves in Figure 7.4, which show the reactivity maximum and minimum obtained from the optimal solutions to Eq. (7.42). The reactivity maximum is seen to be substantially larger than any of the individual P_i and to reach unity at significantly lower energies than any of these solutions. Of greater importance is that the reactivity minimum is, as predicted by the argument

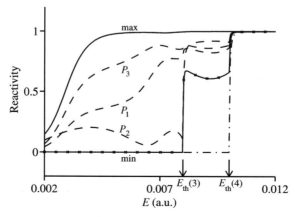

Figure 7.4   Reactivity shown as a function of energy in a model system. Dashed curves labeled P_i correspond to total reactivity from each of three individual initial states in the prepared superposition. Solid curves denote the reactivity obtained by solving Eq. (7.42) for optimal solutions. The two arrows indicate the threshold energies for opening of the third and fourth channels. Dot–dash curve shows minimum reactivity resulting from a separate computation that includes four states in the initial superposition. (From Fig. 2, Ref. [252].)

presented above, exactly zero for $E < E_{th}(3)$ since the total number of states ($N_{pop} = 3$) in the superposition exceeds the number of open-product states ($M = 2$). At $E > E_{th}(3)$ a third product channel opens so that $N_{pop} = M$ and the minimal solution is no longer zero.

Note also that the minimum reactivity curve in Figure 7.4 reflects a variety of different interesting behaviors, depending on the particular energy. Specifically, below the maximum of V_{11} at 0.005 a.u., the zero minimum corresponds to suppression of tunneling through that barrier. Above 0.005 a.u. the zero minimum corresponds to suppression of the reactive scattering that occurs *above* the barrier.

Thus it is clear that the ability to superimpose degenerate scattering states affords great potential to control scattering processes. Note also that, as an obvious extension, similar results hold for tunneling in bound systems if the total number of initial degenerate states at the energy of interest exceeds the number of accessible final states at that energy. This approach has also been extended [253] to the design of decoherence free subspaces, that is, subspaces, as mentioned in Chapter 5, within which the dynamics is free of decoherence effects. In brief, one takes a subspace α that undergoes transitions to a subspace β due to coupling to the environment. In accord with the discussion in this section, these transitions can be suppressed if β is sufficiently larger than α. If this is not the case, then α can be augmented with states from another subspace γ to aid in suppressing transitions to β from α.

7.3.2 Sculpted Imploding Waves

Since, in principle, control can be achieved by superposing any degenerate set of eigenstates, rather than using a superposition of internal diatomic-molecule states, as above, it is possible to affect bimolecular control by using a superposition of equal energy translational wave functions [252]. To see how this is done note that the relative motion of two particles (say in a molecular beam) is generally well described by a plane wave $\exp(i\mathbf{k} \cdot \mathbf{r})$. The character of the plane wave can be exposed by a partial-wave decomposition. That is, assuming the wave is directed along the z axis we can write

$$\exp(i\mathbf{k} \cdot \mathbf{r}) = \exp(ikz) = \exp(ikr\cos\theta) = \sum_l a_l j_l(kr) P_l(\cos\theta), \qquad (7.47)$$

where $a_l = i^l(2l + 1)$ and $j_l(k_m r)$, $P_l(\cos\theta)$ are the spherical Bessel function and Legendre polynomials, respectively. We see that each incoming plane wave is, in fact, a superposition of energetically degenerate states with fixed coefficients a_l. This suggests the possibility of altering the a_l to produce modified states, that is, a sculpted incoming wave packet, $\langle \mathbf{r}|\mathbf{k}_{mod}\rangle$, which will display different quantum interferences, hence altering the product cross sections.

This approach is discussed in detail in Ref. [254]. As an example of the control afforded by this scenario, consider rotational excitation in a model of the Ar + $H_2(j, m) \rightarrow$ Ar + $H_2(j', m')$ collision. Optimizing the phases χ_l of $a_l = |a_l|\exp(i\chi_l)$ allows a direct study of the effect of varying the interferences between partial-wave components on the outgoing flux into any selected product state. Typically, altering

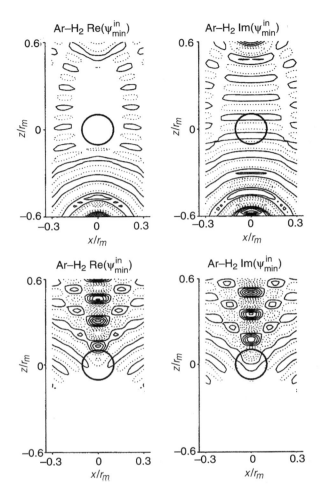

Figure 7.5 Real and imaginary parts of incident wave function leading to maximum and minimum outgoing flux for Ar + H_2 ($v = 0, j = 2, m_j = 0$). (Taken from Fig. 15, Ref. [48].)

χ_l allowed for considerable control. For example, with $j = 2, m = 0$, the outgoing flux into $j' = 0$ could be changed by two orders of magnitude, from 5.1×10^{-4} to 3.8×10^{-2}, just by varying the χ_l. These values are to be compared to a flux of 1×10^{-2} associated with scattering from an incident plane wave. Real and imaginary parts of the incident wave functions leading to these maximum and minimum values of the outgoing flux are shown in Figure 7.5. They are distinctly different from one another, and from a plane wave $\exp(i\mathbf{k} \cdot \mathbf{r})$.

Experimental implementation of this approach, however, requires the development of tools to shape the incoming wave.

CHAPTER 8

COHERENT CONTROL OF SYNTHESIS AND PURIFICATION OF CHIRAL MOLECULES

A molecule is said to be *chiral* if it does not coincide, or cannot be made to coincide via simple rotation, with its mirror image. In such cases, the molecule and its mirror image are called *enantiomers*, with one enantiomer being "right handed" and the other enantiomer being "left handed." A sufficient (though not necessary) condition for chirality is for the molecule to have at least one "asymmetric" carbon atom, that is, a carbon atom bonded to four different groups of atoms. Two enantiomers can be distinguished experimentally, for example, by their ability to rotate linearly polarized light in opposite directions. A mixture of the two enantiomers is called a *racemic mixture* or a *racemate*.

The existence of enantiomers is one of the fundamental broken symmetries in nature [255–258]. It is also one of great practical importance because biological processes are often stereospecific (i.e. sensitive to the particular enantiomer), motivating a long-standing interest in *asymmetric synthesis*, that is, molecular processes that preferentially produce one of the enantiomeric pairs. In this chapter we explain how to use coherent control techniques to perform asymmetric synthesis [136, 259–261] using the strong electric dipole–electric field interaction. This is in sharp contrast with previous techniques [255, 262] where efforts were made to use the far weaker magnetic–dipole interaction terms.

Section 8.1 introduces general rules under which this type of control is possible. Then, in Section 8.2, we consider the two-photon dissociation of a single quantum state of a B–A–B′ molecule to yield BA+B′ and B+AB′, where B and B′ are enantiomers. We demonstrate two results: (1) ordinary photodissociation of the nonchiral BAB′ molecule with linearly polarized light yields identical cross sections

167

for the production of the right-handed (B) and the left-handed (B′) fragment; (2) symmetry breaking can be induced by coherently controlling an interference term. The "control knobs" are the usual relative phases or the delay time between an excitation and a dissociation pulse.

In Section 8.3, we introduce a coherent control scheme for enhancing the fraction of one desired enantiomer, given a racemic mixture of molecules. This scheme, called *laser distillation*, is of great practical interest because the effect obtained is substantial, even in the presence of decoherence. An alternative scheme, due to Fujimura and co-workers [263, 264] is also briefly described.

8.1 PRINCIPLES OF ELECTRIC-DIPOLE-ALLOWED ENANTIOMERIC CONTROL

In this chapter we discuss a number of scenarios for manipulating enantiomer populations via the electric dipole interaction. However, these scenarios are, presumably, a small subset of an entire class of scenarios capable of achieving this goal. The key issue then is to establish the general conditions under which the electric-dipole electromagnetic field interaction may be used to attain selective control over the population of a desired enantiomer. These rules [261] are established in this section.

Consider a molecule, described by the Hamiltonian (*including electrons and nuclei*) H_{MT}. This Hamiltonian has eigenstates describing the L and D enantiomers, denoted $|L_i\rangle$ and $|D_i\rangle$ ($i = 1, 2, 3 \ldots$.) that satisfy

$$\mathcal{I}|L_i\rangle = -|D_i\rangle; \qquad \mathcal{I}|D_i\rangle = -|L_i\rangle, \tag{8.1}$$

where $\mathcal{I}$ is the operator that inverts all space-fixed coordinates through the origin. Note that the choice of phase (here minus one) in Eq. (8.1) is arbitrary, and that neither $|L_i\rangle$ nor $|D_i\rangle$ are of well-defined parity since they are not eigenstates of $\mathcal{I}$.

The dipole interaction of this molecule with an incident time-dependent electric field $\mathbf{E}(t)$ is described by the total Hamiltonian:

$$H(\mathbf{E}) = \mathbf{H}_{MT} - \mathbf{d} \cdot \mathbf{E}. \tag{8.2}$$

Here $\mathbf{d}$ is the total dipole operator, including both electron and nuclear contributions, and we have explicitly indicated the dependence of the Hamiltonian on the electric field. Consider now the effect of inversion on H. Noting that $\mathcal{I}$ operates on the coordinates of the molecule, that $\mathcal{I}^{\dagger} = \mathcal{I}$ and that $[H_{MT}, \mathcal{I}] = 0$, we have [265] that

$$\mathcal{I}H(\mathbf{E})\mathcal{I} = \mathbf{H}(-\mathbf{E}), \tag{8.3}$$

where $H(-\mathbf{E}) = H_{MT} + \mathbf{d} \cdot \mathbf{E}$. Further, if we define $U(\mathbf{E})$ and $U(-\mathbf{E})$ as the propagators corresponding to dynamics under $H(\mathbf{E})$ and $H(-\mathbf{E})$, respectively, then

$$U(\mathbf{E})\mathcal{I} = \mathcal{I}U(-\mathbf{E}). \tag{8.4}$$

To expose the underlying principles allowing achiral light-induced asymmetric synthesis, consider irradiating a racemic mixture of D and L in its ground electronic state with an electric field $\mathbf{E}$ and examine the difference δ between the amount of D and L formed. We consider first the coherent process using transform limited light in the absence of collisions. Then, the difference δ is given by

$$\delta = \sum_i P_i \sum_j [|\langle D_j | U(\mathbf{E}) | D_i \rangle|^2 + |\langle D_j | U(\mathbf{E}) | L_i \rangle|^2]$$
$$- [|\langle L_j | U(\mathbf{E}) | D_i \rangle|^2 + |\langle L_j | U(\mathbf{E}) | L_i \rangle|^2], \tag{8.5}$$

where P_i is the probability of state $|L_i\rangle$ and $|D_i\rangle$ in the initial mixed state. (Since the initial state is a racemic mixture, the states $|L_i\rangle$ and $|D_i\rangle$ appear with equal probability.) If $\delta = 0$, then there is no control over the chirality in the scenario defined by $U(\mathbf{E})$.

To determine the conditions under which δ is nonzero, we rewrite Eq. (8.5) as

$$\delta = \sum_i P_i \sum_j [|\langle D_j | U(\mathbf{E}) | D_i \rangle|^2 - |\langle L_j | U(\mathbf{E}) | L_i \rangle|^2]$$
$$+ [|\langle D_j | U(\mathbf{E}) | L_i \rangle|^2 - |\langle L_j | U(\mathbf{E}) | D_i \rangle|^2] \tag{8.6}$$

and recast the second and third terms using:

$$|\langle L_j | U(\mathbf{E}) | L_i \rangle|^2 = |\langle D_j | \mathcal{I}^\dagger U(\mathbf{E}) \mathcal{I} | D_i \rangle|^2 = |\langle D_j | U(-\mathbf{E}) | D_i \rangle|^2,$$
$$|\langle D_j | U(\mathbf{E}) | L_i \rangle|^2 = |\langle D_j | U(\mathbf{E}) \mathcal{I} | D_i \rangle|^2 = |\langle D_j | \mathcal{I} U(-\mathbf{E}) | D_i \rangle|^2 = |\langle L_j | U(-\mathbf{E}) | D_i \rangle|^2 \tag{8.7}$$

giving

$$\delta = \sum_i P_i \sum_k [|\langle D_k | U(\mathbf{E}) | D_i \rangle|^2 - |\langle D_k | U(-\mathbf{E}) | D_i \rangle|^2]$$
$$+ [|\langle L_k | U(-\mathbf{E}) | D_i \rangle|^2 - |\langle L_k | U(\mathbf{E}) | D_i \rangle|^2]. \tag{8.8}$$

Equation (8.8), the essential result of this section, provides the general condition under which electric fields, assuming a dipole interaction, can break the right–left symmetry of the initial state and result in enhanced production of a desired enantiomer. Specifically, the difference between the amount of D and L formed is seen to depend entirely on the difference between the molecular dynamics when irradiated by $\mathbf{E}$ and by $-\mathbf{E}$. Hence, we can state that *a necessary condition for nonzero enantiomeric excess, and the breaking of the left–right symmetry, is that the dynamics depend on the sign of the electric field*. Note that the fact that molecular dynamics can depend on the phase of the incident electric field is well substantiated [266, 267], but its utility for asymmetric synthesis is only evident from this result. Finally, note that the result is completely consistent with symmetry-based arguments that can usefully provide conditions under which δ must equal zero. For example, a racemic mixture of thermally equilibrated molecules is rotationally invariant. Hence any rotation that converts $\mathbf{E}$ to $-\mathbf{E}$ could not, in this case, result in enantiomeric control. In particular, in this case summing over M_J (where M_J is the component of the reactant's total angular momentum along the direction of laser polarization) implicit in the sum over P_i in Eq. (8.8) would result in $\delta = 0$. By contrast, as discussed below, a racemic mixture of M_J polarized molecules irradiated with linearly polarized light [268] gives nonzero δ. New $\delta \neq 0$ examples emanating from Eq. (8.8) are also expected to display similar nontraditional characteristics.

Both qualitative and quantitative applications of Eq. (8.8) are possible. Qualitatively, for example, a traditional scheme where the ground electronic state of L and D are incoherently excited to bound levels of an excited state, gives $\delta = 0$. This is because all processes connecting the initial and final $|L_i\rangle$ and $|D_i\rangle$ states, that is, contributions to the matrix elements in Eq. (8.8), are even in the amplitude of the electric field. Hence, propagation under $\mathbf{E}$ and $-\mathbf{E}$ are identical. By contrast, consider the four-level model scheme in Figure 8.1, and discussed in detail in Section 8.2. When $\varepsilon_0(t) \neq 0$ there exist processes connecting the initial and final $|L\rangle$ and $|D\rangle$ states that are of the form $|L\rangle \rightarrow |1\rangle \rightarrow |2\rangle \rightarrow |D\rangle$, and hence there are terms in Eq. (8.8) that are odd in the amplitude of the electric field. One therefore anticipates the possibility of altering the enantiomeric excess using this combination of pulses, providing the basis for the control results reported on later below. Further, if $\varepsilon_0 = 0$, then the situation reverts to the case discussed above, where only processes

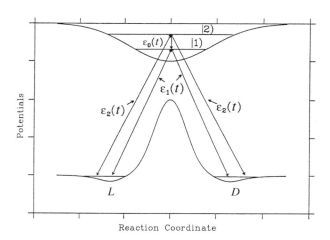

Reaction Coordinate

Fig. 8.1 "Laser distillation" control scenario discussed in detail in Section 8.3. Two lasers, with pulse envelopes $\varepsilon_1(t)$ and $\varepsilon_2(t)$ couple, by virtue of the dipole operator, the states of the D and L enantiomers to two vibrotational states $|1\rangle$ and $|2\rangle$ (denoted $|E_1\rangle$ and $|E_2\rangle$ in the text) in the excited electronic manifold. A third laser pulse with envelope $\varepsilon_0(t)$ couples the excited $|E_1\rangle$ and $|E_2\rangle$ states to one another. The system is allowed to absorb a photon and relax back to the ground state. After many such "excitation–relaxation" cycles, a significant enantiomeric excess is obtained, as explained in Section 8.3.

even in the electric field contribute to transitions between the initial $|D\rangle$, $|L\rangle$ and final $|D\rangle$, $|L\rangle$ transitions, and hence control over the enantiomeric excess is lost. For this reason, the $\varepsilon_0(t)$ coupling laser is crucial to enantiomeric control. This qualitative picture is substantiated quantitatively in Section 8.3.

What is required experimentally to achieve this kind of control is the ability to manipulate the phase of the electric field. One possible approach is to use ultrashort pulses [269, 270] that allow defining the overall electric field phase. Specific applications that are somewhat less experimentally demanding are discussed below.

8.2 SYMMETRY BREAKING IN TWO-PHOTON DISSOCIATION OF PURE STATES

Consider a molecule of the type BAB' where B and B' are enantiomers. This molecule possesses a plane of symmetry σ perpendicular to the B–A–B' axis. The operator corresponding to reflection across this plane is denoted σ_h. In order to coherently control the dissociation of this system, we take advantage of the existence of degenerate continuum states, which do not possess this reflection symmetry. That is, these molecules possess degenerate continuum states $|E, \mathbf{n}, D^-\rangle$ and $|E, \mathbf{n}, L^-\rangle$ that correlate asymptotically with the dissociation of the right B' group and left B group, respectively. The collective quantum index $\mathbf{n}$ in the states $|E, \mathbf{n}, D^-\rangle$ and $|E, \mathbf{n}, L^-\rangle$ includes m, the magnetic quantum number of the B or B' fragment. These states are neither symmetric nor antisymmetric with respect to the reflection operator σ_h, although linear combinations of these states might possess this symmetry.

We now consider using the pump–dump scenario described in Section 3.5, for BAB' photodissociation. The application of this scenario to the chiral synthesis case is depicted schematically in Figure 8.2. Our aim is to control the relative yield of the two product arrangement channels $B - A + B'$ and $B + AB'$. That is, we consider $P_{q,\mathbf{n}}(E)$, with q labeling either the right- ($q = D$) or left- ($q = L$) handed product. As in Section 3.5, the product ratio $R_{DL;\mathbf{n}} = P_{D,\mathbf{n}}(E)/P_{L,\mathbf{n}}(E)$ is a function of the delay time $\Delta_d = (t_d - t_x)$ between pulses and the ratio $x = |c_2/c_3|$, the latter by varying the energy of the initial excitation pulse. Active control over the products $B + AB'$ vs. $B' + AB$, that is, a variation of $R_{DL;\mathbf{n}}$ with Δ_d and x, and hence control over left- vs. right-handed products, will result only if $P_{D,\mathbf{n}}(E)$ and $P_{L,\mathbf{n}}(E)$ have different functional dependences on the control parameters x and Δ_d.

To show that $P_{D,\mathbf{n}}(E)$ may differ from $P_{L,\mathbf{n}}(E)$ for the B'AB case, note first that this molecule belongs to the C_s point group. This group possesses only one (hyper) plane of symmetry, denoted σ above, which is defined as the collection of points satisfying the requirement that the B–A distance equals the A–B' distance. Furthermore, we focus upon transitions between electronic states of the same representations, for example, A' to A' or A'' to A'' (where A' denotes the symmetric representation and A'' the antisymmetric representation of the C_s group). We further assume that the ground vibronic state belongs to the A' representation.

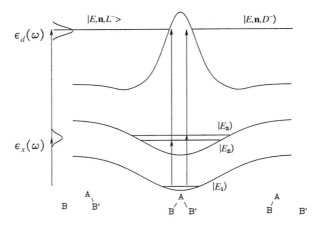

Fig. 8.2 Schematic showing controlled dissociation of molecule B–A–B′ to yield B–A+B′ or B + A–B′ products, where B and B′ are two enantiomers. A molecule is excited from an initial state $|E_1\rangle$ to a superposition of antisymmetric ($|E_2\rangle$) and symmetric ($|E_3\rangle$) vibrational states belonging to an excited electronic state, by excitation pulse $\epsilon_x(\omega)$. After an appropriate delay time, the molecule is dissociated by second pulse $\epsilon_d(\omega)$, to the $|E, \mathbf{n}, D^-\rangle$ or $|E, \mathbf{n}, L^-\rangle$ continuum state.

To obtain control, we choose the intermediate state $|E_3\rangle$ to be *symmetric* and the intermediate state $|E_2\rangle$ to be *antisymmetric*, with respect to reflection in the σ hyperplane. Hence we must first demonstrate that it is possible to optically excite, simultaneously, both the symmetric $|E_3\rangle$ and antisymmetric $|E_2\rangle$ states from the ground state $|E_1\rangle$. Using Eq. (3.75) we see that this requires the existence of both a symmetric dipole component, denoted $\mathbf{d}_s$, and an antisymmetric component, denoted $\mathbf{d}_a$, with respect to reflection in the σ hyperplane because, by the symmetry properties of $|E_3\rangle$ and $|E_2\rangle$,

$$\langle E_3|\mathbf{d} \cdot \hat{\varepsilon}|E_1\rangle = \langle E_3|\mathbf{d}_s \cdot \hat{\varepsilon}|E_1\rangle; \qquad \langle E_2|\mathbf{d} \cdot \hat{\varepsilon}|E_1\rangle = \langle E_2|\mathbf{d}_a \cdot \hat{\varepsilon}|E_1\rangle. \tag{8.9}$$

We note that the coexistence of symmetric and antisymmetric components of the dipole moment is with respect to σ_h. Since the σ plane rotates with the molecule, the σ_h operation is said to be "body-fixed" (or "molecule-fixed"). Both the body-fixed symmetric $\mathbf{d}_s$ and the body-fixed antisymmetric $\mathbf{d}_a$ dipole-moment components do occur in $A' \rightarrow A'$ electronic transitions whenever the geometry of a bent B′–A–B molecule deviates considerably from the points on the σ hyperplane, characterized by the points of equidistance (C_{2v}) geometries (where $\mathbf{d}_a = 0$) (see Fig. 8.3). The deviation of $\mathbf{d}_a$ from zero on the σ plane necessitates going beyond the Franck–Condon approximation, which assumes that the electronic dipole moment does not change as the molecule vibrates. (In the terminology of the theory of vibronic

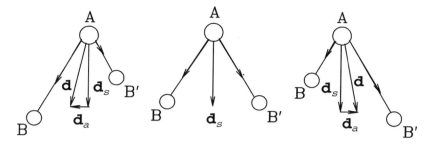

Fig. 8.3 Emergence of antisymmetric dipole component $\mathbf{d}_a$ in addition to symmetric component $\mathbf{d}_s$ in a bent BAB' triatomic molecule as a result of an asymmetric stretching vibration, assuming that the dipole is a vectorial sum of bond dipoles, which are proportional to bond lengths.

transitions, both symmetric and antisymmetric components can be nonzero due to a Herzberg–Teller intensity borrowing [271] mechanism.)

Note also that the dipole-moment operator, being a vector, must invert its sign under inversion $\mathcal{I}$. Hence, with respect to $\mathcal{I}$, the dipole moment is always *antisymmetric*. Thus, for the integrals in Eq. (8.9) to be nonzero also requires that $|E_3\rangle$ and $|E_1\rangle$ be of opposite symmetry with respect to inversion. Given the extant conditions on the behavior of $|E_3\rangle$ and $|E_1\rangle$ with respect to the reflection σ_h, the symmetry requirements with respect to $\mathcal{I}$ are most easily accommodated through the rotational components of the $|E_3\rangle$ and $|E_1\rangle$ states.

Thus, the excitation pulse can create a superposition of $|E_2\rangle$, $|E_3\rangle$ consisting of two states of different reflection symmetry. The resultant superposition possesses no symmetry properties with respect to reflection [272].

We now show that the nonsymmetry created by this excitation of *nondegenerate* bound states translates into a nonsymmetry in the probability of populating the *degenerate* $|E, \mathbf{n}, D^-\rangle$, $|E, \mathbf{n}, L^-\rangle$ continuum states upon subsequent excitation. To do so we examine the properties of the bound-free transition matrix elements $\langle E, \mathbf{n}, q^- | d_{e,g} | E_k \rangle$ that enter into the probability of dissociation [Eq. (3.77)]. Note first that although the continuum states $|E, \mathbf{n}, q^-\rangle$ are nonsymmetric with respect to reflection, we can define symmetric and antisymmetric continuum eigenfunctions $|E, \mathbf{n}, s^-\rangle$ and $|E, \mathbf{n}, a^-\rangle$ via the relations

$$|E, \mathbf{n}, D^-\rangle \equiv [|E, \mathbf{n}, s^-\rangle + |E, \mathbf{n}, a^-\rangle]/\sqrt{2}, \tag{8.10}$$

$$|E, \mathbf{n}, L^-\rangle \equiv [|E, \mathbf{n}, s^-\rangle - |E, \mathbf{n}, a^-\rangle]/\sqrt{2}, \tag{8.11}$$

using the fact that $\sigma_h |E, \mathbf{n}, D^-\rangle = |E, \mathbf{n}, L^-\rangle$.

Consider first the nature of the $d_q(ij)$ that enter Eq. (3.79), prior to averaging over product scattering angles. We denote this as $d_q(ij; \hat{\mathbf{k}})$, where $\hat{\mathbf{k}}$ is the scattering direction. Since $|E_3\rangle$ is symmetric and $|E_2\rangle$ is antisymmetric, and adopting the notation $A_{s2} \equiv \langle E, \mathbf{n}, s^- | d_a | E_2 \rangle$, $S_{a3} \equiv \langle E, \mathbf{n}, a^- | d_s | E_3 \rangle$, and so on, we have [see Eq. (3.79)]

$$d_q(33; \hat{\mathbf{k}}) = \sum{}'' \left[|S_{s3}|^2 + |A_{a3}|^2 \pm 2 \ \mathrm{Re}(A_{a3} S_{s3}{}^*) \right],$$
$$d_q(22; \hat{\mathbf{k}}) = \sum{}'' \left[|A_{s2}|^2 + |S_{a2}|^2 \pm 2 \ \mathrm{Re}(A_{s2} S_{a2}{}^*) \right], \qquad (8.12)$$
$$d_q(32; \hat{\mathbf{k}}) = \sum{}'' \left[S_{s3} A_{s2}{}^* + A_{a3} S_{a2}{}^* \pm S_{s3} S_{a2}{}^* \pm A_{a3} A_{s2}{}^* \right],$$

where the plus sign applies for $q = D$, the minus sign applies for $q = L$, and $d_q(23; \hat{\mathbf{k}}) = d_q{}^*(32; \hat{\mathbf{k}})$. The double prime on the sum denotes a summation over all $q, \mathbf{n}$ other than the scattering angles and the product m, where m denotes the projection of the product angular momentum along the axis of laser polarization.

Equation (8.12) takes on a simpler form after angular averaging. The reason for this is that the overall parity of a state with respect to the inversion operation, $\mathcal{I}$, must change upon photon absorption since a photon has odd parity. As a result, if we have a single photon absorption process in which the parity of a vibrational state is unchanged, then the parity of the rotational states must change, and vice versa. Close examination of Eq. (8.12) reveals that the $S_{s3}{}^*$ term does not involve a change in the parity of the vibrational state, whereas the A_{a3} term does. As a result, the rotational wave functions associated with each term must have opposing parities and the angular integral of the product must vanish. (Appendix 8A discusses the form of these matrix elements and their products.) The same goes for the $A_{s2} S_{a2}{}^*$ term. In a similar manner the $S_{s3} A_{s2}{}^* + A_{a3} S_{a2}{}^*$ term vanishes in the $d_q(32)$ interference term. By contrast, the $\pm S_{s3} S_{a2}{}^* \pm A_{a3} A_{s2}{}^*$ terms do not vanish upon angular integration since they correspond to final rotational states that have the same parity.

As a consequence, the net result is that, after angular averaging, Eq. (8.12) becomes

$$d_q(33) = \sum{}' \left[|S_{s3}|^2 + |A_{a3}|^2 \right],$$
$$d_q(22) = \sum{}' \left[|A_{s2}|^2 + |S_{a2}|^2 \right], \qquad (8.13)$$
$$d_q(32) = \sum{}' \pm \left[S_{s3} S_{a2}{}^* + A_{a3} A_{s2}{}^* \right],$$

where the single prime on the sum indicates that the sum over product m is not carried out.

These equations display two noteworthy features:

1. $d_L(jj) = d_D(jj)$, $j = 2, 3$, that is, lacking interference, no discrimination between the left-handed and right-handed products is possible.
2. $d_L(23) \neq d_D(23)$, that is, laser-controlled symmetry breaking, which depends upon $d_q(23)$ in accordance with Eq. (3.78), is possible. As noted below, this type of discrimination is possible only if we select the direction of the angular momentum of the products (m polarization).

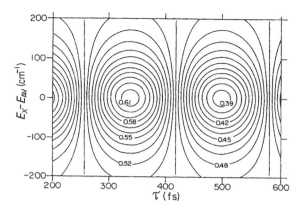

Fig. 8.4 Contour plot of percent $H_aO + H_b$ (as distinct from $H_a + OH_b$) in H_aOH_b photodissociation. Ordinate is the detuning from $E_{av} = (E_3 - E_2)/2$, and the abscissa is the time delay between pulses. (Taken from Fig. 9, Ref. [48].)

To demonstrate the extent of expected control, as well as the effect of m summation, we considered a model of enantiomer selectivity, that is, HOH photodissociation in three dimensions, where the two hydrogens are assumed distinguishable:

$$H_aO + H_b \leftarrow H_aOH_b \rightarrow H_a + OH_b.$$

The computation is done using the formulation and computational methodology of Ref. [273]. In Appendix C we briefly summarize the angular momentum algebra and some other details involved in performing this three-dimensional quantum calculation on triatomic photodissociation. We also discuss the loss of control, in this scenario, with m averaging.

Equation (8.13) in conjunction with the pump–dump control scenario and the artificial channel method [14] for computing the t matrix elements, was used to compute the ratio R_{DL} of the $H_aO + H$ (as distinct from the $H + OH_b$) product in a fixed m state. Figure 8.4 shows the result of first exciting the superposition of symmetric plus asymmetric vibrational modes [(1, 0, 0) + (0, 0, 1)] with $J_i = J_k = 0$ in the ground electronic state, followed by dissociation at $70{,}700$ cm^{-1} to the B state using a pulse width of 200 cm^{-1}. The results show that varying the time delay between pulses allows for controlled variation of P_D from 61 to 39%. This variation is significant, revealing the symmetry breaking arising within this scenario.

8.3 PURIFICATION OF RACEMIC MIXTURES BY "LASER DISTILLATION"

In Section 8.2 we showed that coherent control techniques can be used to direct the photodissociation of a BAB$'$ molecule to yield an excess of a desired B or B$'$

enantiomer. Throughout the treatment we assumed that the BAB' molecule exists in the single quantum state $|E_i, J_i, p_i\rangle$ prior to dissociation.

While the above process is of great scientific interest, practically speaking we usually want to separate a *racemic mixture* of the B and B' enantiomers, that is, our typical initial state is a racemate. If we were to use the scenario of Section 8.2 to accomplish this separation one would have first to prepare the BAB' adduct in a pure state. Since the preparation of the BAB' adduct, and especially its separation from the BAB and B'AB' adducts that would inevitably accompany its preparation, is not a trivial task, it is preferable to find control methods that could separate the B and B' racemic mixture directly. In this section we outline a method that can achieve this much more ambitious task. The essential principles of this method remain the same as in Section 8.2, that is, excitation of a superposition of symmetric and antisymmetric states with respect to σ_h, the reflection operation.

Consider then a molecular system composed of a pair of stable nuclear configurations, denoted L and D, with L being the (distinguishable) mirror image of D. Note that the electronic Hamiltonian H_e commutes with σ_h, hence, the potential energy surfaces, which are the eigenvalues of the electronic Schrödinger equation at all nuclear configurations, must be symmetric with respect to σ_h.

Since L and D are assumed stable, it follows that the ground potential energy surface must possess a sufficiently high barrier at nuclear coordinates separating L and D such that the rate of interconversion between them by tunneling is negligible. By contrast, L and D need not be stable on an excited potential energy surface. To this end, we assume that there is at least one excited potential surface, denoted G, which possesses a potential well midway between the L and the D geometries (see Fig. 8.1). (A number of molecules expected to be of this type are tabulated in Ref. [258], and a number of examples are discussed below.) Hence, the interconversion between L and D on the excited surface G is expected to be very facile.

A direct consequence of the potential well midway between the L and the D geometries on surface G is the existence of stable vibrational eigenstates. Because of the symmetry of G, the vibrational eigenstates must be either symmetric or antisymmetric with respect to σ_h.

The procedure that we propose to enhance the concentration of a particular enantiomer when starting with a racemic mixture, that is, to "purify" the mixture, is as follows [259]. The mixture of statistical (racemic) mixture of L and D is irradiated with a specific sequence of three coherent laser pulses, as described below. These pulses excite a coherent superposition of symmetric and antisymmetric vibrational states of G. After each pulse the excited system is allowed to relax back to the ground electronic state by spontaneous emission or by any other nonradiative process. By allowing the system to go through many irradiation and relaxation cycles, we show below that the concentration of the selected enantiomer L or D can be enhanced, depending on the laser characteristics. We call this scenario *laser distillation* of chiral enantiomers.

We note at the outset that detailed angular momentum considerations show that if the three incident lasers are of the same polarization then control results only if we do not average over M_J, the projection of the total angular momentum of the reactant along the z axis (chosen as the direction of laser polarization). In particular, in this case enantiomeric enhancement of one enantiomer from molecules in state M_J is exactly counterbalanced by enantiomeric enhancement of the other enantiomer by molecules in state $-|M_J|$. Hence, enantiomeric control in this scenario requires prior M_J selection of the molecules. This scenario is discussed below, but results are also provided for the case of three lasers of perpendicular polarization, where M_J averaging is nondestructive and enantiomeric control persists.

Consider then a chiral molecule with Hamiltonian H_M, in the presence of a series of laser pulses. (In general we may deal with lasers that are not fully coherent, but for simplicity we focus here on transform-limited pulses of linearly polarized light.) The treatment is in accord with Chapter 1, Eq. (1.51), where the interaction between the molecule and radiation is given by

$$H_{MR}(t) = -\mathbf{d} \cdot \mathbf{E}(t) = -2\mathbf{d} \cdot \sum_k \mathrm{Re}[\hat{\epsilon}_k \varepsilon_k(t) \exp(-i\omega_k t)]. \tag{8.14}$$

Here $\varepsilon_k(t)$ is the pulse envelope, ω_k is the central laser frequency, and $\hat{\epsilon}_k$ is the polarization direction. Expanding $|\Psi(t)\rangle$ in eigenstates $|E_j\rangle$ of the molecular Hamiltonian (i.e., $H_M|E_j\rangle = E_j|E_j\rangle$):

$$|\Psi(t)\rangle = \sum_j b_j \exp\left(-\frac{iE_j t}{\hbar}\right)|E_j\rangle, \tag{8.15}$$

and substituting Eq. (8.15) into the time-dependent Schrödinger equation gives the standard set of coupled equations:

$$\dot{b}_i = \frac{-i}{\hbar} \sum_{jk} b_j \exp(-i\omega_{ji}t)\langle E_i|H_{MR}(t)|E_j\rangle, \tag{8.16}$$

where $\omega_{ji} = (E_j - E_i)/\hbar$.

As an example of an effective control scenario, consider the molecules D and L in their ground electronic states and in vibrotational states $|E_D\rangle$ and $|E_L\rangle$, of energy $E_D = E_L$. We choose $\mathbf{E}(t)$ so as to excite the system to two eigenstates $|E_1\rangle$ and $|E_2\rangle$ of the electronically excited potential surface G. The states $|E_1\rangle$ and $|E_2\rangle$ are also coupled by an additional laser field (see Fig. 8.1).

Specifically, we choose $\mathbf{E}(t)$ to be composed of three linearly polarized light pulses (all of the same polarization):

$$\mathbf{E}(t) = \sum_{k=0,1,2} 2\,\mathrm{Re}[\varepsilon_k(t)\exp(-i\omega_k t)]\hat{\epsilon}_k, \tag{8.17}$$

with ω_0 in near resonance with $\omega_{2,1} \equiv (E_2 - E_1)/\hbar$, ω_1 is chosen to be near resonant with $\omega_{1,D} \equiv (E_1 - E_D)/\hbar$, and ω_2 near resonant with $\omega_{2,D} \equiv (E_2 - E_D)/\hbar$ (see Fig. 8.1). In this case, only four molecular states are relevant and Eq. (8.15) becomes

$$|\Psi\rangle = b_D(t)\exp(-iE_Dt/\hbar)|E_D\rangle + b_L(t)\exp(-iE_Lt/\hbar)|E_L\rangle$$
$$+ b_1(t)\exp(-iE_1t/\hbar)|E_1\rangle + b_2\exp(-iE_2t/\hbar)|E_2\rangle. \tag{8.18}$$

Equation (8.16), in the rotating wave approximation, is then given by

$$\dot{b}_1 = i\exp(i\Delta_1 t)[\Omega_{D,1}^* b_D + \Omega_{L,1}^* b_L] + i\exp(-i\Delta_0 t)\Omega_0^* b_2,$$
$$\dot{b}_2 = i\exp(i\Delta_2 t)[\Omega_{D,2}^* b_D + \Omega_{L,2}^* b_L] + i\exp(i\Delta_0 t)\Omega_0 b_1,$$
$$\dot{b}_D = i\exp(-i\Delta_1 t)\Omega_{D,1}b_1 + i\exp(-i\Delta_2 t)\Omega_{D,2}b_2, \tag{8.19}$$
$$\dot{b}_L = i\exp(-i\Delta_1 t)\Omega_{L,1}b_1 + i\exp(-i\Delta_2 t)\Omega_{L,2}b_2,$$

where $\Omega_{i,j}(t) \equiv d_{i,j}^{(j)}\varepsilon_1(t)/\hbar$, $\Omega_0 \equiv d_{2,1}^{(0)}\varepsilon_0(t)/\hbar$, $\Delta_j \equiv \omega_{j,D} - \omega_1$, $\Delta_0 \equiv \omega_{2,1} - \omega_0$, where $d_{i,j}^{(k)} \equiv \langle E_i|\mathbf{d}\cdot\hat{\varepsilon}_k|E_j\rangle$, with $i = D, L$; $k = 0, 1, 2$ and $j = 1, 2$.

The essence of the laser distillation process lies in choosing the laser of central frequency ω_1 so that it excites the system to a state $|E_1\rangle$, which is *symmetric* with respect to the reflection operation σ_h, and to a state $|E_2\rangle$, which is *antisymmetric* with respect to σ_h and coupling these states with an additional ω_0 laser. By contrast, $|E_D\rangle$ and $|E_L\rangle$ do not share these symmetries but are related to one another through reflection [i.e., $\sigma_h|E_D\rangle = |E_L\rangle$, $\sigma_h|E_L\rangle = |E_D\rangle$ whereas $\sigma_h|E_1\rangle = |E_1\rangle$, $\sigma_h|E_2\rangle = -|E_2\rangle$].

To consider the nature of the "Rabi frequencies" Ω in Eq. (8.19), we rewrite $|E_D\rangle$ and $|E_L\rangle$ in terms of a symmetric state $|S\rangle$ and an antisymmetric state $|A\rangle$:

$$|E_D\rangle = |A\rangle + |S\rangle,$$
$$|E_L\rangle = |A\rangle - |S\rangle. \tag{8.20}$$

In addition to their symmetry properties with respect to σ_h, we choose the $|S\rangle$ and $|A\rangle$ states to be, respectively, symmetric and antisymmetric under the inversion operation $\mathcal{I}$. Coupled with the fact that the dipole operator must be antisymmetric with respect to $\mathcal{I}$, the relevant matrix elements satisfy the following relations:

$$\langle 1|d^{(1)}|D\rangle = \langle 1|d^{(1)}|A + S\rangle = \langle 1|d^{(1)}|A\rangle,$$
$$\langle 1|d^{(1)}|L\rangle = \langle 1|d^{(1)}|A - S\rangle = \langle 1|d^{(1)}|A\rangle,$$
$$\langle 2|d^{(2)}|D\rangle = \langle 2|d^{(2)}|A + S\rangle = \langle 2|d^{(2)}|S\rangle, \tag{8.21}$$
$$\langle 2|d^{(2)}|L\rangle = \langle 2|d^{(2)}|A - S\rangle = -\langle 2|d^{(2)}|S\rangle.$$

That is,

$$\Omega_{D,1} = \Omega_{L,1}, \ \Omega_{D,2} = -\Omega_{L,2}. \tag{8.22}$$

Given Eq. (8.22), Eq. (8.19) becomes

$$
\begin{aligned}
\dot{b}_1 &= i\exp(i\Delta_1 t)\Omega_{D,1}^*[b_D + b_L] + i\exp(-i\Delta_0 t)\Omega_0^* b_2,\\
\dot{b}_2 &= i\exp(i\Delta_2 t)\Omega_{D,2}^*[b_D = b_L] + i\exp(i\Delta_0 t)\Omega_0 b_1,\\
\dot{b}_D &= i\exp(-i\Delta_1 t)\Omega_{D,1} b_1 + i\exp(-i\Delta_2 t)\Omega_{D,2} b_2,\\
\dot{b}_L &= i\exp(-i\Delta_1 t)\Omega_{D,1} b_1 - i\exp(-i\Delta_2 t)\Omega_{D,2} b_2.
\end{aligned}
\tag{8.23}
$$

The essence of optically controlled enantioselectivity in this scenario lies in Eq. (8.22) and the effect of these relationships on the dynamical equations for the level populations [Eq. (8.23)]. Note specifically that the equation for $\dot{b}_D(t)$ is different than the equation for $\dot{b}_L(t)$, due to the sign difference in the last term in Eq. (8.23). Although not sufficient to ensure enantiomeric selectivity, the ultimate consequence of this difference is that populations of $|E_D\rangle$ and $|E_L\rangle$ after laser excitation are different when there is radiative coupling between levels $|E_1\rangle$ and $|E_2\rangle$.

Note, in accord with Section 8.1, the behavior of Eq. (8.23) under the transformation $\mathbf{E} \to -\mathbf{E}$. Specifically, changing $\mathbf{E}$ to $-\mathbf{E}$ means changing all $\varepsilon_j(t)$ to $-\varepsilon_j(t)$. Doing so, and defining $b_1' = -b_1$ and $b_2' = -b_2$ converts Eq. (8.23) into

$$
\begin{aligned}
\dot{b}_1' &= i\exp(i\Delta_1 t)\Omega_{D,1}^*[b_D + b_L] - i\exp(i\Delta_0 t)\Omega_0^* b_2',\\
\dot{b}_2' &= i\exp(i\Delta_2 t)\Omega_{D,2}^*[b_D - b_L] - i\exp(-i\Delta_0 t)\Omega_0 b_1',\\
\dot{b}_D &= i\exp(-i\Delta_1 t)\Omega_{D,1} b_1' + i\exp(-i\Delta_2 t)\Omega_{D,2} b_2',\\
\dot{b}_L &= i\exp(-i\Delta_1 t)\Omega_{D,1} b_1' - i\exp(-i\Delta_2 t)\Omega_{D,2} b_2'.
\end{aligned}
\tag{8.24}
$$

Clearly, Eq. (8.24) is the same as Eq. (8.23) barring the change of sign in the Ω_0 terms. Thus, the solution to Eq. (8.23) depends on the sign of $\mathbf{E}$ when $\varepsilon_0 \neq 0$. Hence, by the argument in Section 8.1, this scenario allows for chirality control when $\varepsilon_0(t) \neq 0$. For $\varepsilon_0(t) = 0$ Eq. (8.24) is the same as Eq. (8.23) so that enantiomer control is not possible.

To obtain quantitative estimates for the extent of obtainable control, we consider results for model cases assuming Gaussian pulses

$$\varepsilon_k(t) = \epsilon_k \exp[-((t - t_k)/\alpha_k)^2], \qquad (k = 0, 1, 2), \tag{8.25}$$

and system parameters $\langle 1|d^{(1)}|D\rangle = \langle 1|d^{(1)}|L\rangle = \langle 2|d^{(2)}|L\rangle = -\langle 2|d^{(2)}|D\rangle = 1$ a.u., $\langle 1|d^{(0)}|2\rangle = 1$ a.u., $\omega_{2,1} = 100 \text{ cm}^{-1}$, and $\Delta_0 = 0$. Figure 8.5 displays the probabilities $P_D = |b_D(\infty)|^2$, $P_L = |b_L(\infty)|^2$ of population in $|E_D\rangle$ and $|E_L\rangle$, after a single pulse, for a variety of pulse parameters. Results are shown for various values of

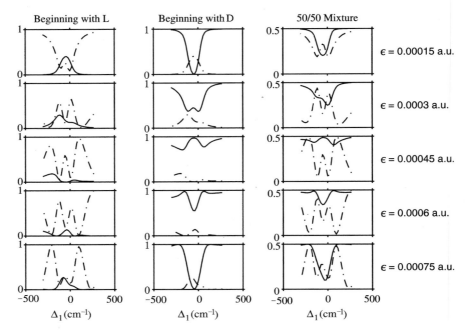

Fig. 8.5 Probabilities of populating the $|E_D\rangle$ (solid lines) and $|E_L\rangle$ (dot–dash lines) after laser excitation, but prior to relaxation, as function of detuning Δ_1. Three different cases are shown, corresponding to three different initial conditions: (1) only state $|E_L\rangle$ is initially populated; (2) only state $|E_D\rangle$ is initially populated; (3) a statistical mixture made up of equal shares of $|E_D\rangle$ and $|E_L\rangle$ states is initially populated. Results are shown for five different $\epsilon_2 = \epsilon_1 = \epsilon_0 \equiv \epsilon$ laser peak electric fields, where Gaussian pulses are assumed with $\alpha_0 = \alpha_1 = 0.15$ ps, and $t_0 = t_1$.

$\Delta_1 = \omega_{1,0} - \omega_1$ at various different pulse powers assuming that one starts solely with D, solely with L, or with a racemic mixture of both enantiomers. Clearly, for particular parameters, one can significantly enhance the population of one chiral enantiomer over the other. For example, for $\Delta_1 = -115$ cm^{-1}, $\epsilon_0 = \epsilon_1 = 4.5 \times 10^{-4}$, a racemic mixture of D and L can be converted, after a single pulse, to a enantiomerically enriched mixture with predominantly D.

Control is strongly affected by the relative phase θ of the ε_1 and ε_0 fields, as shown in Figure 8.6. Here it is clear that changing θ by π interchanges the dynamical evolution of the L and D enantiomers.

Although not immediately obvious, this control scenario relies entirely upon quantum interference effects. To see this note that in the absence of an $\varepsilon_0(t)$ pulse, excitation from $|D\rangle$ or $|L\rangle$ to level $|E_i\rangle$, for example, occurs via one photon excitation with $\varepsilon_i(t)$, $i = 1, 2$. In this case, as noted above, there is no chiral control. By contrast, with nonzero $\varepsilon_0(t)$, there is an additional (interfering) route to $|E_i\rangle$, that is, a two-photon route using $\varepsilon_j(t)$ excitation to level $|E_j\rangle, j \neq i$, followed by an $\varepsilon_0(t)$

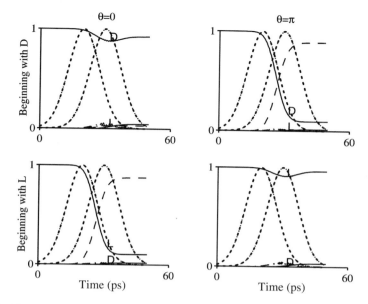

Fig. 8.6 Time evolution of enantiomeric populations for two different relative phases θ between the ϵ_1 and ϵ_0 beams. (_____) population in the D or L enantiomer; (......) the ϵ_1 and ϵ_0 laser pulses; (– – –) excited state population in levels $|E_1\rangle + |E_2\rangle$.

induced transition from $|E_j\rangle$ to $|E_i\rangle$. The one- and two-photon routes interfere and, as implied in Section 3.3.3, allow for symmetry-breaking transitions.

The computation, which results in Figure 8.5, which gives the result of a single pulse, provides input into a calculation of the overall result. In the overall process we begin with an incoherent mixture of N_D molecules of type D and N_L molecules of type L. In the first step the system is excited, as above, with a laser pulse sequence. In the second step, the system collisionally and radiatively relaxes so that all the population returns to the ground state to produce an incoherent mixture of $|E_L\rangle$ and $|E_D\rangle$. This pair of steps is then repeated until the populations of $|E_L\rangle$ and $|E_D\rangle$ reach convergence.

To obtain the result computationally note that the population after laser excitation, but before relaxation, consists of the weighted sum of the results of two computations: N_D times the results of laser excitation starting solely with molecules in $|E_D\rangle$, plus N_L times the results of laser excitation starting solely with molecules in $|E_L\rangle$. If $P_{D\leftarrow D}$ and $P_{L\leftarrow D}$ denote the probabilities of $|E_D\rangle$ and $|E_L\rangle$ resulting from laser excitation assuming the first of these initial conditions, and $P_{D\leftarrow L}$ and $P_{L\leftarrow L}$ for the results of excitation following from the second of these initial conditions, then the populations of $|E_D\rangle$ and $|E_L\rangle$ after laser excitation of the mixture are $N_D P_{D\leftarrow D} + N_L P_{D\leftarrow L}$ and $N_D P_{L\leftarrow D} + N_L P_{L\leftarrow L}$, respectively. The remainder of the population, $N_D[1 - P_{D\leftarrow D} - P_{L\leftarrow D}] + N_L[1 - P_{D\leftarrow L} - P_{L\leftarrow L}]$, is in the upper two levels $|E_1 >$ and $|E_2 >$. Relaxation from levels $|E_1\rangle$ and $|E_2\rangle$ then follows, with the excited population dividing itself equally between $|E_D\rangle$ and $|E_L\rangle$. The resultant populations $\mathcal{N}_D$ and $\mathcal{N}_L$ in ground state $|E_D\rangle$ and $|E_L\rangle$ is then:

$$\mathcal{N}_D = 0.5N_D[1 + P_{D\leftarrow D} - P_{L\leftarrow D}] + 0.5N_L[1 + P_{D\leftarrow L} - P_{L\leftarrow L}],$$
$$\mathcal{N}_L = 0.5N_D[1 + P_{L\leftarrow D} - P_{D\leftarrow D}] + 0.5N_L[1 + P_{L\leftarrow L} - P_{D\leftarrow L}]. \tag{8.26}$$

The sequence of laser excitation followed by collisional relaxation and radiative emission is then iterated to convergence. In the second step, for example, the populations in Eq. (8.26) are taken as the initial populations for two independent computations, one assuming a population of $\mathcal{N}_D$ in $|E_D\rangle$, with $|E_L\rangle$ unpopulated, and the second assuming a population of $\mathcal{N}_L$ in $|E_L\rangle$, with $|E_D\rangle$ unpopulated.

Clearly, convergence is obtained when the populations, postrelaxation, are the same as those prior to laser excitation, that is, when $\mathcal{N}_D = N_D$, and $\mathcal{N}_L = N_L$. These conditions reduce to

$$N_D(1 - P_{D\leftarrow D} + P_{L\leftarrow D}) = N_L(1 + P_{D\leftarrow L} - P_{L\leftarrow L}). \tag{8.27}$$

If the total population is chosen to be normalized ($N_D + N_L = 1$), then the final probabilities $\mathcal{P}_D, \mathcal{P}_L$ of populating states $|E_D\rangle$ and $|E_L\rangle$ are

$$\mathcal{P}_D = \frac{1 + P_{D\leftarrow L} - P_{L\leftarrow L}}{2 - P_{D\leftarrow D} + P_{L\leftarrow D} + P_{D\leftarrow L} - P_{L\leftarrow L}},$$
$$\mathcal{P}_L = \frac{1 - P_{D\leftarrow D} + P_{L\leftarrow D}}{2 - P_{D\leftarrow D} + P_{L\leftarrow D} + P'_{D\leftarrow D} - P_{L\leftarrow L}}, \tag{8.28}$$

and the equilibrium enantiomeric branching ratio is simply

$$R_{D,L} \equiv \frac{\mathcal{P}_D}{\mathcal{P}_L} = \frac{1 + P_{D\leftarrow L} - P_{L\leftarrow L}}{1 - P_{D\leftarrow D} + P_{L\leftarrow D}}. \tag{8.29}$$

Results for the converged probabilities for the cases depicted in Figure 8.5 are shown in Figure 8.7. The results clearly show substantially enhanced enantiomeric ratios at various choices of control parameters. For example, at $\epsilon_0 = \epsilon_1 = 1.5 \times 10^{-3}$, tuning Δ_1 to 50 cm^{-1} gives a preponderance of L whereas tuning to the $\Delta_1 = -125$ cm^{-1} gives more D.

Numerous other parameters in this system, such as the pulse shape, time delay between pulses, pulse frequencies, and pulse powers can be varied to affect the final L to D ratio [259], resulting in a very versatile approach to asymmetric synthesis.

Finally, note that although we have only included two ground state levels, the method applies equally well when a large number of ground state levels are included. In this case, relaxation will be among all of these ground state levels, but the proposed scenario, tuned to the above set of transitions, will "bleed" population from one M_J level of the desired enantiomer. As relaxation refills this level, it will

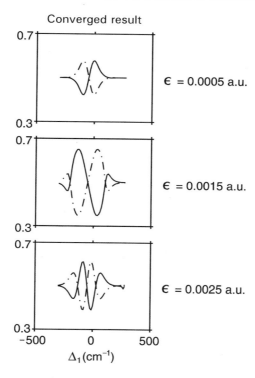

Fig. 8.7 Results for laser distillation after a convergent series of steps comprised of radiative excitation, and collisional and radiative relaxation. Shown are the results at three different field strengths $\epsilon_1 = \epsilon_2 = \epsilon_0 = \epsilon$.

continue to be pumped over to the other enantiomer, with the overall effect that the major amount of the population will be transferred from one enantiomer to the other.

As a realization of the above scheme we now examine [274] the case of enantiomer control in dimethylallene, a molecule shown in Figure 8.8. Note that, at equilibrium in the ground state, the H—C—CH$_3$ groups at both ends of the molecule lie on planes that are perpendicular to one another, resulting in a molecule that is chiral. By contrast, in the excited state, the C=C double bond breaks, allowing for rotation of one plane relative to the other. Cuts through the ground and first two excited state potential energy surfaces for this molecule along the α and θ coordinates (see Fig. 8.8) are shown in Figure 8.9. The potentials show the features required for control in this scenario, that is, a minimum in the excited state potential surface at the geometry corresponding to the potential energy maximum on the ground state potential.

The results of a computation [260] on the control of L vs. D 1,3-dimethylallene are shown in Figure 8.10. Outstanding enantiomeric control over the dimethylallene enantiomers is evident for a wide variety of powers. For example, a most impressive result is achieved for $\Delta_1 = 0.0986$ cm^{-1} and $\epsilon_0 = 1.5 \times 10^{-4}$ a.u., $\epsilon_1 = \epsilon_2 = 4.31 \times 10^{-5}$ a.u., corresponding to laser powers of 7.90×10^8 W/cm^2 and

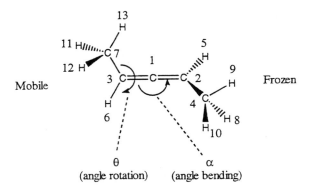

Fig. 8.8 Geometry of 1,3-dimethylallene and two angles θ and α that were varied to scan the potential energy surface. Here θ is the dihedral angle between the H_3C—C=C and the C=C—CH_3 planes and α is the C—C—C bending angle, here shown by an arrow that brings the H_3C—C—H out of the plane of the paper. (From Fig. 2, Ref. [274].)

6.52×10^7 W/cm^2, respectively. Here a racemic mixture of dimethylallene in a specific J, M_J, λ state can be converted, after a series of pulses, to a mixture of dimethylallene, containing 92.7% of the D-dimethylallene in this state. (Here λ is the projection of the total angular momentum J along an axis fixed in the molecule.) Similarly, detuning to $\Delta_1 = -0.0986$ cm^{-1} results in a similar enhancement of L-dimethylallene. Slightly lower extremes of control are seen to be achievable for the two other laser powers shown. Further, control was achievable to field strengths down to 10^4 W/cm^2. Note, however, that this computation neglects the competitive process of internal conversion, discussed below.

It is of some interest to note the character of the eigenstates $|E_1\rangle$ and $|E_2\rangle$ that contribute to these results; they are shown in Figure 8.11. Clearly they are states with considerable vibrational energy, so that they are broad enough in configuration space to overlap the ground electronic state, ground vibrational state wave functions. If this is not the case, then the dipole matrix elements are too small to allow control at reasonable laser intensities.

The primary experimental difficulty associated with this scenario is the need to isolate a particular subset of M_J levels, in order to avoid cancellation of M_J and $-|M_J|$ control. That is, from the viewpoint of the M_J structure, this scenario is associated with the level structure shown in Figure 8.12.

To remove this restriction we introduced another scenario [268] where all of the three laser polarizations, $\hat{\varepsilon}_0$, $\hat{\varepsilon}_1$, and $\hat{\varepsilon}_2$, are perpendicular to one another. This laser arrangement now allows for transitions between different M_J levels. The first few of these levels are shown in Figure 8.13. Under these circumstances, control survives averaging over M_J levels [268].

Sample results for the three laser cases with perpendicular polarizations are shown in Figure 8.14, first row, where extensive control is evident. Here, even with M_J averaging, one can choose to convert the racemic mixture to over 90%

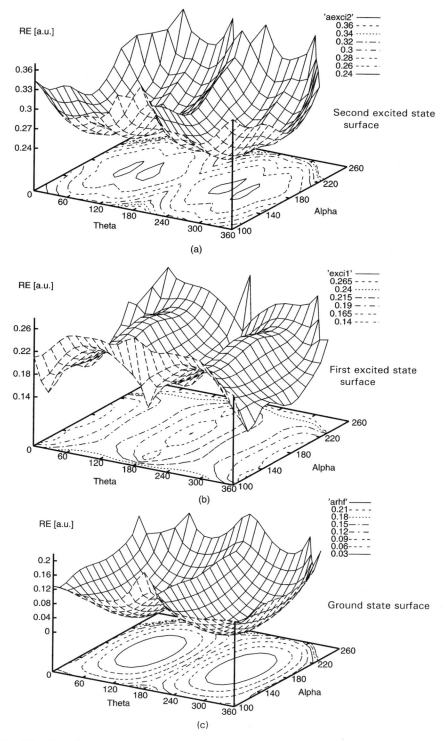

Fig. 8.9 Potential energy surfaces for 1,3-dimethylallene. Here we show in-plane surfaces for the ground and first two excited electronic states. (From Fig. 4, Ref. [274].)

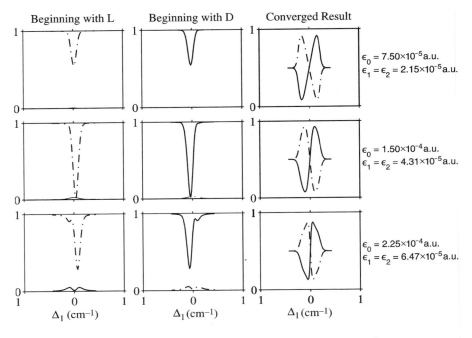

Fig. 8.10   Control over dimethylallene enantiomer populations as a function of detuning Δ_1 for various laser powers. First column corresponds to probabilities of L (dot–dash curves) and D (solid curves) after a single laser pulse, assuming that the initial state is all L. Second column is similar, but for an initial state, which is all D. Rightmost column corresponds to probabilities of L and D after repeated excitation–relaxation cycles, as described in the text. (This is a corrected version of Fig. 2, Ref. [260].)

of the L enantiomer, or of the D enantiomer, depending on the detuning. In this case the 1,3-dimethylallene was treated as an asymmetric top and averaging was carried out over all M_J levels.

A realistic model of dimethylallene control must also recognize the possibility of internal conversion to the ground state. In this process the electronically excited molecule undergoes a radiationless transition to the ground electronic state, leaving a highly vibrationally excited species. Only a few estimates or measurements of the internal conversion time scales for molecules are available [275, 276], and dimethyl-allene has not been explored. Further, after internal conversion one expects, in the dimethylallene case, that the excited molecule subsequently dissociates, leaving molecular fragments that no longer participate in the control scenario. Hence, the process of internal conversion serves as a decoherence mechanism that can reduce control. Further decoherence effects, but on a slower time scale, would arise, for example, if the control was carried out in solution.

The second row in Figure 8.14 shows control with similar parameters as in the first row, but in the presence of a T_2 associated with decoherence chosen arbitrarily

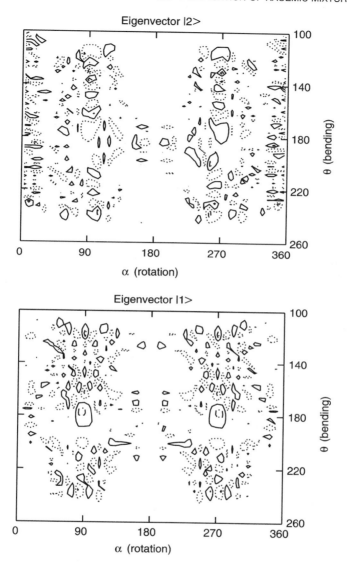

Fig. 8.11 Contour plots of $|E_1\rangle$ and $|E_2\rangle$ where dash–dash lines $= 0.012$ a.u., dot–dot lines $= 0.0004$ a.u., solid lines $= -0.004$ a.u., and dot–dash lines $= -0.012$ a.u. Note that $|E_1\rangle$ is symmetric with respect to reflection and $|E_2\rangle$ is antisymmetric. Reflection here corresponds to changing α to $(360° - \alpha)$ (From Fig. 1, Ref. [260].)

as 10 ps. Clearly, almost all of the control is lost. However, if the laser parameters are changed to those shown on the right-hand side of the figure, bottommost column, then significant control is restored once again. In this case, however, the process occurs with the loss of considerable reactant population to dissociated dimethyl-

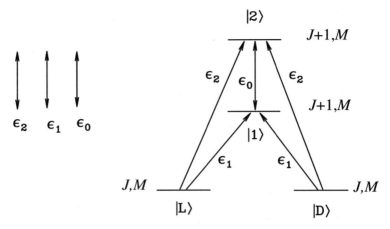

Fig. 8.12 Schematic level diagram emphasizing the $M \equiv M_J$ features of the four-level scheme in Fig. 8.1.

allene. Additional studies designed to establish the relationship between the laser requirements for control, and the internal conversion rates, are in progress as of the writing of this book [277]. Results are very encouraging showing significant control with minimal dissociation. The possibility of alternate substituents to replace the hydrogens is also of interest, as is the effect of changes to molecular structure to alter the radiative lifetime, the internal conversion rates, and so forth.

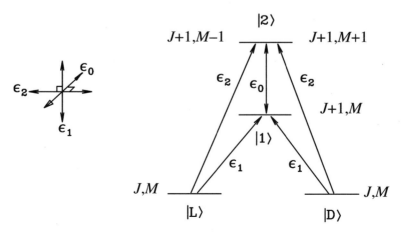

Fig. 8.13 Schematic level diagram emphasizing the $M \equiv M_J$ coupling where three lasers of perpendicular polarization irradiate D and L enantiomers. Only the first five levels coupled by these lasers are shown.

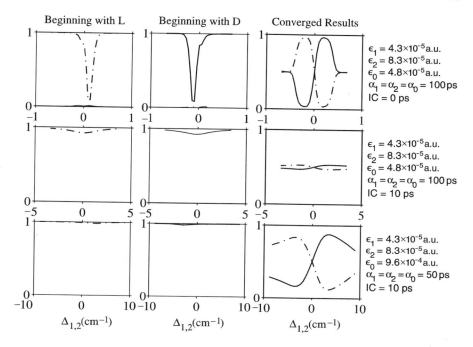

Fig. 8.14 Control over dimethylallene enantiomer populations as a function of the detuning Δ_1 for various laser powers. First column corresponds to probabilities of L (dot–dash curves) and D (solid curves) after a single laser pulse, assuming that the initial state is all L. Second column is similar, but for an initial state that is all D. Rightmost column corresponds to the probabilities L and D after repeated excitation–relaxation cycles, as described in the text. First row corresponds to control using laser parameters on the extreme right, in which there is no internal conversion; second row uses the same laser parameters as does the first row, but with an internal conversion time of 10 ps; bottom row shows results for an internal conversion time of 10 ps, but with modified laser parameters shown.

Enantiomeric control is more difficult if the excited molecular potential energy surfaces do not posses an appropriate minimum at the σ_h hyperplane configurations (see Fig. 8.1). In this case the method introduced in this section is not applicable. One may, however, be able to apply the laser distillation procedure by adding molecule B to the initial L, D mixture to form weakly bound L–B and B–D, which are themselves right- and left-handed enantiomeric pairs [278]. Molecule B is chosen so that electronic excitation of B–D and L–B forms an excited species G, which has stationary rovibrational states that are either symmetric or antisymmetric with respect to reflection through σ_h. The species L–B and B–D now serve as the L and D enantiomers in the general scenario above, and the laser distillation procedure described above then applies. Further, molecule B serves as a catalyst that may be removed from the final product by traditional chemical means.

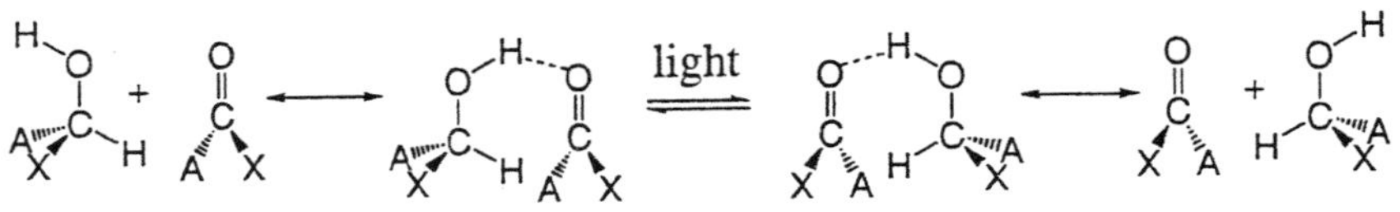

Fig. 8.15 Sample scenario for enhanced enantiomeric selectivity in a racemic mixture of two chiral alcohols related by inversion. An alcohol and ketone exchange two hydrogen atoms so as to produce the ketone, but with an alcohol of reverse handedness. Here A and X are distinct organic groups and dashes denote, in the upper panel, hydrogen bonds. The electronically excited species G, which is formed upon excitation with light, is postulated to be given by the structure at the bottom of the figure. In this case the topmost and bottommost hydrogens are attached to the oxygens and carbons, respectively, by "half-bonds." (From Fig. 4, Ref. [259].)

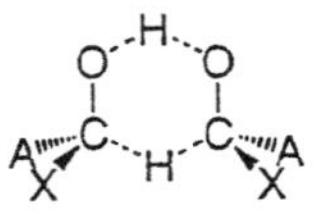

For example, L and D might be the left- and right-handed enantiomers of a chiral alcohol, and B is the ketone derived from this alcohol (see Fig. 8.15). In this case, studies [278] of the electronic structure of the alcohol–ketone system show that there are weakly bound chiral alcohol–ketone minima in the ground electronic state, as desired. The particular advantage of using the ketone–alcohol complex is that the ketone, which is "recycled" after the conversion of one enantiomer to another, serves as a catalyst for the process.

The results in this chapter make clear that a chiral outcome, the enhancement of a particular enantiomer, can arise by coherently encoding quantum interference information in the laser excitation of a racemic mixture. The fact that the initial state displays a broken symmetry and that the excited state has states that are either symmetric or antisymmetric with respect to σ_h allows for the creation of a superposition state that does not have these symmetry properties. Radiatively coupling the states in the superposition then allows for the transition probabilities from L and D to differ, allowing for depletion of the desired enantiomer.

8.4 ENANTIOMER CONTROL: ORIENTED MOLECULES

An alternative way to introduce chirality into the interaction of matter with light using linearly polarized light has been introduced by Fujimura et al. [263, 264]. In this approach one first preorients the racemic mixture of D and L along some axis. Under these circumstances, there is a difference in the direction of the transition dipole moments of the left- and right-handed enantiomers. That is, matrix elements such as $\langle E_i | \mathbf{d} \cdot \hat{\varepsilon}_k | E_j \rangle$ are different for the two enantiomers $i =$ D or L. This distinction between L and D suffices to allow for the possibility of control over enantiomers.

As a simple example, consider the model [264] shown in Figure 8.16. The system is initially in a mixture of the ground vibrational state of D and L. The pump laser carries the system to a single excited vibrational state of the excited electronic state, and the dump laser returns the system to an excited vibrational state of the ground electronic state. This then is a pump–dump scenario, but transitions are solely between bound states.

To appreciate the essence of this control scenario, recall the results of applying a laser pulse $\varepsilon(t)$ to induce a transition between two bound states $|E_i\rangle$ and $|E_j\rangle$. We denote the dipole transition matrix element between these two states by $d_{i,j}$ and define $\kappa_{i,j} = 2d_{i,j}/\hbar$. Then it is well known [279] that complete population transfer between these levels can be accomplished by using a π pulse, that is, a pulse of duration t satisfying

$$\int_{-\infty}^{t} \kappa_{i,j}\, \varepsilon(t')dt' = \pi. \tag{8.30}$$

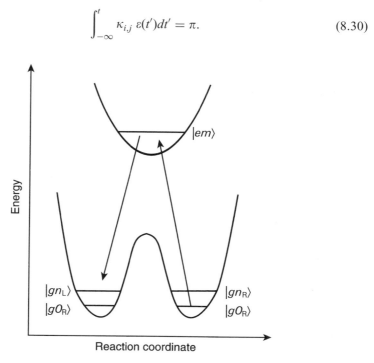

Fig. 8.16 Model system for enantiomer control in accord with Ref. [264]. Notation is such that $|gn_R\rangle$ denotes the D enantiomer on the ground electronic state in vibrational state n, $|gn_L\rangle$ is the analogous state of the L enantiomer and $|em\rangle$ is the mth level of excited electronic state. (From Fig. 1, Ref. [264].)

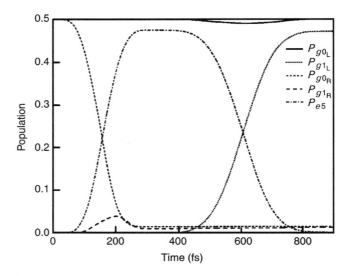

Fig. 8.17 Sample computation of control in the pump–dump scenario for controlling chirality. Populations shown are defined in the upper right-hand corner of the figure. For example, P_{g0_R} denotes the population of $|g0_R\rangle$, etc. (From Fig. 4, Ref. [264].)

Consider then the scenario in Figure 8.16, and suppose that we wish to transfer population from L to D. Then, since $\langle E_i|\mathbf{d}\cdot\hat{\varepsilon}_k|E_j\rangle$ differ for the L and D states, we can choose a laser polarization such that this matrix element is zero for excitation of the ground state of D, but not L. Application of a π pulse at this polarization will then transfer the ground state L population to the excited state. Application of a second π pulse of different polarization can then transfer this population to the excited vibrational state of the ground electronic state of D, by now choosing a polarization that does not couple the excited state to L.

Sample results for the control over the oriented enantiomers of H_2POSH are shown in Figure 8.17. Clearly, as proposed, the method is very effective. In this case the primary experimental challenge is to orient the system prior to irradiation.

APPENDIX 8A: COMPUTATION OF B–A–B′ ENANTIOMER SELECTIVITY

Section 8.2 requires the computation of the B–A–B′ photodissociation matrix elements in Eq. (8.12). To carry out this computation, and to understand the origin of the loss of control with m averaging, we sketch the details of the formalism for the model H_aOH_b system.

First, it is necessary to specify the relevant $\mathbf{n}$ and i quantum numbers that enter the bound-free matrix elements $\langle E, \mathbf{n}, q^-|d_{e,g}|E_i\rangle$. For the continuum states, $\mathbf{n} = \{\hat{\mathbf{k}}, v, j, m\}$ where $\hat{\mathbf{k}}$ is the scattering direction, v and j are the vibrational and rotational product quantum numbers and m is the space-fixed z projection of j. For the bound states, $|E_i\rangle$ actually denotes $|E_i, M_i, J_i, p_i\rangle$, where J_i, M_i, and p_i are, respectively, the bound state angular momentum, its space-fixed z projection, and its parity. The full (six-dimensional) bound-free matrix element can be written as a product of analytic functions involving $\hat{\mathbf{k}}$ and (three-dimensional) radial matrix elements:

$$\langle E, \hat{\mathbf{k}}, v, j, m, q^- | \mathrm{d}_{e,g} | E_i, M_i, J_i, p_i \rangle$$

$$= \left(\frac{\mu k_{vj}}{2\pi^2 \hbar^2} \right)^{1/2} \sum_{J\lambda} (2J+1)^{1/2} (-1)^{M_i - j - m} D_{\lambda M_i}^J (\phi_k, \theta_k, 0)$$

$$\times D_{-\lambda - m}^j (\phi_k, \theta_k, 0) t^{(q)} (E, J, v, j, \lambda | E_i J_i p_i). \tag{8.31}$$

Here the $D_{\lambda M_i}^J (\phi_k, \theta_k, 0)$ are the rotation matrices [248, 323], ϕ_k, θ_k are the scattering angles, μ is the reduced mass, k_{vj} is the momentum of the products, and $t^{(q)} (E, J, v, j, \lambda | E_i J_i p_i)$ is proportional to the radial partial-wave matrix element [273] $\langle E, J, M, p, v, j, \lambda, q^- | \mathrm{d}_{e,g} | E_i, M_i, J_i, p_i \rangle$. Here λ is the projection of J along the body fixed axis of the H–OH (c.m.) product separation.

The product of the bound-free matrix elements of Eq. (8.31), which enter Eq. (3.79), integrated over scattering angles and averaged over the initial [280] $M_k (= M_i)$ quantum numbers, is

$$(2J_i + 1)^{-1} \sum_{M_i} \int d\hat{\mathbf{k}} \langle E_k, M_i, J_k, p_k | \mathrm{d}_{e,g} | E, \hat{\mathbf{k}}, v, j, m, q'^- \rangle$$

$$\langle E, \hat{\mathbf{k}}, v, j, m, q^- | \mathrm{d}_{e,g} | E_i, M_i, J_i, p_i \rangle$$

$$= (-1)^m \frac{8\pi\mu}{\hbar^2 (2J_i + 1)} \sum_{vj} k_{vj} \sum_{J\lambda J'\lambda'} [(2J+1)(2J'+1)]^{1/2} (-1)^{\{\lambda - \lambda' + J + J' + J_i\}}$$

$$\times \sum_{\ell=0,2} (2\ell + 1) \begin{pmatrix} J & J' & \ell \\ \lambda & -\lambda' & \lambda' - \lambda \end{pmatrix} \begin{pmatrix} j & j & \ell \\ -\lambda & \lambda' & \lambda - \lambda' \end{pmatrix} \begin{pmatrix} 1 & 1 & \ell \\ 0 & 0 & 0 \end{pmatrix}$$

$$\times \begin{pmatrix} j & j & \ell \\ -m & m & 0 \end{pmatrix} \begin{Bmatrix} 1 & 1 & \ell \\ J & J' & J_i \end{Bmatrix} t^{(q)} (EJvj\lambda p | E_i J_i p_i) t^{(q')*} (EJ'vj\lambda' p | E_k J_k p_k). \tag{8.32}$$

Here $J_i = J_k$ has been assumed for simplicity. The $t^{(q)}$ matrix elements are computed with the artificial channel method [14].

Finally, we sketch the effect of a summation over product m states on symmetry breaking and chirality control. In this regard the three-body model is particularly informative. Specifically, note that Eq. (8.12) provides $\mathrm{d}_q(32)$ in terms of products of matrix elements involving $|E, \mathbf{n}, a^-\rangle$ and $|E, \mathbf{n}, s^-\rangle$. Focus attention on those products that involve both of these wave functions, for example, terms like $S_{s3} S_{a2}*$. These matrix element products can be written in the form of Eq. (8.32) where q and q' now refer to the antisymmetric or symmetric continuum states, rather than to channels D or L. Thus, for example, $S_{s3} S_{a2}*$ results from using $|E, \mathbf{m}, s^-\rangle$ in

Eq. (8.31) to form S_{s3} and $|E, \mathbf{n}, a^-\rangle$ to form S_{a2}^*. The resultant $S_{s3}S_{a2}^*$ has the form of Eq. (8.32) with $t^{(q)}$ and $t^{(q')}$ associated with the symmetric and antisymmetric continuum wave functions, respectively. Consider now the effect of summing over m. Standard formulas [273, 281] imply that this summation introduces a $\delta_{\ell,0}$, which, in turn, forces $\lambda = \lambda'$ via the first and second $3j$ symbol in Eq. (8.32). However, it is possible to show that t-matrix elements associated with symmetric continuum eigenfunctions and those associated with antisymmetric continuum eigenfunctions must have λ of different parities. Hence summing over m eliminates all contributions to Eq. (8.13) that involve both $|E, \mathbf{n}, a^-\rangle$ and $|E, \mathbf{n}, s^-\rangle$. Thus, we find after m summation:

$$\sum_m d_L(23) = \sum_m d_D(23) = 0. \tag{8.33}$$

That is, control over the enantiomer ratio is lost upon m summation, both channels $q = D$ and $q = L$ having equal photodissociation probabilities.

CHAPTER 9

COHERENT CONTROL BEYOND THE WEAK-FIELD REGIME: BOUND STATES AND RESONANCES

In Chapters 2 and 3 we treated n-photon molecular dissociation and coherent control in the weak-field regime and showed that control arises through quantum interference effects. The moderately strong field regime treated in this chapter is far more complicated because the higher laser powers imply that nth-order perturbation theory is no longer valid. The "moderately strong" field regime discussed here, in contradistinction to the "strong" field regime discussed in Chapter 12, is taken here to mean that, although nth-order perturbation theory is inappropriate, the (radiation plus matter) wave functions can still be expressed in terms of a moderate number of molecular states. In the moderately strong field regime the use of the "field-dressed" basis of states, discussed in Chapter 12, though an option, is not an absolute necessity.

The formalism is best divided into a discussion of control involving bound states, treated in this section, and control involving the continuum, discussed in Chapters 10 and 11.

9.1 ADIABATIC POPULATION TRANSFER

In this section we show how strong fields can be used to adiabatically control population transfer between bound states and, especially, how to achieve *complete* population transfer between such states. In doing so we describe realistic methods for control, introduce a number of useful methods in strong-field control, and pave

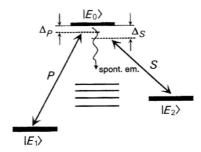

Fig. 9.1 The Λ configuration associated with adiabatic passage. (Taken from Fig. 1, Ref. [245].)

the way for a discussion of adiabatic population transfer in problems involving the continuum.

The ability to induce complete population transfer between states is intimately linked to the concept of a "trapped" state, that is, a state that remains invariant under the action of cw irradiation. These states, which only change when the field changes, often enable one to guide a quantum system from one state to another, a phenomenon known as adiabatic passage (AP), first introduced in the context of magnetic resonance [282] and described in detail below.

Adiabatic passage was most commonly exploited in two- and three-level systems [245, 279, 283–296]. In particular, the three-level "Λ system" (see Fig. 9.1), in which one (initially unpopulated) level is higher in energy than the two other levels, was extensively investigated theoretically [245, 246, 288, 289, 295] and experimentally [293–295]. Gaubatz et al. [294] were the first to experimentally demonstrate, in a process called stimulated raman adiabatic passage (STIRAP), that adiabatic passage in a Λ system enables the *complete* transfer of population from one level to another under certain conditions. A related phenomenon, called electromagnetically induced transparency (EIT), in which a medium is made transparent at a certain transition frequency, was investigated by Harris et al. [297, 298, 302] and others [299, 300]. Moreover, the possibility of using trapped states to cause lasing without inversion (LWI) was shown by Harris [301], Scully et al. [303], Kocharovskaya [304], and others [305, 306]. These phenomena are described below and are seen to share a common basis insofar as they rely upon interference effects associated with the preparation and evolution of specific superposition states.

9.1.1 Adiabatic States, Trapping, and Adiabatic Following

To understand these phenomena we first consider a three-level Λ system, composed of a lowest energy state $|E_1\rangle$, coupled radiatively to an intermediate state $|E_0\rangle$, which in turn is coupled radiatively to a third state $|E_2\rangle$ where $E_0 > E_2 > E_1$. The coupling is due to the combined action of two laser pulses of central frequencies ω_1 and ω_2. We assume (see Fig. 9.1) that ω_1, "the pump pulse" (labeled P), is in near resonance with a transition from $|E_1\rangle$ to the bound state $|E_0\rangle$ and that ω_2, "the dump pulse" or

Stokes pulse (labeled S), is in near resonance with the transition from $|E_0\rangle$ to $|E_2\rangle$. In most applications one chooses the two frequencies to fulfill the *two-photon* resonance condition,

$$\omega_1 - \omega_2 = (E_2 - E_1)/\hbar. \tag{9.1}$$

Writing the total Hamiltonian in the dipole approximation as

$$H = H_M - 2\mathbf{d}_1 \cdot \hat{\varepsilon}_1 \mathcal{E}_1(t) \cos(\omega_1 t) - 2\mathbf{d}_2 \cdot \hat{\varepsilon}_2 \mathcal{E}_2(t) \cos(\omega_2 t), \tag{9.2}$$

where $\hat{\varepsilon}_1$ and $\hat{\varepsilon}_2$ are the polarization directions, $\mathcal{E}_1(t)$ and $\mathcal{E}_2(t)$ are "slowly varying" electric field amplitudes and $\mathbf{d}_1$ and $\mathbf{d}_2$ are the electronic transition dipole operators, we can solve the time-dependent Schrödinger equation, $i\hbar\partial|\Psi(t)\rangle/\partial\, dt = H|\Psi(t)\rangle$, by expanding the total wave function as

$$|\Psi(t)\rangle = b_1(t)|E_1\rangle e^{-iE_1 t/\hbar} + b_0(t)|E_0\rangle e^{-iE_0 t/\hbar} + b_2(t)|E_2\rangle e^{-iE_2 t/\hbar}, \tag{9.3}$$

where

$$[E_i - H_M]|E_i\rangle = 0, \qquad (i = 0, 1, 2). \tag{9.4}$$

Doing so gives the three-state version of Eq. (2.3):

$$\frac{db_1}{dt} = i\Omega_1^*(t)e^{-i\Delta_1 t}b_0(t),$$

$$\frac{db_0}{dt} = i\Omega_1(t)e^{i\Delta_1 t}b_1(t) + i\Omega_2^*(t)e^{i\Delta_2 t}b_2(t),$$

$$\frac{db_2}{dt} = i\Omega_2(t)e^{-i\Delta_2 t}b_0(t),$$

where the Rabi frequencies Ω_i and detunings Δ_i are

$$\Omega_i(t) \equiv \langle E_0|\mathbf{d}_i \cdot \hat{\varepsilon}_i|E_i\rangle \mathcal{E}_i(t)/\hbar, \qquad \Delta_i \equiv (E_0 - E_i)/\hbar - \omega_i. \tag{9.6}$$

The two-photon resonance condition [Eq. (9.1)] implies that

$$\Delta_2 = \Delta_1. \tag{9.7}$$

Defining a vector of coefficients $\mathbf{b} \equiv (b_1,\ b_0,\ b_2)^{\mathbf{T}}$, where $\mathbf{T}$ denotes the transpose, we write Eq. (9.5) in matrix notation as

$$\frac{d}{dt}\mathbf{b}(t) = i\underline{\underline{\mathrm{H}}} \cdot \mathbf{b}(t), \tag{9.8}$$

where

$$\underline{\underline{H}} = \begin{pmatrix} 0 & \Omega_1^*(t)e^{-i\Delta_1 t} & 0 \\ \Omega_1(t)e^{i\Delta_1 t} & 0 & \Omega_2^*(t)e^{i\Delta_2 t} \\ 0 & \Omega_2(t)e^{-i\Delta_2 t} & 0 \end{pmatrix} \tag{9.9}$$

We first derive the adiabatic approximation to Eq. (9.8). The derivation begins by diagonalizing the $\underline{\underline{H}}$ matrix,

$$\underline{\underline{H}} \cdot \underline{\underline{U}} = \underline{\underline{U}} \cdot \underline{\underline{\lambda}}, \tag{9.10}$$

where $\underline{\underline{\lambda}}$ is a diagonal eigenvalue matrix. We then transform the Schrödinger equation to the adiabatic representation by operating on Eq. (9.8) with $\underline{\underline{U}}^\dagger(t)$, the Hermitian adjoint of $\underline{\underline{U}}$. Defining

$$\mathbf{a}(t) = \underline{\underline{U}}^\dagger(t) \cdot \mathbf{b}(t), \tag{9.11}$$

and using the unitarity property of $\underline{\underline{U}}$:

$$\underline{\underline{U}} \cdot \underline{\underline{U}}^\dagger = \underline{\underline{U}}^\dagger \cdot \underline{\underline{U}} = \underline{\underline{I}}, \tag{9.12}$$

we have that

$$\frac{d\mathbf{a}}{dt} = \left\{ i\underline{\underline{\lambda}}(t) + \underline{\underline{A}} \right\} \cdot \mathbf{a}, \tag{9.13}$$

where

$$\underline{\underline{A}} \equiv \frac{d\underline{\underline{U}}^\dagger(t)}{dt} \cdot \underline{\underline{U}}, \tag{9.14}$$

is the *nonadiabatic coupling matrix*.

In the adiabatic approximation one neglects $\underline{\underline{A}}$. This can be done whenever the rate of change of $\underline{\underline{U}}$ with time is slow, or more specifically, whenever

$$\left(\underline{\underline{A}} \right)_{i,j} \ll |\lambda_i - \lambda_j|. \tag{9.15}$$

A discussion of the range of validity of the adiabatic approximation, and methods of improving upon it, is given in Section 11.1.

When the adiabatic approximation is adopted, Eq. (9.13) becomes

$$\frac{d}{dt}\mathbf{a} = i\underline{\underline{\lambda}}(t) \cdot \mathbf{a}(t), \tag{9.16}$$

yielding the adiabatic solutions

$$\mathbf{a}(t) = \exp\left\{ i \int_0^t \underline{\underline{\lambda}}(t')\, dt' \right\} \cdot \mathbf{a}(0). \tag{9.17}$$

Given $\mathbf{a}(t)$ and $\underline{U}(t)$ one can generate the desired $\mathbf{b}(t)$ via Eq. (9.11).

For the case in Eq. (9.9) it is easy to show that $\underline{\underline{\lambda}}$ comprises the following eigenvalues on the diagonal:

$$\begin{aligned} \lambda_1 &= 0, \\ \lambda_{2,3}(t) &= \pm[|\Omega_1(t)|^2 + |\Omega_2(t)|^2]^{1/2}, \end{aligned} \tag{9.18}$$

where the plus sign applies to λ_2 and the minus sign applies to λ_3. Because of their time dependence, these adiabatic eigenvalues are also called *quasi-energies*.

We now seek the adiabatic states, denoted $|\lambda_i(t)\rangle$, which are the $\mathbf{b}(t)$ vectors obtained from the $\mathbf{a}(t)$ eigenvectors using Eq. (9.11), that is, $\mathbf{b}(t) = \underline{U}(t) \cdot \mathbf{a}(t)$ and reinserted into Eq. (9.3).

Solving explicitly for $\mathbf{U}^{(1)}(t)$, the eigenvector corresponding to $\lambda_1 = 0$, that is,

$$\underline{\underline{H}} \cdot \mathbf{U}^{(1)} = 0, \tag{9.19}$$

together with Eq. (9.9), yields

$$\mathbf{U}^{(1)} = \frac{1}{[|\Omega_1|^2 + |\Omega_2|^2]^{1/2}} \begin{pmatrix} \Omega_2^* \\ 0 \\ -\Omega_1 e^{-i(\Delta_2 - \Delta_1)t} \end{pmatrix} = \begin{pmatrix} \cos\theta(t) \\ 0 \\ -e^{i\chi(t)} \sin\theta(t) \end{pmatrix}. \tag{9.20}$$

Equation (9.20) involves both the *mixing angle* $\theta(t)$, where

$$\theta(t) = \arctan\left(\frac{|\Omega_1(t)|}{|\Omega_2(t)|} \right), \tag{9.21}$$

and the *azimuthal angle* $\chi(t)$, where

$$\chi(t) \equiv (\Delta_1 - \Delta_2)t - \phi_2(t) + \phi_1(t) = \phi_1(t) - \phi_2(t), \tag{9.22}$$

and where $\phi_i(t)$ are the phases of $\Omega_i(t)$:

$$\Omega_i(t) \equiv |\Omega_i(t)| e^{i\phi_i(t)}, \qquad i = 1, 2. \tag{9.23}$$

The last equality in Eq. (9.22) is a result of the two-photon resonance condition [Eq. (9.7)].

Thus, the adiabatic state associated with the $\lambda_1 = 0$ eigenvector is

$$|\lambda_1(t)\rangle = \cos\theta(t)e^{-iE_1t/\hbar}|E_1\rangle - e^{i\chi(t)}\sin\theta(t)e^{-iE_2t/\hbar}|E_2\rangle. \qquad (9.24)$$

It follows from Eqs. (9.6), (9.20) and (9.21) that $\mathbf{U}^{(1)}(t)$ does not change with time for cw fields. Rather, it evolves in time, in accord with Eq. (9.20), only when $\mathcal{E}_1(t)$ and $\mathcal{E}_2(t)$ are themselves time dependent. The adiabatic eigenvectors with zero eigenvalues are called *trapped* (or *null*) states, since, for cw fields, the population stays trapped in the two states $|E_1\rangle$ and $|E_2\rangle$. It is possible to show that the *trapping* is a consequence of quantum interference established between the two routes to level $|E_0\rangle$. However, if the pulse envelopes do vary in time, the trapped states will also vary in time, following the time evolution in Eq. (9.24). This phenomenon is called *adiabatic following*.

Now focus attention on the dynamics of $|\lambda_1(t)\rangle$. It follows from Eqs. (9.20), (9.21), and (9.24) that if the pulse sequence is arranged in the so-called counter-intuitive order [245], that is, where the $\mathcal{E}_2(t)$ dump pulse is applied *before* the $\mathcal{E}_1(t)$ pump pulse, then the trapped state $\lambda_1(t)$ will smoothly pass from state $|E_1\rangle$ at $t = 0$ to state $|E_2\rangle$ at $t = \infty$. This phenomenon, known as *adiabatic passage* occurs because $\theta(t = 0) = 0$, which means that [see Eq. (9.24)] $|\lambda_1(t = 0)\rangle = |E_1\rangle$, [corresponding to $\mathbf{U}^{(1)}(t = 0) = (1 \quad 0 \quad 0)$], whereas $\theta(t \to \infty) = \pi/2$ and $|\lambda_1(t \to \infty)\rangle = e^{-i\chi(t)}e^{-iE_2t/\hbar}|E_2\rangle$. Thus, the trapped state starts as $|E_1\rangle$ and goes over to the $|E_2\rangle$ state while never populating the intermediate state $|E_0\rangle$. In this way full population transfer from $|E_1\rangle$ to $|E_2\rangle$ results. Further, there are no losses due to spontaneous emission since $|E_0\rangle$ is never populated and selection rules prevent spontaneous emission from $|E_2\rangle$ to $|E_1\rangle$.

Note that because $|E_0\rangle$ does not contribute to the trapped state, the adiabatic passage process can be described by the nutation of a *Bloch vector* [307], shown in Figure 9.2, which is parameterized by an azimuthal angle $\chi(t)$ and a polar angle $2\theta(t)$. During an adiabatic passage the Bloch vector nutates [245] from the $|b_1|^2 - |b_2|^2 = 1$ up-vertical position to the $|b_1|^2 - |b_2|^2 = -1$ down-vertical position while undergoing a precession at a frequency $\omega_{2,1}$ about the z axis.

Figure 9.3 shows computed results for the time evolution of the Ω associated with the dump and pump pulse as well as the mixing angle θ, the dressed state eigenvalues and the population of $|E_1\rangle$ and $|E_2\rangle$. The smooth and complete transfer of population between the initial and final state is evident. The dependence of the transfer efficiency, gleaned from experimental studies on Ne, is shown as a function of the pulse ordering in Figure 9.4. Specifically, the transfer of population between the $^3P_0$ and $^2P_2$ states of Ne is seen to be small when the dump and the pump pulses do not overlap, and to maximize at 100% population transfer when the dump laser of frequency ω_2 slightly precedes the pump laser of frequency ω_1.

The two remaining eigenvectors $\mathbf{U}^{(2)}$, $\mathbf{U}^{(3)}$ are obtained by solving the second and third eigenvalue equations:

$$\underline{\underline{\mathbf{H}}} \cdot \left(\mathbf{U}^{(2)}, \ \mathbf{U}^{(3)}\right) = \left(\lambda_2\mathbf{U}^{(2)}, \ \lambda_3\mathbf{U}^{(3)}\right) = \left(\lambda_2\mathbf{U}^{(2)}, \ -\lambda_2\mathbf{U}^{(3)}\right). \qquad (9.25)$$

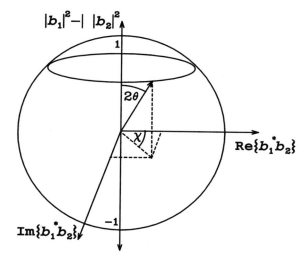

Fig. 9.2 Bloch vector view of time evolution of $b_1(t)$ and $b_2(t)$ complex coefficients of the $b_1(t)\exp(-iE_1t/\hbar)|E_1\rangle + b_2(t)\exp(-iE_2t/\hbar)|E_2\rangle$ superposition state. Two complex coefficients are constrained by normalization and therefore can be represented as a point on a three-dimensional sphere, shown here as a function of twice the mixing angle θ and the azimuthal angle χ of Eq. (9.22). In this space the z axis corresponds to the $|b_1|^2 - |b_2|^2$ population difference, the x axis to $\mathrm{Re}\{b_1^*b_2\}$, and the y axis to $\mathrm{Im}\{b_1^*b_2\}$. In case of two-photon resonance $(\Delta_1 - \Delta_2 = 0)$ and when phases of two lasers fields are the same $[\phi_2(t) = \phi_1(t)]$, it follows from Eq. (9.22) that the azimuthal angle is constant and the Bloch vector executes a pure nutational motion from the north pole of the sphere to the south pole during a complete adiabatic population transfer.

Using Eq. (9.9) the solutions are

$$
\mathbf{U}^{(2,3)} = \frac{1}{\sqrt{2\lambda_2}} \begin{pmatrix} \Omega_1^* e^{-i\Delta_1 t} \\ \pm\lambda_2 \\ \Omega_2 e^{-i\Delta_2 t} \end{pmatrix} = \frac{1}{\sqrt{2}} \begin{pmatrix} e^{-i\phi_1(t) - i\Delta_1 t}\sin\theta \\ \pm 1 \\ e^{i\phi_2(t) - i\Delta_2 t}\cos\theta \end{pmatrix}, \qquad (9.26)
$$

where the plus sign applies to $\mathbf{U}^{(2)}$ and the minus sign to $\mathbf{U}^{(3)}$. The associated adiabatic eigenvectors are

$$
\begin{aligned}
|\lambda_2(t)\rangle &= \frac{1}{\sqrt{2}}\exp\left[\int_0^t i\lambda_2(t')\,dt'\right]\big[e^{-i\phi_1(t) - i\Delta_1 t}\sin\theta e^{-iE_1 t/\hbar}|E_1\rangle + |E_0\rangle \\
&\quad + e^{-i\phi_2(t) - i\Delta_2 t}\cos\theta e^{-iE_2 t/\hbar}|E_2\rangle\big], \\
|\lambda_3(t)\rangle &= \frac{1}{\sqrt{2}}\exp\left[\int_0^t -i\lambda_2(t')\,dt'\right]\big[e^{-i\phi_1(t) - i\Delta_1 t}\sin\theta e^{-iE_1 t/\hbar}|E_1\rangle - |E_0\rangle \\
&\quad + e^{-i\phi_2(t) - i\Delta_2 t}\cos\theta e^{-iE_2 t/\hbar}|E_2\rangle\big].
\end{aligned} \qquad (9.27)
$$

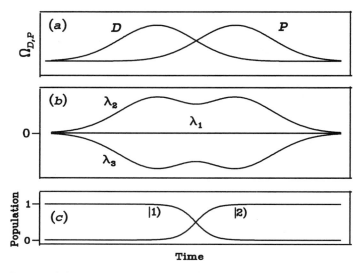

Fig. 9.3   Computed time evolution of Rabi frequencies, adiabatic eigenvalues, and population of states $|1\rangle$ and $|2\rangle$. (After Fig. 3, Ref. [245].)

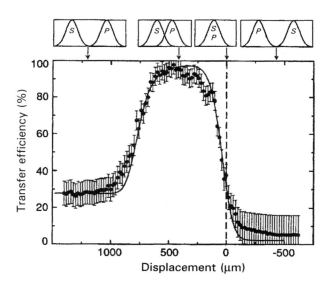

Fig. 9.4   Experimental results on population transfer between two states of Ne as a function of laser pulse ordering. The ordering of the two laser pulses is shown at the top and is controlled by the displacement between the two cw laser beams transversing the path of the Ne beam, thereby introducing an effective time delay between pulses. S of the figure denotes the Stokes pulse (the dump pulse of the text), and P of the figure denotes the pump pulse of the text. (From Fig. 9, Ref. [245].)

Assuming that $\phi_1(t) = \phi_2(t) = 0$, we have that in the counterintuitive pulse ordering

$$\mathbf{U}^{(2,3)}(t=0) = \frac{1}{\sqrt{2}} \begin{pmatrix} 0 \\ \pm 1 \\ 1 \end{pmatrix}, \qquad \mathbf{U}^{(2,3)}(t \to \infty) = \frac{1}{\sqrt{2}} \begin{pmatrix} e^{-i\Delta_1 t} \\ \pm 1 \\ 0 \end{pmatrix}. \qquad (9.28)$$

Since at $t = 0$ the first element of $\mathbf{U}^{(2)}$ and of $\mathbf{U}^{(3)}$ is zero, then the two "nontrapped" adiabatic states are orthogonal to the initial state $|E_1\rangle$ at the beginning of the process. Hence, the only adiabatic state populated initially is the trapped state $|\lambda_1(t)\rangle$. If there is no coupling between the three adiabatic states, the system will continue to evolve as the trapped state $|\lambda_1(t)\rangle$, executing an adiabatic passage to the $|E_2\rangle$ state as $t \to \infty$. Thus we can achieve the control objective of complete population transfer between two bound states.

A variety of STIRAP pulse configurations and generalizations of STIRAP have been studied. For example, and of interest below, is when one or more of the levels is either part of the continuum [292, 308] or a resonant state [309, 310] (i.e., a bound state coupled to the continuum).

9.1.2 Electromagnetically Induced Transparency

Thus, STIRAP uses interference effects generated by irradiation with two lasers to control characteristics of the time-dependent level populations. An alternative phenomenon, also associated with the existence of trapped states, utilizes interferences to control features of the absorption and emission of radiation. If, instead of applying the STIRAP pulse ordering, in which $\mathcal{E}_2(t)$ is applied before $\mathcal{E}_1(t)$ and allowed to decay as $\mathcal{E}_1(t)$ reaches its peak value, we maintain $\mathcal{E}_2(t) \gg \mathcal{E}_1(t)$ at all times, we produce a phenomenon termed *electromagnetically induced transparency* (EIT). In this phenomenon the absorption of radiation by the system at a specific frequency is suppressed. The suppression of the absorption whenever $\mathcal{E}_2(t) \gg \mathcal{E}_1(t)$, follows from Eq. (9.21), since in that case $\theta \approx 0$ at all times. Under these circumstances $b_1(t) \approx 1$, $b_0(t) \approx 0$, $b_2(t) \approx 0$. Hence the trapped state $|\lambda_1(t)\rangle \approx |E_1\rangle$ does not react to the light at all, becoming a so-called dark state. Qualitatively, this arises because the two fields $\mathcal{E}_1(t)$ (termed the probe laser) and $\mathcal{E}_2(t)$ (termed the coupling laser) have created competitive absorption pathways that destructively interfere with one another, canceling the absorption of light.

The formation of a dark state may be detected by allowing two beams of light of frequencies $\omega_1 = [E_0 - E_1]/\hbar$ and an $\omega_2 = [E_0 - E_2]/\hbar$ to transverse a dense medium of atoms or molecules. In the absence of the ω_2 beam, the ω_1 beam is readily absorbed due to the $\omega_1 = [E_0 - E_1]/\hbar$ resonance condition. By contrast, in the presence of the ω_2 beam, as long as $\mathcal{E}_2(t) \gg \mathcal{E}_1(t)$, neither beam will be absorbed [302].

9.1.3 EIT: A Resonance Perspective

Electromagnetically induced transparency is, however, a much more dramatic effect than the general description given above would lead us to believe. Not only is the absorption small, as the choice of the pulse ordering derived above implies, but the absorption actually goes to zero as a very sharp function of the frequency. To understand this phenomenon it is necessary to utilize a new type of interference, that between *overlapping resonances*. To understand how such interference is associated with the strong field effects in EIT, we simplify the problem by noting that since $\mathcal{E}_2(t) \gg \mathcal{E}_1(t)$, we can treat $\mathcal{E}_1(t)$ as a small perturbation and first obtain the adiabatic eigenstates resulting from the $\mathcal{E}_2(t)$-induced interaction between $|E_0\rangle$ and $|E_2\rangle$.

Consider again Eqs. (9.8) and (9.9). Temporarily neglecting $\mathcal{E}_1(t)$, we focus on the component of the Hamiltonian matrix in the $(|E_0\rangle, |E_2\rangle)$ space, which is

$$\underline{\underline{H}} = \begin{pmatrix} 0 & \Omega_2^*(t)e^{i\Delta_2 t} \\ \Omega_2(t)e^{-i\Delta_2 t} & 0 \end{pmatrix}. \tag{9.29}$$

To utilize the adiabatic condition, we transform $\underline{\underline{H}}$ to a form that does not contain oscillatory terms that might cause rapid variations of $d\underline{\underline{U}}/dt$ and hence invalidate the adiabatic condition.

Multiplying Eq. (9.8) by a diagonal matrix, $e^{i\underline{\underline{\Delta}}t/2}$, with $\underline{\underline{\Delta}}$ given by

$$\underline{\underline{\Delta}} \equiv \begin{pmatrix} -\Delta_2 & 0 \\ 0 & \Delta_2 \end{pmatrix}, \tag{9.30}$$

we obtain,

$$e^{i\underline{\underline{\Delta}}t/2} \cdot \frac{d\,\mathbf{b}}{dt} = \frac{de^{i\underline{\underline{\Delta}}t/2}}{dt} \cdot \mathbf{b} - i\underline{\underline{\Delta}}e^{i\underline{\underline{\Delta}}t/2} \cdot \frac{\mathbf{b}}{2} = ie^{i\underline{\underline{\Delta}}t/2} \cdot \underline{\underline{H}} \cdot e^{-i\underline{\underline{\Delta}}t/2} \cdot e^{i\underline{\underline{\Delta}}t/2} \cdot \mathbf{b}(t). \tag{9.31}$$

The oscillatory terms can be eliminated by defining

$$\mathbf{c} \equiv e^{i\underline{\underline{\Delta}}t/2}\mathbf{b}, \tag{9.32}$$

and we obtain from Eq. (9.31) that

$$\frac{d\,\mathbf{c}}{dt} = i\underline{\underline{H}}' \cdot \mathbf{c}(t), \tag{9.33}$$

where

$$\underline{\underline{H}}' = \begin{pmatrix} -\Delta_2/2 & \Omega_2^*(t) \\ \Omega_2(t) & \Delta_2/2 \end{pmatrix}. \tag{9.34}$$

Having removed the oscillatory $e^{\pm i\Delta_2 t}$ terms, we now build adiabatic solutions by diagonalizing Eq. (9.34) [as in Eq. (9.10)] to obtain a 2×2 unitary eigenvector matrix:

$$\underline{\underline{U}} = \begin{pmatrix} \cos\theta & e^{-i\phi_2}\sin\theta \\ -e^{i\phi_2}\sin\theta & \cos\theta \end{pmatrix}. \tag{9.35}$$

The corresponding eigenvalue matrix $\underline{\underline{\lambda}}$ is composed of the two roots

$$\lambda_{1,2}(t) = \pm\lambda(t) = \pm\left[\Delta_2^2/4 + |\Omega_2(t)|^2\right]^{1/2}, \tag{9.36}$$

where θ and ϕ_2 in $\underline{\underline{U}}$ are given as

$$\tan\theta = \left[\frac{|\Omega_2(t)|}{(\Delta_2^2/4 + |\Omega_2(t)|^2)^{1/2} + \Delta_2/2}\right]^{-1}, \tag{9.37}$$

with ϕ_2 being the argument of Ω_2 [Eq. (9.23)].

Operating with $\underline{\underline{U}}^\dagger$ on Eq. (9.33), defining $\mathbf{a} = \underline{\underline{U}}^\dagger \cdot \mathbf{c} \equiv (a_1\; a_2)^\dagger$ and neglecting $d\underline{\underline{U}}^\dagger/dt$, we obtain an equation analogous to Eq. (9.13) for $\mathbf{a}$, with $\underline{\underline{H}}'$ replacing $\underline{\underline{H}}$. The solutions to this equation in the adiabatic approximation are

$$a_{1,2}(t) = \exp\left\{i\int_0^t \pm\lambda(t')\,dt'\right\}a_{1,2}(0) = \exp\left\{i\int_0^t \pm\left[\frac{\Delta_2^2}{4} + |\Omega_2(t')|^2\right]^{1/2}dt'\right\}a_{1,2}(0).$$

$$\tag{9.38}$$

To gain insight into requirements for the validity of the adiabatic approximation consider the case of $\phi_2 = 0$. Here the nonadiabatic coupling matrix $\underline{\underline{A}}$ is given by

$$\underline{\underline{A}} = \begin{pmatrix} 0 & \dot{\theta} \\ -\dot{\theta} & 0 \end{pmatrix}. \tag{9.39}$$

Examining Eq. (9.13) shows that in order for $\underline{\underline{A}}$ to be a small correction, the time derivative of the mixing angle must be small with respect to the gap between the eigenvalues:

$$|\dot{\theta}(t)| \ll |\lambda_2(t) - \lambda_1(t)|. \tag{9.40}$$

If Eq. (9.40) holds, then adiabaticity is a good approximation.

We now introduce the (weak) $\mathcal{E}_1(t)$ pulse. Since $|E_1\rangle$ is the initially populated state, neither the $|E_0\rangle$ or $|E_2\rangle$ states, nor the $|\lambda_1\rangle$ and $|\lambda_2\rangle$ adiabatic states, can ever be populated in the absence of $\mathcal{E}_1(t)$. Therefore, in the absence of the $\mathcal{E}_1(t)$ pulse, the only noticeable effect of $\mathcal{E}_2(t)$ is to change the spectrum of the Hamiltonian. Assuming that the adiabatic condition [Eq. (9.40)] indeed holds, the states seen by the $\mathcal{E}_1(t)$

pulse with $\mathcal{E}_2(t)$ on, are the adiabatic states $|\lambda_1\rangle$ and $|\lambda_2\rangle$, rather than the $|E_0\rangle$ and $|E_2\rangle$ material states.

Using the definition of $\mathbf{a}$ (i.e., $\mathbf{a} = \underline{U}^\dagger \cdot \mathbf{c}$), along with Eq. (9.32), gives the adiabatic states as $\mathbf{b} = e^{-i\underline{\Delta}t/2} \cdot \underline{U} \cdot \mathbf{a}$. That is,

$$|\lambda_1(t)\rangle = e^{i\int_0^t \lambda(t')dt' + i\Delta_2 t/2}\left\{\cos\theta|E_0\rangle e^{-iE_0t/\hbar} + \sin\theta e^{-i\phi_2(t)}e^{-i\Delta_2 t}e^{-iE_2t/\hbar}|E_2\rangle\right\}$$

$$|\lambda_2(t)\rangle = e^{-i\int_0^t \lambda(t')dt' + i\Delta_2 t/2}\left\{-\sin\theta e^{i\phi_2(t)}e^{-iE_0t/\hbar}|E_0\rangle + \cos\theta e^{-i\Delta_2 t}e^{-iE_2t/\hbar}|E_2\rangle\right\}.$$

$$(9.41)$$

Here $|\lambda_1(t)\rangle$ and $|\lambda_2(t)\rangle$ are obtained using $\mathbf{a} = (a_1\ 0)^\dagger$ and $\mathbf{a} = (0\ a_2)^\dagger$, respectively. When $\Delta_2 = 0$ (i.e. when ω_2 is exactly resonant with the $|E_2\rangle$ to $|E_0\rangle$ transition), it follows from Eq. (9.37) that $\theta = \pi/4$. If, in addition, we assume that the pulse has no chirp (i.e. that the phase of $\mathcal{E}_2(t)$, $\phi_2(t) = 0$), we have that

$$|\lambda_1(t)\rangle = \frac{1}{\sqrt{2}}\left[\exp i\int_0^t |\Omega_2(t')|\,dt'\right]\left\{e^{-iE_0t/\hbar}|E_0\rangle + e^{-iE_2t/\hbar}|E_2\rangle\right\}$$

$$|\lambda_2(t)\rangle = \frac{1}{\sqrt{2}}\exp\left[-i\int_0^t |\Omega_2(t')|\,dt'\right]\left\{-e^{-iE_0t/\hbar}|E_0\rangle + e^{-iE_2t/\hbar}|E_2\rangle\right\}.\qquad(9.42)$$

Thus the time evolution of the $|E_0\rangle$ component of $|\lambda_1(t)\rangle$ is governed by a "quasi-energy" of $E_0 - |\Omega_2(t)|$, whereas the time evolution of the $|E_0\rangle$ component of $|\lambda_2(t)\rangle$ is governed by a quasi-energy of $E_0 + |\Omega_2(t)|$.

We now switch on the $\mathcal{E}_1(t)$ pulse and simultaneously excite the two adiabatic eigenstates. The simultaneous excitation occurs since the two adiabatic levels are broadened (e.g., by spontaneous emission) by an amount comparable to, or exceeding, the $2|\Omega_2(t)|$ gap. Below we show that EIT occurs due to interference between the two excited adiabatic states. Figure 9.5 shows the original levels and adiabatic states for the case of Sr.

The fact that these adiabatic levels are broadened implies that they are coupled to the continuum, that is, that they can be regarded as resonance states. Hence we utilize the resonance theory introduced in Section 6.3.1. In the first instance this approach implies that whenever broadening occurs, excitation is to the fully interacting scattering state $|E, \mathbf{n}^-\rangle$.

Consider then the probability of the one-photon absorption to this scattering state, which is given by Eq. (6.5.1) (with $\bar{\epsilon}_1 = 0$) as

$$P_\mathbf{n}(E) = \left[\frac{2\pi\bar{\epsilon}(\omega_{E,1})}{\hbar}\right]^2 |A_\mathbf{n}(E)|^2 = \left[\frac{2\pi\bar{\epsilon}(\omega_{E,1})}{\hbar}\right]^2 \left|\sum_s \langle E_1|d_{e,g}|\phi_s\rangle\langle\phi_s|E, \mathbf{n}^-\rangle\right|^2,$$

$$(9.43)$$

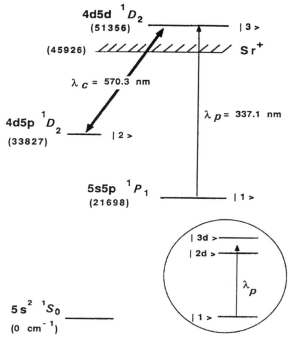

Fig. 9.5 Scenario associated with electromagnetically induced transparency (i.e., old levels, new levels $|\lambda_1\rangle$ and $|\lambda_2\rangle$, and associated fields). Note that the inset shows excitation to two adiabatic states, as discussed in the text. Here our levels $|E_0\rangle$, $|E_1\rangle$, and $|E_2\rangle$ are denoted $|3\rangle$, $|1\rangle$, and $|2\rangle$, respectively. The laser wavelengths are denoted λ_p and λ_c. (Taken from Fig. 1, Ref. [297].)

where we have assumed for simplicity that there is no direct transition to the continuum and where the amplitude $A_\mathbf{n}(E)$ is defined via Eq. (9.43). If we now identify the $|\phi_s\rangle$ bound states with the adiabatic states $|\lambda_s\rangle$, then we can write that

$$A_\mathbf{n}(E) = \sum_{s=1,2} \langle E_1|\mathrm{d}_{e,g}|\lambda_s\rangle\langle\lambda_s|E, \mathbf{n}^-\rangle, \tag{9.44}$$

where we include only the two resonances. Given Eq. (9.42) and the fact that $\langle E_2|\hat\varepsilon\cdot\mathbf{d}|E_1\rangle = 0$, we have that

$$A_\mathbf{n}(E) = \frac{1}{\sqrt{2}}e^{-iE_0 t/\hbar}\langle E_1|\mathrm{d}_{e,g}|E_0\rangle$$

$$\times \left\{\langle\lambda_1|E,\ \mathbf{n}^-\rangle \exp\left[i\int_0^t |\Omega_2(t')|\,dt'\right] - \langle\lambda_2|E,\ \mathbf{n}^-\rangle \exp\left[-i\int_0^t |\Omega_2(t')|dt'\right]\right\}$$

$$\tag{9.45}$$

Equation (9.45) requires an expression for $\langle \lambda_2 | E, \ \mathbf{n}^- \rangle$ that is given via Eq. (6.36) as

$$\langle \lambda_s | E, \ \mathbf{n}^- \rangle = \sum_{s'} \langle \lambda_s | [E - i\epsilon - Q\mathcal{H}Q]^{-1} | \lambda_{s'} \rangle \langle \lambda_{s'} | H | E, \ \mathbf{n}^-; 1 \rangle. \tag{9.46}$$

Here Q projects onto the adiabatic states.

Following Ref. [14] [see also Eq. (6.41)] we can write the $E - Q\mathcal{H}Q$ matrix in the two overlapping resonances case as

$$E - Q\mathcal{H}Q = \begin{pmatrix} E - E_0 - \lambda_1 - D_1 - \dfrac{i\Gamma_1}{2} & -D_{1,2} - \dfrac{i\Gamma_{1,2}}{2} \\[2ex] -D_{2,1} - \dfrac{i\Gamma_{2,1}}{2} & E - E_0 - \lambda_2 - D_2 - \dfrac{i\Gamma_2}{2} \end{pmatrix}, \tag{9.47}$$

where

$$\Gamma_{s,s'} = 2\pi \sum_{\mathbf{n}} \langle \lambda_s | H | E, \ \mathbf{n}^-; 1 \rangle \langle E, \ \mathbf{n}^-; 1 | H | \lambda_{s'} \rangle, \tag{9.48}$$

and

$$D_{s,s'} = \mathbf{P}_v \int dE' \sum_{\mathbf{n}} \frac{\langle \lambda_s | H | E', \ \mathbf{n}^-; 1 \rangle \langle E', \ \mathbf{n}^-; 1 | H | \lambda_{s'} \rangle}{E - E'}. \tag{9.49}$$

We have denoted $\Gamma_{s,s}$ by Γ_s and $D_{s,s}$ by D_s.

The inverse matrix can be calculated and is given by

$$[E - Q\mathcal{H}Q]^{-1} = \frac{1}{\text{Det}} \times \begin{pmatrix} E - E_0 - \lambda_2 - D_2 - \dfrac{i\Gamma_2}{2} & D_{1,2} + \dfrac{i\Gamma_{1,2}}{2} \\[2ex] D_{2,1} + \dfrac{i\Gamma_{2,1}}{2} & E - E_0 - \lambda_1 - D_1 - \dfrac{i\Gamma_1}{2} \end{pmatrix}, \tag{9.50}$$

where

$$\text{Det} = \left(E - E_0 - \lambda_1 - D_1 - \frac{i\Gamma_1}{2} \right) \left(E - E_0 - \lambda_2 - D_2 - \frac{i\Gamma_2}{2} \right)$$

$$- \left(D_{1,2} + \frac{i\Gamma_{1,2}}{2} \right) \left(D_{2,1} + \frac{i\Gamma_{2,1}}{2} \right). \tag{9.51}$$

Using Eq. (9.46) we obtain that

$$\langle \lambda_1 | E, \mathbf{n}^- \rangle = \frac{1}{\text{Det}} \left[\left(E - E_0 - \lambda_2 - D_2 - \frac{i\Gamma_2}{2} \right) \langle \lambda_1 | H | E, \mathbf{n}^-; 1 \rangle \right.$$
$$\left. + \left(D_{1,2} + \frac{i\Gamma_{1,2}}{2} \right) \langle \lambda_2 | H | E, \mathbf{n}^-; 1 \rangle \right]$$

(9.52)

and

$$\langle \lambda_2 | E, \mathbf{n}^- \rangle = \frac{1}{\text{Det}} \left[\left(D_{2,1} + \frac{i\Gamma_{2,1}}{2} \right) \langle \lambda_1 | H | E, \mathbf{n}^-; 1 \rangle \right.$$
$$\left. + \left(E - E_0 - \lambda_1 - D_1 - \frac{i\Gamma_1}{2} \right) \langle \lambda_2 | H | E, \mathbf{n}^-; 1 \rangle \right].$$

(9.53)

If there is only one open product channel $\mathbf{n}$ we have, following Eq. (9.48), that

$$\Gamma_2 \langle \lambda_1 | H | E, \mathbf{n}^-; 1 \rangle = \Gamma_{1,2} \langle \lambda_2 | H | E, \mathbf{n}^-; 1 \rangle$$
$$\Gamma_{2,1} \langle \lambda_1 | H | E, \mathbf{n}^-; 1 \rangle = \Gamma_1 \langle \lambda_2 | H | E, \mathbf{n}^-; 1 \rangle,$$

(9.54)

and hence that

$$\langle \lambda_1 | E, \mathbf{n}^- \rangle = \frac{1}{\text{Det}} \left[(E - E_0 - \lambda_2 - D_2) \langle \lambda_1 | H | E, \mathbf{n}^-; 1 \rangle + D_{1,2} \langle \lambda_2 | H | E, \mathbf{n}^-; 1 \rangle \right],$$

$$\langle \lambda_2 | E, \mathbf{n}^- \rangle = \frac{1}{\text{Det}} \left[D_{2,1} \langle \lambda_1 | H | E, \mathbf{n}^-; 1 \rangle + (E - E_0 - \lambda_1 - D_1) \langle \lambda_2 | H | E, \mathbf{n}^-; 1 \rangle \right].$$

(9.55)

Therefore, from Eq. (9.45)

$$A_{\mathbf{n}}(E) = \frac{1}{\sqrt{2}\text{Det}} e^{-iE_0 t/\hbar} \langle E_1 | d_{e,g} | E_0 \rangle$$

$$\times \left\{ \left[(E - E_0 - \lambda_2 - D_2) \langle \lambda_1 | H | E, \mathbf{n}^-; 1 \rangle + D_{1,2} \langle \lambda_2 | H | E, \mathbf{n}^-; 1 \rangle \right] \right.$$

$$\times \exp \left[-i \int_0^t |\Omega_2(t')| \, dt' \right] - \left[D_{2,1} \langle \lambda_1 | H | E, \mathbf{n}^-; 1 \rangle \right.$$

$$\left. + (E - E_0 - \lambda_1 - D_1) \langle \lambda_2 | H | E, \mathbf{n}^-; 1 \rangle \right] \exp \left[-i \int_0^t |\Omega_2(t')| \, dt' \right] \right\}.$$

(9.56)

Assuming that only $|E_0\rangle$ decays, that is, $|E_0\rangle$ is coupled to $|E, \mathbf{n}^-; 1\rangle$ while $|E_2\rangle$ is not, then

$$\langle E_2 | H | E, \mathbf{n}^-; 1 \rangle = 0. \tag{9.57}$$

We have, using Eq. (9.42), that

$$\langle \lambda_1 | H | E, \mathbf{n}^-; 1 \rangle = \exp\left[\frac{iE_0 t}{\hbar} - i \int_0^t |\Omega_2(t')| \, dt'\right] \langle E_0 | H | E, \mathbf{n}^-; 1 \rangle / \sqrt{2}$$

and

$$\langle \lambda_2 | H | E, \mathbf{n}^-; 1 \rangle = -\exp\left[\frac{iE_0 t}{\hbar} + i \int_0^t |\Omega_2(t')| \, dt'\right] \langle E_0 | H | E, \mathbf{n}^-; 1 \rangle / \sqrt{2}. \tag{9.58}$$

Therefore, from Eq. (9.56)

$$A_{\mathbf{n}}(E) = \frac{1}{2\mathrm{Det}} \langle E_1 | d_{e,g} | E_0 \rangle \langle E_0 | H | E, \mathbf{n}^-; 1 \rangle$$

$$\times \left\{ \left[(E - E_0 - \lambda_2 - D_2) - D_{1,2} \exp\left[2i \int_0^t |\Omega_2(t')| \, dt' \right] \right] \right.$$

$$\left. - \left[D_{2,1} \exp\left[-2i \int_0^t |\Omega_2(t')| \, dt' \right] - (E - E_0 - \lambda_1 - D_1) \right] \right\}. \tag{9.59}$$

It follows from Eqs. (9.57), (9.42), and (9.49) that

$$\tilde{D}_{1,2} \equiv D_{1,2} \exp\left[2i \int_0^t |\Omega_2(t')| \, dt' \right] = D_{2,1} \exp\left[-2i \int_0^t |\Omega_2(t')| \, dt' \right] = -D_1 = -D_2 \tag{9.60}$$

is a real number. Hence there is a real E value, satisfying

$$(E - E_0 - \lambda_2 - D_2) - \tilde{D}_{1,2} - \tilde{D}_{1,2} + (E - E_0 - \lambda_1 - D_1) = 0, \tag{9.61}$$

where $A_{\mathbf{n}}(E)$ is zero. Using Eq. (9.36) and Eq. (9.60), we see that the solution to Eq. (9.61) is the unshifted bound state energy $E = E_0$. At this point the absorption vanishes *exactly* and EIT results.

Thus, we have shown that there is one energy point ($E = E_0$) where the adiabatic states contribute two equal absorption amplitudes of opposite signs, thereby entirely canceling the absorption process. The location of this point is independent of the strength of the $\mathcal{E}_2(t)$ pulse envelope *or* of its time dependence, *or* of the strength of the $\langle E_0 | H | E, \mathbf{n}^-; 1 \rangle$ interaction that confers an energetic width to the resonance.

The vanishing of the line shapes at certain points due to the interference between resonance, first discovered in the context of overlapping resonances of van der Waals complexes [14], is shown in Figure 9.6.

In Figure 9.7 we illustrate the formation of the analogous EIT dark state, as described by Eq. (9.59), as a function of time. We see that when the pulse is weak ($t = 1.7$) the (Autler–Townes) splitting between the two field-dressed states is small and the EIT dark state resembles a very narrow "hole." As the pulse gets

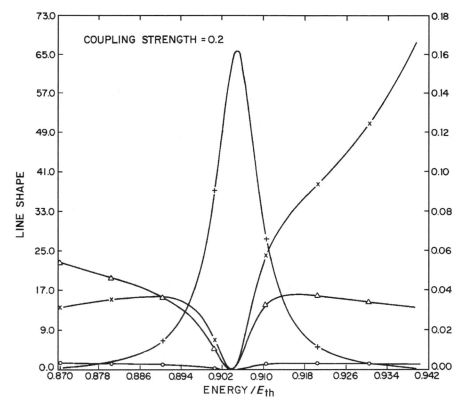

Fig. 9.6 Resonance line shapes in model He–H_2 complex. The Q space composed of the He–H_2 complex in which the H_2 fragment is confined to the $j = 2$ rotational state is coupled by potential anisotropy to the P space composed of the He–H_2 ($j = 0$) manifold. Shown are four ($v = 0, 1, 2, 3$) vibrational resonances, with $v = 0$ marked as O, $v = 1$ marked as Δ, $v = 2$ marked as $+$, and $v = 3$ marked as X, in the vicinity of the center of the $v = 2$ resonance. Left-hand scale pertains to $v = 2$ and right-hand scale to all other resonances. We see that although the resonances do not overlap appreciably (note difference between the $v = 2$ scale on left-hand side and right-hand scale pertaining to the "tails" of the other resonances), each of the $v = 0, 1, 3$ resonance exhibits a hole at the exact position of $v = 2$ maximum.

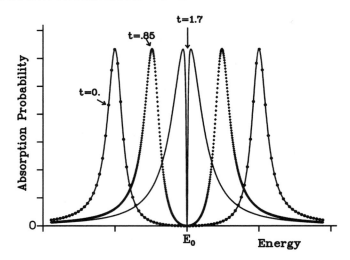

Fig. 9.7 Formation of the EIT hole as a result of an Autler–Townes splitting of a resonance according to Eq. (9.59). Shown is the line shape at three different times, at the peak of the pulse $t = 0$, as the pulse begins to wane, $t = 0.85$, and at the tail of the pulse, $t = 1.7$. A simple Gaussian pulse of the form $\mathcal{E}_2(t) = \mathcal{E}_0 \exp(-t^2)$ was assumed.

stronger ($t = 0.85$ and $t = 0$), the splitting increases and the hole widens. It stays centered at $E = E_0$ irrespective of the intensity of the pulse.

Note that in the multichannel case (i.e., when the quasi-energy states are broadened due to the interaction with more than one n continuum) Eq. (9.54) does not necessarily hold, and the numerator of Eq. (9.53) does not become a purely real function. Hence there is no real E value that would cause it to vanish. In this case EIT will not occur, although a substantial dip in the absorption cross section may still be observed.

Experimental demonstrations of EIT abound. For example, Harris' group [297] showed that a gas of Sr atoms, which is normally opaque when irradiated with a laser operating at the ω_1 transition frequency, becomes transparent when accompanied by a strong laser operating at the ω_2 frequency. An example is shown in Figure 9.8. Here, applying a 570.3-nm laser beam resonant with the 4d 5p $^1D_2$ to 4d 5d $^1D_2$ transition increases the transmission of light at 337.1 nm (a wavelength that is near resonant with 5s 5p $^1D_2$ to 4d 5d $^1D_2$) by a factor of e^{19}, with nearly all of the Sr atoms remaining in their ground state. Experimental results (solid line) are compared to computations (dashed line) in Figure 9.8 where the significant increase in transmission with changes in the central frequency of the $\mathcal{E}_1(t)$ pulse is shown. Note the enormous increase in transmission at a detuning close to zero. Related results have been obtained by Hakuta et al. [311] in atomic hydrogen. Also of interest is the use of EIT, by Kasapi [312], who was able to identify the presence of 0.3% ^{207}Pb in 99.7% ^{208}Pb by eliminating the absorption of the ^{208}Pb species, revealing the underlying traces of ^{207}Pb that still absorbed the incident radiation. The ability to distin-

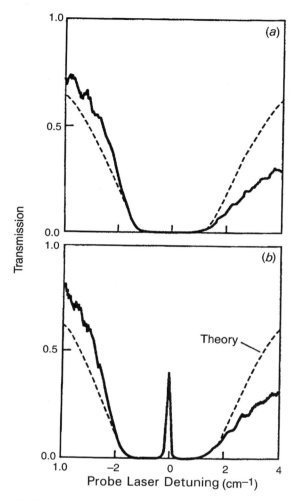

Fig. 9.8   Transmission vs. probe laser detuning in Sr at density of 5×10^{17} atoms/cm^3. Upper panel is without the ω_2 laser. Lower panel with the Rabi frequency $\Omega_{0,2} = 1.5$ cm^{-1}. (Taken from Fig. 3, Ref. [297].)

guish the two isotopes of lead by this technique is based on the small energy differences in the spectra of the two systems.

Since these initial EIT experiments in dilute gases, EIT has also been observed in solids where the decoherence rates are considerably higher. Specifically, EIT has been seen in Y_2SiO_5 doped with Pr^{3+} [313, 314] and in solid ruby [315]. The experimental results suggest that EIT persists as long as the inhomogeneous width associated with the transition between levels $|E_1\rangle$ and $|E_2\rangle$ is less than the Rabi frequencies coupling $|E_1\rangle$ with $|E_0\rangle$ and $|E_0\rangle$ with $|E_2\rangle$.

9.1.4   Lasing without Inversion

If we reverse the energetic ordering of the levels (see Fig. 9.9), making $E_1 > E_0$, we can use trapped states to give rise to a phenomenon known as lasing without inversion (LWI). Usually, laser action requires that the population in the upper levels, from which there is radiative emission, is larger than that in the lower levels. However, this is not the case in LWI. Here we consider a strong pulse $\varepsilon_2(t)$ coupling two low-lying states $|E_0\rangle$ and $|E_2\rangle$. As in the EIT case, two adiabatic states, $|\lambda_1\rangle$ and $|\lambda_2\rangle$, with energy separation $2|\Omega_2|$ are formed as a result. A higher lying state $|E_1\rangle$ can spontaneously emit radiation to both the adiabatic states, but due the $2|\Omega_2(t)|$ separation between these two states there is no interference between the amplitudes for (spontaneous or stimulated) emission to these states. That is, the energy separation between these two states is sufficiently large to allow us to distinguish the state to which emission had occurred.

For zero detuning, however, the two frequencies thus emitted *cannot* be reabsorbed. That is, contrary to the emission process, the reabsorption process would lead to the same final state $|E_1\rangle$, and there would be no way we can tell which pathway was chosen by the system (see Fig. 9.9). Thus, the resultant cancellation of reabsorption of the two emitted frequencies in the LWI case is seen to be a special case of the *bichromatic* control scenario (Section 3.1.1).

The net result of the phenomenon just described is the possibility of having lasing without the need for population inversion [303–306]. We can envision a pumping process that transfers some population to state $|E_1\rangle$. This state can emit to the two

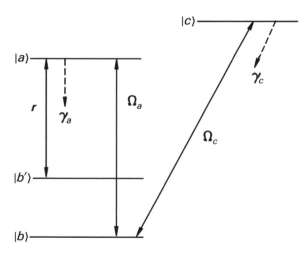

Fig. 9.9   Simplified four-level model for lasing without inversion. Shown is a model of Rb[87] in which $|a\rangle$ corresponds to sublevels $|P_{1/2}, F = 2\rangle$, $|c\rangle = |P_{3/2}, F = 2\rangle$, and the two sublevels of ground state $|b\rangle = |S_{1/2}, F = 1\rangle$ and $|b'\rangle = |S_{1/2}, F = 2\rangle$. Driving field ω_c is tuned to the D_2 resonance from the $F = 1$ ground state to the $F = 2$ of excited state, the probe field ω_a couples the same sublevels of the ground state with levels having $F = 2$ of the $P_{1/2}$ excited state. Polarizations of fields are described in text. (After Fig. 1(a), Ref. [316].)

adiabatic states $|\lambda_1\rangle$ and $|\lambda_2\rangle$ created by the $\varepsilon_2(t)$ photon, but the light thus emitted cannot be reabsorbed due to a destructive interference effect similar to the EIT described above. As a result, the basic conditions for lasing, namely higher probability for emission than for absorption, can be met without ever creating population inversion between the $|E_1\rangle$ and the $|E_0\rangle$ states. Experimental confirmation of this idea in a Λ system of the Na atom has been reported [317].

9.2 ANALYTIC SOLUTION OF NONDEGENERATE QUANTUM CONTROL PROBLEM

The most general "quantum control" problem can be phrased as the problem of finding ways to completely transfer population from an arbitrary initial state to a desired "target" state, under the guidance of external fields (e.g., laser pulses). As discussed in Chapter 4, its general solution can only be attained using "brute-force" numerical optimization schemes. However, there is a more restricted problem, namely achieving population transfer between superpositions of *nondegenerate* energy eigenstates, which we term the *nondegenerate quantum control problem*, that can be solved analytically using the concept of trapped states discussed above. This solution does not help in the control of chemical reactions and photodissociation, dealt with in previous chapters, because in those cases the continuum is characterized by exact degeneracies. It is nevertheless useful when we encounter a nondegenerate manifold composed of bound (vibrational) states. This problem has also received much attention (see, e.g., Refs. [318–321]) and has usually been solved numerically. Here [322] we present an analytic solution of the problem that generalizes STIRAP (Section 9.1) to the multilevel case.

Consider population transfer between an arbitrary initial state $|\Psi\rangle = \sum_k c_k e^{-iE_k t/\hbar}|k\rangle$ to an arbitrary target state $|\Psi'\rangle = \sum_l c'_l e^{-iE_l t/\hbar}|l\rangle$, where both spectra E_k and E_l are nondegenerate and differ from one another.

Figure 9.10 illustrates the proposed method. The transfer process is induced by two pulses, represented by a multimode electric field:

$$\varepsilon(t) = \mathrm{Re} \sum_{k=1}^{n+m} \mathcal{E}_{0,k}(t) e^{-i\omega_{0,k}t}, \qquad (9.62)$$

where $\omega_{i,j} \equiv (E_i - E_j)/\hbar$, and $\mathcal{E}_{0,k}(t)$ are the slowly varying amplitudes of each $\omega_{0,k}$ mode. By choosing a different time dependence for the $\mathcal{E}_{0,k}(t)$ amplitudes of the dump process, connecting the $|k\rangle = |n+1\rangle, |n+2\rangle, \ldots, |n+m\rangle$ states to the $|0\rangle$ state, and of the pump process, connecting the $|0\rangle$ state to the $|k\rangle = |1\rangle, |2\rangle, \ldots, |n\rangle$ states, we in effect construct two temporally distinct (dump and pump) pulses. The intensity and phase of each $\mathcal{E}_{0,k}(t)$ amplitude is adjusted, as explained below, to yield the desired transfer. In the counterintuitive scheme, the pump pulse with the $\mathcal{E}_{1,0}, \ldots, \mathcal{E}_{n,0}$ components follows the dump pulse with the $\mathcal{E}_{n+1,0}, \ldots, \mathcal{E}_{n+m,0}$ components.

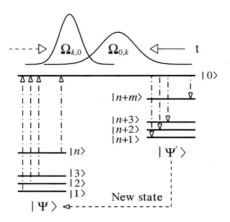

Fig. 9.10   Transfer scheme between states $|\Psi\rangle \to |0\rangle \to |\Psi'\rangle$. The first (dump) laser pulse, with Rabi frequencies $\Omega_{0,k'}(t)$, couples all $|k'\rangle$ final (empty) states to state $|0\rangle$, while the second (pump) pulse, with Rabi frequency $\Omega_{k,0}(t)$, couples state $|0\rangle$ to the initial (populated) states $|k\rangle$. (Taken from Fig. 1, Ref. [322].)

The system evolution is described by the wave function $|\Psi(t)\rangle = \sum_{k=0}^{n+m} c_k(t) e^{-iE_k t/\hbar}|k\rangle$. Below, the vector $\mathbf{c}(t)$ denotes the $c_k(t)$ coefficients:

$$\mathbf{c}(t) = (c_0, \ c_1, \ldots, c_n, \ c_{n+1}, \ldots, c_{n+m}). \tag{9.63}$$

In order to solve for $\mathbf{c}(t)$ we consider the dynamics in the rotating-wave approximation. One convenient way to do this is to write the Hamiltonian directly in this approximation, and neglect off-resonance terms. This corresponds to a molecule–field interaction of

$$H_{\mathrm{MR}} = \hbar \sum_{k=1}^{n+m} \left[\Omega_{0,k}(t) e^{-i\omega_{0,k}t}|0\rangle\langle k| + \Omega_{k,0}(t) e^{i\omega_{0,k}t}|k\rangle\langle 0| \right]. \tag{9.64}$$

Here, $\Omega_{i,j}(t)$, the time-dependent Rabi frequencies, are given by

$$\Omega_{i,j}(t) \equiv \mathcal{O}_{i,j} f_{D(P)}(t) \equiv \hat{\varepsilon} \cdot \mathbf{d}_{i,j}\mathcal{E}_{i,j}(t)/\hbar, \tag{9.65}$$

where $\hat{\varepsilon} \cdot \mathbf{d}_{0,k}$ are the electric-dipole matrix elements, projected along the field polarization and $0 < f_D(t) < 1$ and $0 < f_P(t) < 1$ describe the pulse envelopes of the dump and the pump pulses, respectively.

Inserting $\Psi(t)$ into the time-dependent Schrödinger equation and using Eq. (9.64) gives

$$\dot{\mathbf{c}}^{\mathrm{T}}(t) = -i\underline{\underline{H}}(t) \cdot \mathbf{c}^{\mathrm{T}}(t), \tag{9.66}$$

where $\mathbf{T}$ designates the matrix transpose, and the effective Hamiltonian matrix is

$$
\underline{\underline{H}}(t) =
\begin{bmatrix}
0 & \Omega_{0,1} & \cdots & \Omega_{0,n} & \Omega_{0,n+1} & \cdots & \Omega_{0,n+m} \\
\Omega_{1,0} & 0 & \cdots & 0 & 0 & \cdots & 0 \\
\cdots & \cdots & \cdots & \cdots & \cdots & \cdots & \cdots \\
\Omega_{n,0} & 0 & \cdots & 0 & 0 & \cdots & 0 \\
\Omega_{n+1,0} & 0 & \cdots & 0 & 0 & \cdots & 0 \\
\cdots & \cdots & \cdots & \cdots & \cdots & \cdots & \cdots \\
\Omega_{n+m,0} & 0 & \cdots & 0 & 0 & \cdots & 0
\end{bmatrix}
$$

Of the $n+m+1$ eigenvalues of $\underline{\underline{H}}(t)$, $n+m-1$ are zero and two are nonzero:

$$
\lambda_{1,2,\ldots,n+m-1} = 0,
$$

$$
\lambda_{n+m} = -\lambda_{n+m+1} = \left(\sum_{k=1}^{n+m} |\Omega_{0,k}(t)|^2 \right)^{1/2}. \tag{9.67}
$$

We can find nm (trapped) eigenvectors of the type

$$
|D_{kl}\rangle = \Omega_{0,l}|k\rangle e^{-iE_k t/\hbar} - \Omega_{0,k}|l\rangle e^{-iE_l t/\hbar}, \quad (k=1,\ldots,n; \; l=n+1,\ldots,n+m), \tag{9.68}
$$

all having zero eigenvalues of H_{MR}. However, at most only $n+m-1$ of these states can be linearly independent. We can use the above trapped eigenvectors to obtain a state that correlates (in the counterintuitive pulse ordering) at $t=0$ with the initial state $|\Psi\rangle = \sum_k c_k^0 |k\rangle$, and at $t=t_{\text{end}}$ with the final state $|\Psi'\rangle = \sum_l c_l^e |l\rangle e^{-iE_l t_{\text{end}}/\hbar}$. The particular combination that satisfies these asymptotic conditions is

$$
|D\rangle = \sum_{k,l} t_{kl}|D_{kl}\rangle = \sum_{k=1}^{n} |k\rangle e^{-iE_k t/\hbar} \sum_{l=n+1}^{n+m} t_{kl}\Omega_{0,l} - \sum_{l=n+1}^{n+m} |l\rangle e^{-iE_l t/\hbar} \sum_{k=1}^{n} t_{kl}\Omega_{0,k}, \tag{9.69}
$$

where the t_{kl} coefficients are chosen such that

$$
\sum_{l=n+1}^{n+m} t_{kl}\Omega_{0,l} = c_k^0, \qquad \sum_{k=1}^{n} t_{kl}\Omega_{0,k} = c_l^e. \tag{9.70}
$$

Equations (9.70) can be satisfied by choosing a *counterintuitive* pulse ordering (in which only the $\mathcal{E}_{0,l}$, $l=n+1,\ldots,n+m$ components exist at $t=0$ while only the $\mathcal{E}_{0,k}$, $k=1,\ldots,n$ components exist at $t=t_{\text{end}}$) and choosing t_{kl} and $\Omega_{k,0}(t) \equiv \mathcal{O}_{k,0}f_P(t)$ such that

$$
t_{kl} \propto \mathcal{O}_{k,0}\mathcal{O}_{l,0}, \quad \text{and} \quad \mathcal{O}_{k,0} = \mathcal{C}c_k^0, \; \mathcal{O}_{l,0} = \mathcal{C}c_l^e, \tag{9.71}
$$

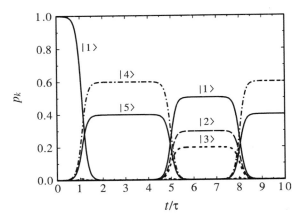

Fig. 9.11 Complete population transfer between various superpositions of $|k\rangle$ $k = 1 - 5$ states. Lines are denoted by symbols $|k\rangle$ give respective populations p_k. (Taken from Fig. 2, Ref. [322].)

where C is an arbitrary complex number. The only limitation on the choice of C is that the (slowly varying) Rabi frequencies should be strong enough to guarantee the adiabaticity of the transfer process. Other than that, the process is quite robust against changes in C.

In Figure 9.11 we test this approach by considering a population transfer chain. This is composed of the transfer from state $|1\rangle$ to a linear combination of states $|4\rangle$ and $|5\rangle$, followed by the transfer from this superposition state to a superposition of the $|1\rangle$, $|2\rangle$, and $|3\rangle$ states, and back to the $|4\rangle$ plus $|5\rangle$ superposition. This transfer chain is performed with Gaussian pulses for which the Rabi frequencies are given as $\Omega_{0,k}(t) = \mathcal{O}_{0,k} \exp[-(t-t_0)^2/\tau^2]$ $(k = 1, \ldots, n)$ and $\Omega_{0,l}(t) = \mathcal{O}_{0,l} \exp[-t^2/\tau^2]$ $(l = n+1, \ldots, n+m)$, with $t_0 = 2\tau$ being the delay between the pulses. The C coefficient of Eq. (9.71) is chosen to be $C = 50/\tau$.

Thus, there is a very simple and analytic pulse shaping recipe for achieving a complete population transfer between two arbitrary superposition states $|\Psi\rangle$ and $|\Psi'\rangle$, composed of nondegenerate energy eigenstates.

CHAPTER 10

PHOTODISSOCIATION BEYOND THE WEAK-FIELD REGIME

In this chapter we continue our discussion of control in moderately strong fields, focusing on the photodissociation of molecules by moderately strong pulses using approximate analytical approaches. The treatment of the dissociation of a molecule by a strong cw light is postponed to Section 12.3, where we show that the "dressed-state" picture emerging from the quantum description of light is the most natural and useful way of thinking about the interaction of molecules with truly strong cw fields.

10.1 ONE-PHOTON DISSOCIATION WITH LASER PULSES

We consider now the case of dissociation by the net absorption of just one photon, to be termed *one-photon dissociation*. As in the weak-field domain (Chapter 3) the molecule is assumed to dissociate into two fragments as a result of the interaction with a laser pulse. It is convenient to parametrize the incident electric field [Eq. (1.35)] as

$$\mathbf{E}(t) = 2\hat{\varepsilon}\mathcal{E}(t)\cos(\omega_1 t), \qquad (10.1)$$

where we suppress the spatial z variable.

Assuming that the field is in near resonance or on resonance with transitions from the initial bound state $|E_1\rangle$ to the continuum (see Fig. 10.1a), we expand the full time-dependent wave function as:

$$|\Psi(t)\rangle = b_1(t)|E_1\rangle \exp\left(\frac{-iE_1 t}{\hbar}\right) + \sum_{\mathbf{n}} \int dE \, b_{E,\mathbf{n}}(t)|E, \mathbf{n}^-\rangle \exp\left(\frac{-iEt}{\hbar}\right). \qquad (10.2)$$

Here we have suppressed the channel (q) index for convenience, assuming that it is contained in the $\mathbf{n}$ index. As usual, we insert Eq. (10.2) into the time-dependent Schrödinger equation, $i\hbar\partial\Psi/\partial t = H\Psi(t)$, and use the orthogonality of the eigenfunctions of H_M, to obtain an indenumerable set of first-order differential equations that are analogous to Eq. (2.3):

$$\frac{db_1}{dt} = i \int dE \sum_{\mathbf{n}} \Omega_{1,E,\mathbf{n}}(t) b_{E,\mathbf{n}}(t) \exp(-i\Delta_E t), \qquad (10.3a)$$

$$\frac{db_{E,\mathbf{n}}}{dt} = i\Omega^*_{1,E,\mathbf{n}}(t) \exp(i\Delta_E t) b_1(t), \quad \text{for each } E \text{ and } \mathbf{n}. \qquad (10.3b)$$

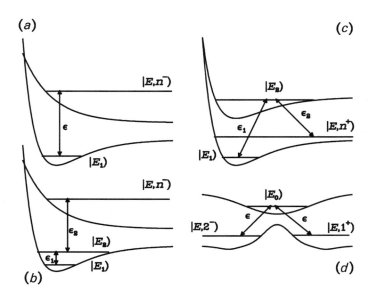

Fig. 10.1   Energy levels and pulses pertaining to: (a) One-photon dissociation, (b) resonantly enhanced two-photon dissociation, (c) resonantly enhanced two-photon association, and (d) laser catalysis.

Here we have retained the rotating-wave terms [see Eq. (2.13)] only. The detuning, Δ_E, is given by

$$\Delta_E \equiv \omega_{E,1} - \omega_1, \quad \text{with } \omega_{E,1} \equiv (E - E_1)/\hbar, \tag{10.4}$$

and $\Omega_{1,E,\mathbf{n}}(t)$ is the (time-varying) Rabi frequency, defined as

$$\Omega_{1,E,\mathbf{n}}(t) \equiv \langle E_1|\hat{\varepsilon} \cdot \mathbf{d}|E, \mathbf{n}^-\rangle \mathcal{E}(t)/\hbar. \tag{10.5}$$

Unlike the treatment in the weak-field domain, we do not assume that $b_1(t) \approx 1$ at all times. Rather, we integrate the $b_{E,\mathbf{n}}$ continuum coefficients of Eq. (10.3) over time, while imposing the boundary condition that the continuum states are empty at the start of the process [i.e., $b_{E,\mathbf{n}}(t \to -\infty) = 0$], to obtain

$$b_{E,\mathbf{n}}(t) = i \int_{-\infty}^{t} dt' \, \Omega_{1,E,\mathbf{n}}^*(t')b_1(t') \exp(i\Delta_E t'). \tag{10.6}$$

The state-specific photodissociation probability, $P_{\mathbf{n}}(E)$ [Eq. (2.74)], is the long-time probability, at fixed energy E, of observing a particular internal state $|\mathbf{n}\rangle$ of the dissociated fragments. Hence, using Eqs. (10.5) and (10.6), we have that

$$P_{\mathbf{n}}(E) = |b_{E,\mathbf{n}}(t \to \infty)|^2 = \left| \frac{1}{\hbar} \langle E_1|\hat{\varepsilon} \cdot \mathbf{d}|E, \mathbf{n}^-\rangle \int_{-\infty}^{\infty} dt' \, \mathcal{E}^*(t')b_1(t') \exp(i\Delta_E t') \right|^2. \tag{10.7}$$

We can obtain a closed-form solution for the bound part of the problem by substituting Eq. (10.6) into Eq. (10.3a), giving a first-order integro-differential equation for b_1:

$$\frac{db_1}{dt} = \frac{-1}{\hbar^2} \int dE \sum_{\mathbf{n}} |\langle E, \mathbf{n}^-|\hat{\varepsilon} \cdot \mathbf{d}|E_1\rangle|^2 \mathcal{E}^*(t) \int_{-\infty}^{t} dt' \, \mathcal{E}^*(t') \exp[-i\Delta_E(t - t')]b_1(t'). \tag{10.8}$$

Equation (10.8) can be solved numerically in a straightforward fashion.

Nevertheless, it is instructive to analyze it in terms of $F_1(t - t')$, the *spectral autocorrelation function* [27, 29, 191, 324], defined as the Fourier transform of the absorption spectrum:

$$F_1(t - t') = \int dE \, A_1(E) \exp[-i\omega_{E,1}(t - t')]. \tag{10.9}$$

Here $A_i(E)$ is the absorption spectrum from the ith state, where

$$A_i(E) \equiv \sum_{\mathbf{n}} |\langle E, \mathbf{n}^-|\hat{\varepsilon} \cdot \mathbf{d}|E_i\rangle|^2. \tag{10.10}$$

Using this definition for $F_1(t - t')$, we can rewrite Eq. (10.8) as

$$\frac{db_1}{dt} = \frac{-\mathcal{E}(t)}{\hbar^2} \int_{-\infty}^{t} dt' \; \mathcal{E}^*(t') F_1(t - t') \exp[i\omega_1(t - t')] b_1(t'). \tag{10.11}$$

The value of the ground state coefficient at time t is seen to be determined by its past history at $t' < t$ through the *memory kernel* $\mathcal{E}(t)\mathcal{E}^*(t')F_1(t - t')$.

10.1.1 Slowly Varying Continuum

The simplest (though approximate) solution of Eq. (10.11) is obtained by assuming that all the continua are "flat," that is, that the bound-continuum matrix elements vary slowly with energy and can be replaced by their value at some average energy, say $E_L = E_1 + \hbar\omega_1$,

$$\sum_n |\langle E, \mathbf{n}^- | \hat{\varepsilon} \cdot \mathbf{d} | E_1 \rangle|^2 \approx \sum_n |\langle E_L, \mathbf{n}^- | \hat{\varepsilon} \cdot \mathbf{d} | E_1 \rangle|^2. \tag{10.12}$$

This approximation, called the *flat continuum* or *slowly varying continuum approximation* (SVCA) [191, 325, 326] localizes the autocorrelation function in time, since by Eqs. (10.9) and (10.12)

$$F_1(t - t') = 2\pi\hbar A_1(E_L)\delta(t - t'). \tag{10.13}$$

Substituting Eq. (10.13) into Eq. (10.11) and integrating over E and t', we obtain that

$$\frac{db_1}{dt} = -\Omega_1^I(t)b_1(t). \tag{10.14}$$

Hence,

$$b_1(t) = b_1(-\infty) \exp\left[-\int_{-\infty}^{t} \Omega_1^I(t') \, dt'\right], \tag{10.15}$$

where $\Omega_1^I(t)$, the *imaginary Rabi frequency*, is defined as

$$\Omega_1^I(t) \equiv \frac{\pi A_1(E_L)|\mathcal{E}(t)|^2}{\hbar} = \pi \sum_n \frac{|\langle E_L, \mathbf{n}^- | \hat{\varepsilon} \cdot \mathbf{d} | E_1 \rangle \mathcal{E}(t)|^2}{\hbar}. \tag{10.16}$$

The factor of $\frac{1}{2}$ relative to Eq. (10.13) arises because the integration over t' in Eq. (10.11) is carried out over the $[-\infty, t]$ range and not over the usual $[-\infty, +\infty]$ range.

It follows from Eq. (10.15) that a "slowly varying" continuum acts as an irreversible "perfect absorber" since in this approximation $b_1(t)$ decreases monotonically (though not necessarily as a simple exponential) with time.

In many cases the continuum may have structures that are narrower than the bandwidth of the pulse. Such structures may be due to either the natural spectrum of the molecular Hamiltonian [327, 328] or to the interaction with the strong external field [195, 197–199, 329]. Under such circumstances we expect the SVCA approximation to break down, yielding nonmonotonic decay dynamics.

Given Eqs. (10.6) and (10.15) the amplitude $b_{E,\mathbf{n}}(t)$ is given by

$$b_{E,\mathbf{n}}(t) = i \int_{-\infty}^{t} dt'\, \Omega^*_{1,E,\mathbf{n}}(t') b_1(-\infty) \exp\left[-\int_{-\infty}^{t'} \Omega_1^I(t'')\, dt'' + i\Delta_E t' \right]. \quad (10.17)$$

10.1.2 Bichromatic Control

We now generalize this result by considering photoexcitation from a superposition state $b_1|E_1\rangle + b_2|E_2\rangle$ where $b_i \equiv b_i(-\infty)$. Assuming no transitions between levels 1 and 2, and using the SVCA, we can write an analytic formula for *bichromatic control* that goes beyond perturbation theory. Allowing the coefficients b_1 and b_2 to decay according to Eq. (10.15) we obtain that

$$
\begin{aligned}
b_{E,\mathbf{n}}(t \to \infty) = \frac{i}{\hbar} \Bigg\{ & \langle E, \mathbf{n}^-|\hat{\varepsilon} \cdot \mathbf{d}|E_1\rangle b_1 \int_{-\infty}^{\infty} dt'\, \mathcal{E}_1(t') \\
& \times \exp\left[i\Delta_{E,1} t' - \frac{\pi}{\hbar} A_1(E_L) \int_{-\infty}^{t'} |\mathcal{E}_1(t'')|2\, dt'' \right] \\
& + \langle E, \mathbf{n}^-|\hat{\varepsilon} \cdot \mathbf{d}|E_2\rangle b_2 \int_{-\infty}^{\infty} dt'\, \mathcal{E}_2(t') \\
& \times \exp\left[i\Delta_{E,2} t' - \frac{\pi}{\hbar} A_2(E_L) \int_{-\infty}^{t'} |\mathcal{E}_2(t'')|^2\, dt'' \right] \Bigg\},
\end{aligned}
\quad (10.18)
$$

where $\Delta_{E,i} = \omega_{E,i} - \omega$, $i = 1,\ 2$. Therefore, the probability of observing a particular channel $\mathbf{n}$ is given as

$$P_{\mathbf{n}}(E) = \frac{4\pi^2}{\hbar^2} \left| \langle E, \mathbf{n}^-|\hat{\varepsilon} \cdot \mathbf{d}|E_1\rangle|\zeta_1(E)|e^{-i\theta_1(E)} b_1 + \langle E, \mathbf{n}^-|\hat{\varepsilon} \cdot \mathbf{d}|E_2\rangle|\zeta_2(E)|e^{-i\theta_2(E)} b_2 \right|^2, \quad (10.19)$$

where

$$\zeta_i(E) \equiv \left(\frac{1}{2\pi} \right) \int_{-\infty}^{\infty} dt \exp(-i\Delta_{E,i} t) \mathcal{E}_i(t) \exp\left[-\frac{\pi}{\hbar} A_i(E_L) \int_{-\infty}^{t} |\mathcal{E}_i(t')\, dt'|^2 \right], \quad (10.20)$$

and $\theta_i(E)$ is the phase of $\zeta_i(E)$.

Thus, we obtain a form, which is correct (within the range of validity of the SVCA) for strong fields, that resembles the weak-field bichromatic control result of Eq. (3.12). The only difference is that instead of the Fourier transform of the electric field of the pulse, Eq. (10.19) depends on the Fourier transform of the product of the pulse electric field and the decaying factor $\exp[-(\pi/\hbar)A_i(E_L)\int_{-\infty}^{t}|\mathcal{E}_i(t')|^2\,dt']$, which describes the depletion of the initial state(s) due to the action of the pulse.

10.1.3 Resonance

Consider once again the case of excitation from a single $|E_1\rangle$ level, but now the limit where the continuum is composed of a *single* resonance at $E = E_s$. This means that the continuous spectrum is described by a single Lorentzian form, positioned at E_s with full width at half maximum Γ_s:

$$A_1(E) = \frac{\overline{d_s^2}\,\Gamma_s^2/4}{(E - E_s)^2 + \Gamma_s^2/4},\tag{10.21}$$

where $\overline{d_s^2}$ is an average dipole-moment square that determines the height of the absorption curve and $2\pi\overline{d_s^2}\Gamma_s/4$ is the area under the absorption curve. In the time-dependent picture, the resonance decays at the rate $\Gamma_s/2\hbar$.

Substituting this form into Eq. (10.9), and using the fact that $t \geq t'$, gives the spectral autocorrelation:

$$\exp\big[i\omega_1(t - t')\big]F_1(t - t') = \frac{\overline{d_s^2}\,\Gamma_s}{4}f_s^+(t)f_s^-(t'),\tag{10.22}$$

where

$$f_s^{\pm}(t) = \sqrt{2\pi}\,\exp[\mp i\chi_s t],\tag{10.23}$$

with

$$\chi_s \equiv \Delta_s - i\frac{\Gamma_s}{2\hbar}, \quad \text{and} \quad \Delta_s \equiv \frac{E_s - E_1}{\hbar} - \omega_1.\tag{10.24}$$

Using Eq. (10.22) we can transform Eq. (10.11) into two coupled first-order differential equations:

$$\frac{db_1}{dt} = \frac{i}{4\hbar}\mathcal{E}(t)\overline{d_s^2}\,\Gamma_s f_s^+(t)B_s(t),$$

$$\frac{dB_s}{dt} = \frac{i}{\hbar}\mathcal{E}^*(t)f_s^-(t)b_1(t).\tag{10.25}$$

These two equations can be solved by reducing them to a form similar to the Schrödinger equation and using the WKB-like approximate solution [330]. Specifically, we first obtain a second-order equation by differentiating Eq. (10.25) to give

$$\frac{d^2 b_1}{dt^2} = \frac{d \ln \mathcal{E}(t)}{dt} \frac{db_1}{dt} - \frac{\pi |\mathcal{E}(t)|^2}{2\hbar^2} \overline{d_s^2} \Gamma_s b_1 - \frac{\mathcal{E}(t)}{4\hbar} \overline{d_s^2} \Gamma_s \chi_s f_s^+(t) B_s(t), \tag{10.26}$$

where we have used the explicit form of f_s^+ [Eq. (10.23)]. We thus obtain from Eqs. (10.25) and (10.26), that

$$\frac{d^2 b_1}{dt^2} = \left(\frac{d \ln \mathcal{E}(t)}{dt} - i\chi(t) \right) \frac{db_1}{dt} - \frac{\pi}{2\hbar^2} |\mathcal{E}(t)|^2 \overline{d_s^2} \Gamma_s b_1, \tag{10.27}$$

where $\chi(t)$ is defined by equating Eqs. (10.26) and (10.27).

This second-order differential equation in time [Eq. (10.27)] is homomorphic to a one-dimensional time-independent Schrödinger equation in a spatial variable. This can be seen by defining

$$g_1(t) = -\frac{d \ln \mathcal{E}(t)}{dt} + i\chi(t), \tag{10.28}$$

$$g_0(t) = \frac{\pi}{2\hbar^2} |\mathcal{E}(t)|^2 \overline{d_s^2} \Gamma_s, \tag{10.29}$$

writing Eq. (10.27) as

$$\frac{d^2 b_1}{dt^2} + g_1(t) \frac{db_1}{dt} + g_0(t) b_1 = 0, \tag{10.30}$$

and transforming $b_1(t)$ according to

$$c(t) = \exp\left[\frac{1}{2} \int_{t*}^t g_1(t') \, dt' \right] b_1(t) = \left[\frac{\mathcal{E}(t)}{\mathcal{E}(t*)} \right]^{-1/2} \exp\left[\frac{i}{2} \int_{t*}^t \chi(t') \, dt' \right] b_1(t). \tag{10.31}$$

Doing so, we obtain a Schrödinger-like equation in $c(t)$,

$$\left[\frac{d^2}{dt^2} - W(t) \right] c(t) = 0, \tag{10.32}$$

where

$$\begin{aligned} W(t) &= \frac{1}{2} g_1'(t) + \frac{1}{4} g_1^2(t) - g_0(t) \\ &= -\frac{d\mathcal{E}(t)}{2dt} \frac{d^2 \ln \mathcal{E}(t)}{dt^2} + \frac{1}{4} \left(\frac{d \ln \mathcal{E}(t)}{dt} - i\chi(t) \right)^2 - \frac{\pi}{2\hbar^2} |\mathcal{E}(t)|^2 \overline{d_s^2} \Gamma_s. \end{aligned} \tag{10.33}$$

The term $W(t)$ is analogous to the term $2m[V(x) - E]/\hbar^2$ in the time-independent Schrödinger equation, with x being the coordinate, $V(x)$ is the potential, E is the energy, and m is the mass of a particle. The quantity t^*, satisfying the $W(t^*) = 0$ equation, is defined, again in analogy with the time-independent Schrödinger equation, as the "turning point" [i.e., the point where $V(x^*) = E$].

If there is only one turning point, the solutions of Eq. (10.32) can be well approximated in terms of the uniform regular and irregular Airy functions A_i and B_i [330, 331],

$$c_{\text{uni}}(t) = \left[\frac{T(t)}{-W(t)}\right]^{1/4} \{C_a A_i[-T(t)] + C_b B_i[-T(t)]\}, \qquad (10.34)$$

where the complex argument T is defined as

$$T(t) = \left[\frac{3}{2}\int_{t^*}^{t}[-W(t')]^{1/2}\,dt'\right]^{2/3}, \qquad (10.35)$$

and C_a and C_b are constants determined by the initial conditions $b_s = 1$, $b_{E,\mathbf{n}} = 0$.

Equation (10.34) is the exact solution of the equation,

$$\left[\frac{d^2}{dt^2} - W(t) - \eta(t)\right]c_{\text{uni}}(t) = 0, \qquad (10.36)$$

where

$$\eta(t) = [T(t)]^{1/2}\frac{d^2}{dt^2}[T(t)]^{-1/2}. \qquad (10.37)$$

Since $|\eta(t)| \ll |W(t)|$, c_{uni} is an excellent approximate solution to Eq. (10.32). If there is more than one turning point, the Airy functions can still be used (provided the turning points do not coalesce), by writing the solutions of Eq. (10.34) for each time interval containing a turning point and matching these solutions and their derivatives across the time intervals. Usually no more than two turning points exist.

The above equations can be simplified when $|T|$ is large because in that case we can use the asymptotic forms of the A_i and B_i functions [4] in Eq. (10.34):

$$\begin{aligned}
A_i(T) &\xrightarrow{|T|\to\infty} \frac{1}{2}\pi^{-1/2}T^{-1/4}\exp(-\zeta), & |\arg T| &< \pi, \\
B_i(T) &\xrightarrow{|T|\to\infty} \pi^{-1/2}T^{-1/4}\exp(\zeta), & |\arg T| &< \frac{1}{3}\pi, \\
A_i(-T) &\xrightarrow{|T|\to\infty} \pi^{-1/2}T^{-1/4}\sin\left(\zeta + \frac{\pi}{4}\right), & |\arg T| &< \frac{2}{3}\pi, \\
B_i(-T) &\xrightarrow{|T|\to\infty} \pi^{-1/2}T^{-1/4}\cos\left(\zeta + \frac{\pi}{4}\right), & |\arg T| &< \pi,
\end{aligned} \qquad (10.38)$$

where $\zeta = \frac{2}{3}T^{3/2} = \int_{t^*}^{t}[W(t')]^{1/2}\,dt'$.

For small enough Γ_s, $|\arg T| < \frac{2}{3}\pi$ and we can use the last two identities in Eq. (10.38) to obtain a (first-order) WKB approximation,

$$
\begin{aligned}
c_{\text{WKB}}(t) = \left[-\pi^2 W(t)\right]^{-1/4} \Bigg\{ &C_a \sin\left[\int_{t*}^{t} (-W(t'))^{1/2} \, dt' + \frac{\pi}{4}\right] \\
&+ C_b \cos\left[\int_{t*}^{t} (-W(t'))^{1/2} \, dt' + \frac{\pi}{4}\right]\Bigg\}.
\end{aligned}
\tag{10.39}
$$

Using the relation Eq. (10.31) we obtain the first-order WKB approximation that

$$
\begin{aligned}
b_1(t) = \left[\frac{\pi\mathcal{E}(t)}{\mathcal{E}(t*)}\right]^{-1/2} &\left[-W(t)\right]^{-1/4} \exp\left[\frac{i}{2}\int_{t*}^{t} \chi(t') \, dt'\right] \\
&\times \left\{ C_a \sin\left[\int_{t*}^{t} (-W(t'))^{1/2} \, dt' + \frac{\pi}{4}\right] + C_b \cos\left[\int_{t*}^{t} (-W(t'))^{1/2} \, dt' + \frac{\pi}{4}\right]\right\}.
\end{aligned}
\tag{10.40}
$$

For cw radiation $\mathcal{E}(t) = \mathcal{E}_0$ is a constant. Hence,

$$
(-W)^{1/2} = \frac{1}{2}\left[\left(\Delta_s - i\frac{\Gamma_s}{2\hbar}\right)^2 + \frac{2\pi}{\hbar^2}\mathcal{E}_0^2\overline{d_s^2}\Gamma_s\right]^{1/2}.
\tag{10.41}
$$

For cw radiation (that is turned on at $t = 0$) we choose C_a and C_b such that $b_1(0) = 1$, and obtain that

$$
b_1(t) = \exp\left(-i\Delta_s t/2 - \Gamma_s t/4\hbar\right) \cos\left[(-W)^{1/2}t\right].
\tag{10.42}
$$

To obtain the cw transition amplitude to a level E in the continuum we use Eq. (10.6), according to which

$$
\begin{aligned}
b_{E,\mathbf{n}}(t) = i\Omega^*_{1,E,\mathbf{n}} \int_0^t dt' \, b_1(t') \exp(i\Delta_E t') = &\frac{i}{\hbar}\langle E_s, \mathbf{n}^-|\hat{\varepsilon}\cdot\mathbf{d}|E_1\rangle \mathcal{E}_0^* \\
&\times \frac{e^{-\eta_{E,s}t}\left\{-\eta_{E,s}\cos\left[(-W)^{1/2}t\right] + (-W)^{1/2}\sin\left[(-W)^{1/2}t\right]\right\} + \eta_{E,s}}{\eta_{E,s}^2 - W},
\end{aligned}
\tag{10.43}
$$

where $\eta_{E,s} \equiv i(\Delta_s/2 - \Delta_E) + \Gamma_s/4\hbar$ and where we have carried out the integral over t' in Eq. (10.43).

When we irradiate at the center of the resonance ($\Delta_s = 0$) and examine the amplitude for populating a continuum energy level at this center [$\Delta_E = 0$, see Eq.

(10.4)], then $\eta_{E,S} = \Gamma_s/4\hbar$, Eq. (10.43) simplifies and the probability of observing the state $|E_s, \mathbf{n}^-\rangle$ is

$$
P_\mathbf{n}(E_s, t) = |b_{E_s,\mathbf{n}}(t)|^2 = \left| \frac{\langle E_s, \mathbf{n}^-|\hat{\varepsilon} \cdot \mathbf{d}|E_1\rangle \mathcal{E}_0}{2\pi\mathcal{E}_0^2 \mathrm{d}_s^2 - \Gamma_s/2} \right|^2
$$

$$
\times \left| e^{-\Gamma_s t/4\hbar} \left\{ -\cos[(-W)^{1/2}t] + \left(\hbar(-W)^{1/2}/\Gamma_s\right) \sin[(-W)^{1/2}t] \right\} + 1 \right|^2.
$$

$$\tag{10.44}$$

Similar formulas were obtained in Refs. [332, 333], providing an analytic expression for the transition from a single level $|E_1\rangle$ to a resonance in the continuum.

We see that $P_\mathbf{n}(E_s, t)$ displays damped Rabi-type oscillations between the initial state and the final continuum states, where the damping is given by resonance *decay rate* $\Gamma_s/2\hbar$. Although the frequency of the Rabi oscillations is a function of the field strength $\mathcal{E}_0$, the branching ratio between channels is independent of the laser parameters, as pointed out, based on general grounds, in Eq. (10.8).

The above equations are only valid when $2\pi\mathcal{E}_0^2 \mathrm{d}_s^2 > \Gamma_s/2$. If $2\pi\mathcal{E}_0^2 \mathrm{d}_s^2 < \Gamma_s/2$, for example, when the field is weak or when the resonance width is very large, as in the flat unstructured continuum situation, another limit applies. Noting that in that case $(-W)^{1/2}$ of Eq. (10.41) becomes imaginary, we make use of the first of Eqs. (10.38), from which it follows that the initial state decays in a monotonic (exponential) fashion. It is easy to show that in that case one obtains the SVCA result of Eq. (10.15).

If there is more than one resonance, we follow the approach in Section 6.3.1, and express the bound-continuum dipole matrix elements $\langle E_1|\hat{\varepsilon} \cdot \mathbf{d}|E, \mathbf{n}^-\rangle$ as a sum of resonances [326]:

$$
\langle E_1|\hat{\varepsilon} \cdot \mathbf{d}|E, \mathbf{n}^-\rangle = \sum_{s=1}^{N} \frac{i\mathrm{d}_{s\mathbf{n}}\Gamma_s/2}{E - E_s + i\Gamma_s/2}.
$$

$$\tag{10.45}$$

The spectrum is now given by

$$
A_1(E) = \sum_\mathbf{n} |\langle E_1|\hat{\varepsilon} \cdot \mathbf{d}|E, \mathbf{n}^-\rangle|^2 = \sum_{s's} \frac{\mathrm{d}_{s's}\Gamma_s\Gamma_{s'}/4}{(E - E_s + i\Gamma_s/2)(E - E_{s'} - i\Gamma_{s'}/2)},
$$

$$\tag{10.46}$$

where $\mathrm{d}_{s's} \equiv \sum_\mathbf{n} \mathrm{d}_{s\mathbf{n}}\mathrm{d}_{s'\mathbf{n}}^*$. If we only keep the diagonal ($s = s'$) terms, $A_1(E)$ becomes a sum of Lorentzians whereas the off-diagonal terms allow for interferences between overlapping resonances. Further, in the large Γ_s limit one obtains the SVCA [191]. An illustration of a typical spectrum obtained from Eq. (10.46) is given in Figure 6.13.

In this multiple resonance case the Fourier transform of $A_1(E)$ [where, in Eq. (10.11), $t \geq t'$], now becomes

$$\exp[i\omega_1(t - t')]F_1(t - t') = 2\pi \sum_s \overline{d_s^2} \exp[-i\chi_s(t - t')] = \sum_s \overline{d_s^2} f_s^+(t) f_s^-(t'),$$

$$(10.47)$$

where

$$\overline{d_s^2} \equiv \sum_{s'} \frac{-i d_{s's} \Gamma_s \Gamma_{s'}/4}{E_s - E_{s'} - i(\Gamma_s + \Gamma_{s'})/2},$$

$$(10.48)$$

and where $f_s^{\pm}$ are defined in Eq. (10.23).

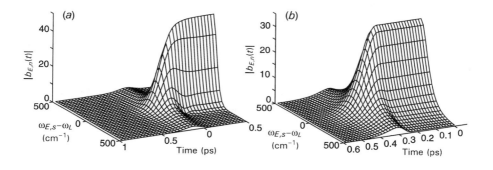

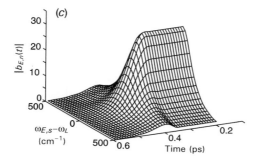

Fig. 10.2 Time dependence of continuum coefficient $b_{E,n}$ for different pulse intensities at center of the absorption spectrum. The spectral width $\Gamma_s = 2000$ cm^{-1}; the laser bandwidth is 120 cm^{-1}. Transition dipole moment is 2.8×10^{-3} a.u. and peak intensity is (a) 0.01 a.u., (b) 0.1 a.u., and (c) 0.5 a.u. (Taken from Fig. 1, Ref. [326].)

Using Eq. (10.47) we can transform Eq. (10.8) into a discrete set of coupled differential equations:

$$\frac{db_1}{dt} = \frac{i}{\hbar}\mathcal{E}(t)\sum_s \overline{d_s^2}f_s^+(t)B_s(t), \tag{10.49a}$$

$$\frac{dB_s}{dt} = \frac{i}{\hbar}\mathcal{E}^*(t)f_s^-(t)b_1(t), \qquad s = 1,\ldots, N, \tag{10.49b}$$

which can be solved in a routine way using a variety of propagation methods. Once $b_1(t)$ is known, the continuum coefficient $b_{E,\mathbf{n}}$ can be computed by straightforward quadrature, using Eq. (10.6).

10.2   COMPUTATIONAL EXAMPLES

Numerical studies allow us to explore aspects of these models for a number of molecular continua and pulse configurations. Consider first the effect of the pulse intensity on transition probabilities to a slowly varying continuum by considering a continuum composed of single broad Lorentzian [Eq. (10.21)] of width $\Gamma_s = 2000$ cm^{-1}, excited by a 120 cm^{-1} wide pulse (i.e., a pulse of ~ 80 fs duration). The central frequency of the pulse is tuned to the center of the continuum ($\Delta_s = 0$) and the pulse peaks at $t = 0$.

In Figure 10.2a we show the $|b_{E,\mathbf{n}}(t)|$ continuum coefficients [Eq. (10.44)] as a function of time, at different intensities. The onset of off-resonance processes is typified by a nonmonotonic behavior: At off-pulse-center energies, the continuum coefficients rise and fall with the pulse, with the effect becoming more pronounced the further away from the line center the continuum energy levels are. In the far wings of the pulse the continuum coefficients are zero at the end of the pulse, giving rise to a pure transient, otherwise known as a *virtual* state. These results should be compared to the weak-field transients discussed in Section 2.1 and shown in Figure 2.2.

As we increase the field strength, the line shapes of the photodissociation amplitudes [given as $|b_{E,\mathbf{n}}(t = \infty)|$] broaden. This broadening is due to saturation of the continuum population, which is greater for continuum states near the pulse center than at the pulse wings. Since a slowly varying continuum is an almost perfect absorber, as the pulse intensity increases the initial state $|E_1\rangle$ empties faster and the dissociation is over before any recurrence can occur. For example, in the 0.01 a.u. peak height case (Fig. 10.2a), the continuum levels reach their final population by the time the pulse peaks (at $t = 0$). This time gets progressively shorter as the field strength is increased.

The situation is quite different for a *structured* continuum. Consider Figure 10.3, where the strong-pulse-induced transition to a narrow continuum ($\Gamma_s = 50$ cm^{-1}) is displayed. The results show behavior that is intermediate between a "flat" continuum and a discrete set of levels. We see that "center-line", $\omega_{E,1} - \omega_1 \approx 0$, continuum levels display recurrences, or Rabi oscillations, similar, though not identical,

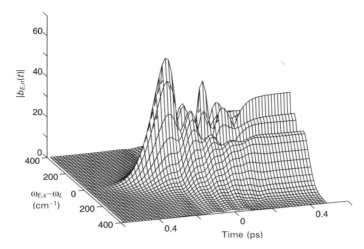

Fig. 10.3 Time dependence of continuum coefficients at center of absorption spectrum for narrow absorption band, $\Gamma_s = 50$ cm^{-1}. Other parameters are: pulse's bandwidth $= 120$ cm^{-1}, peak intensity $= 0.05$ a.u., transition dipole strength 5.7×10^{-5}. (Taken from Fig. 3, Ref. [326].)

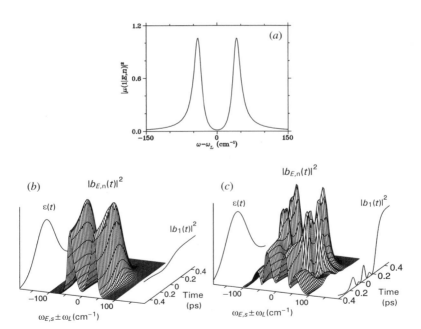

Fig. 10.4 Time dependence of continuum populations for a bound-continuum spectrum comprised of two ($s = 1, 2$) overlapping resonances. (a) The weak-field absorption spectrum. (b) $|b_{E,n}|^2$ as a function of t and E, for laser amplitude $\mathcal{E}_0 = 5 \times 10^{-3}$ a.u. (c) The same as in (b), but for $\mathcal{E}_0 = 5 \times 10^{-2}$ a.u. Also shown are the field envelope $\varepsilon(t)$ as a function of time, as well as $|b_1(t)|^2$. Note the direction of the time axis. (Taken from Fig. 5, Ref. [326].)

to those of a discrete two-level system. By contrast, continuum states at the pulse wings rise and fall smoothly with the pulse, as in the slowly varying continuum case.

Consider now narrow-band continua composed of *two* distinct diffuse features [Eq. (10.45)] (as shown in Fig. 10.4a). In Figures 10.4b and 10.4c as we switch on the pulse, the two initially separated lines begin to slowly merge. For moderate laser powers (Fig. 10.4b), this merging is a signature of saturation: The continuum states at the center of the absorption lines cease to rise while the population of continuum states between the line centers continues to increase. At higher laser intensities (Fig. 10.4c), one sees the effect of the Rabi cycling: The populations of the continuum states at the line centers oscillate at a higher frequency than the populations at the wings of the lines. It may happen, as shown in Figure 10.4c, that continuum states at the line center execute a 2π cycle and are empty at the end of the pulse, whereas continuum states away from the line centers execute only a π cycling and are highly populated. Thus, under the action of the moderately strong laser pulse, the lines can be reversed: The absorption is effectively zero at the resonance (E_s) positions. In addition, the optically induced interference between the lines causes the formation of "dark states" [14, 302, 334, 335], that is, the cancellation of the absorption at the conclusion of the pulse, lying midway between the two line centers. As a result, we see three holes in the continuum populations in Figure 10.4c: Two transparent lines at the center of the resonances due to 2π cycling of these continuum states, accompanied by a third transparent line residing midway between the resonances, due to destructive interference between the resonances. This effect is similar to EIT, discussed in Section 9.1.

CHAPTER 11

COHERENT CONTROL BEYOND THE WEAK-FIELD REGIME: THE CONTINUUM

In this chapter we extend the discussion of control beyond the weak-field regime to consider various transitions involving the continuum. We examine two-photon transitions from a bound state to a continuum (complete population transfer to a continuum), from a continuum to a bound state (photoassociation), and from one continuum to another (laser catalysis).

11.1 CONTROL OVER POPULATION TRANSFER TO THE CONTINUUM BY TWO-PHOTON PROCESSES

In this section we extend the treatment presented in Section 9.1 for three bound states to the case of resonantly enhanced two-photon dissociation. That is, we replace the final state $|E_2\rangle$ by a continuum of states $|E, \mathbf{n}^-\rangle$. The states are renumbered in order of increasing energy, $|E_1\rangle$, $|E_2\rangle$, and $|E, \mathbf{n}^-\rangle$ (see Fig. 10.1b). If the behavior of such a system is similar to that of three-level systems, then complete population transfer to the continuum, hence complete ionization or dissociation, is possible [191].

As in Section 9.1, we consider a molecule, initially ($t = 0$) in state $|E_1\rangle$, being excited to a continuum of states $|E, \mathbf{n}^-\rangle$, due to the combined action of *two* laser pulses of central frequencies ω_1 and ω_2. We assume, as depicted in Figure 10.1b, that ω_1 is in near resonance with the transition from $|E_1\rangle$ to an intermediate bound state $|E_2\rangle$ and that ω_2 is in near resonance with the transition from $|E_2\rangle$ to the continuum.

The total matter–radiation Hamiltonian is given in Eq. (9.2). We then solve the Schrödinger equation by expanding the total wave function as

$$|\Psi(t)\rangle = b_1(t)|E_1\rangle \exp\left(-\frac{iE_1 t}{\hbar}\right) + b_2(t)|E_2\rangle \exp\left(-\frac{iE_2 t}{\hbar}\right)$$
$$+ \sum_{\mathbf{n}} \int dE\, b_{E,\mathbf{n}}(t)|E, \mathbf{n}^-\rangle \exp\left(-\frac{iEt}{\hbar}\right). \tag{11.1}$$

and obtain a set of first-order differential equations, which is now of the form

$$\frac{db_1}{dt} = -\Omega_1^*(t) \exp(-i\Delta_1 t)b_2(t), \tag{11.2a}$$

$$\frac{db_2}{dt} = i\Omega_1(t) \exp(i\Delta_1 t)b_1(t) + i \int dE \sum_{\mathbf{n}} \Omega_{2,E,\mathbf{n}}(t) \exp(-i\Delta_E t)b_{E,\mathbf{n}}(t), \tag{11.2b}$$

$$\frac{db_{E,\mathbf{n}}}{dt} = i\Omega_{2,E,\mathbf{n}}^*(t) \exp(i\Delta_E t)b_2(t), \quad \text{for all } E \text{ and } \mathbf{n}, \tag{11.2c}$$

where

$$\Omega_1(t) \equiv \langle E_2|\mathbf{d}_1 \cdot \hat{\varepsilon}_1|E_1\rangle \mathcal{E}_1(t)/\hbar, \qquad \Omega_{2,E,\mathbf{n}}(t) \equiv \langle E_2|\mathbf{d}_2 \cdot \hat{\varepsilon}_2|E, \mathbf{n}^-\rangle \mathcal{E}_2(t)/\hbar,$$
$$\Delta_1 \equiv (E_2 - E_1)/\hbar - \omega_1, \qquad \Delta_E \equiv (E - E_2)/\hbar - \omega_2. \tag{11.3}$$

We can eliminate the continuum equations by substituting the formal solution of Eq. (11.2c),

$$b_{E,\mathbf{n}}(t) = i \int_0^t dt'\, \Omega_{2,E,\mathbf{n}}^*(t') \exp(i\Delta_E t')b_2(t'), \tag{11.4}$$

into Eq. (11.2b) to obtain

$$\frac{db_2}{dt} = i\Omega_1(t) \exp(i\Delta_1 t)b_1(t)$$
$$- \sum_{\mathbf{n}} \int dE\, \Omega_{2,E,\mathbf{n}}(t) \exp(-i\Delta_E t) \int_0^t dt'\, \Omega_{2,E,\mathbf{n}}^*(t') \exp(i\Delta_E t')b_2(t'). \tag{11.5}$$

By invoking the slowly varying continuum approximation (SVCA) [Eq. (10.13)] we obtain the two-photon analog of Eq. (10.15):

$$\frac{db_2}{dt} = i\Omega_1(t) \exp(i\Delta_1 t)b_1(t) - \Omega_2^I(t)b_2(t), \tag{11.6}$$

where

$$\Omega_2^I(t) = \pi \sum_{\mathbf{n}} |\langle E_L, \mathbf{n}^-|\mathbf{d}_2 \cdot \hat{\varepsilon}_2|E_2\rangle \mathcal{E}_2(t)|^2/\hbar. \tag{11.7}$$

Coupled with Eq. (11.2c), Eq. (11.6) gives a closed set of equations for $b_1(t)$ and $b_2(t)$.

11.1.1 Adiabatic Approximation for Final Continuum Manifold

Equations (11.2a) and (11.6) can be expressed as a 2×2 version of Eq. (9.8) where in this case

$$\mathbf{b} \equiv [\exp(i\Delta_1 t)b_1, b_2]^{\mathrm{T}}, \tag{11.8}$$

and

$$\underline{\underline{H}} = \begin{pmatrix} \Delta_1 & \Omega_1^* \\ \Omega_1 & i\Omega_2^I \end{pmatrix}. \tag{11.9}$$

Assuming that Ω_1 is real, we obtain the adiabatic solutions to Eq. (9.8) by diagonalizing the $\underline{\underline{U}}$ matrix, as in Eq. (9.10).

The presence of the continuum, coupled with the SVCA, results in a complex-symmetric $\underline{\underline{H}}$ matrix. Such matrices are diagonalizable using complex-orthogonal matrices $\underline{\underline{U}}$, satisfying

$$\underline{\underline{U}}(t) \cdot \underline{\underline{U}}^{\mathrm{T}}(t) = \underline{\underline{I}}. \tag{11.10}$$

Note that $\underline{\underline{U}}$ must be nonunitary on physical grounds in order to allow flux loss to the continuum.

In contrast to the three-level case discussed in Section 9.1, the 2×2 complex-orthogonal $\underline{\underline{U}}$ matrix obtained here can be parameterized in terms of a single complex "mixing angle" α, where

$$\underline{\underline{U}} = \begin{pmatrix} \cos \alpha & \sin \alpha \\ -\sin \alpha & \cos \alpha \end{pmatrix} \tag{11.11}$$

and where

$$\alpha(t) = \frac{1}{2}\arctan\left(\frac{2\Omega_1}{i\Omega_2^I - \Delta_1}\right). \tag{11.12}$$

Operating with $\underline{\underline{U}}^{\mathrm{T}}(t)$ on Eq. (9.8), and defining

$$\mathbf{a}(t) = \underline{\underline{U}}^{\mathrm{T}}(t) \cdot \mathbf{b}(t), \tag{11.13}$$

we obtain that

$$\frac{d}{dt}\mathbf{a} = \{i\underline{\underline{\lambda}}(t) + \underline{\underline{A}}\} \cdot \mathbf{a}, \tag{11.14}$$

with

$$\underline{\underline{A}} \equiv \frac{d\underline{\underline{U}}^T(t)}{dt} \cdot \underline{\underline{U}} = \begin{pmatrix} 0 & \dot{\alpha} \\ -\dot{\alpha} & 0 \end{pmatrix}. \tag{11.15}$$

Once again, the adiabatic solutions are given by

$$\mathbf{a}(t) = \exp\left\{i\int_0^t \underline{\underline{\lambda}}(t')\,dt'\right\}\mathbf{a}(0), \tag{11.16}$$

and with the elements of the diagonal eigenvalue matrix, $\underline{\underline{\lambda}}$, given this time as

$$\lambda_{1,2} = \tfrac{1}{2}\{\Delta_1 + i\Omega_2^I \pm [(\Delta_1 - i\Omega_2^I)^2 + 4|\Omega_1|^2\}. \tag{11.17}$$

Using Eqs. (11.8) and (11.13), and imposing the initial condition, $\mathbf{b}(0) = (1, 0)$, we obtain for the $b_1(t)$ and $b_2(t)$ coefficients,

$$b_1(t) = \left\{ U_{1,1}(t)\exp\left[i\int_0^t \lambda_1(t')\,dt'\right]U_{1,1}(0) \right.$$
$$\left. + U_{2,1}(t)\exp\left[i\int_0^t \lambda_2(t')\,dt'\right]U_{2,1}(0) \right\}\exp(-i\Delta_1 t),$$
$$b_2(t) = U_{1,2}(t)\exp\left[i\int_0^t \lambda_1(t')\,dt'\right]U_{1,1}(0) + U_{2,2}(t)\exp\left[i\int_0^t \lambda_2(t')\,dt'\right]U_{2,1}(0). \tag{11.18}$$

If both lasers are assumed to be off initially, that is, $\mathcal{E}_1(0) = \mathcal{E}_2(0) = 0$, we have that $\alpha(0) = 0$. Hence $U_{1,1}(0) = 1$, $U_{2,1}(0) = 0$, and

$$\begin{pmatrix} b_1(t) \\ b_2(t) \end{pmatrix} = \begin{pmatrix} \exp(-i\Delta_1 t)\cos\alpha(t) \\ \sin\alpha(t) \end{pmatrix}\exp\left\{i\int_0^t \lambda_1(t')\,dt'\right\}. \tag{11.19}$$

Once $b_2(t)$ is known, the continuum coefficients $b_{E,\mathbf{n}}(t)$ are obtained directly via Eq. (11.4).

As an example of this process, we consider the resonantly enhanced two-photon dissociation, by two laser pulses, of a molecule (e.g., Na_2) where the energy gap between the two bound states, $E_2 - E_1 = 20{,}000$ cm^{-1}, and the gap between E_2 and the continuum is $19{,}000$ cm^{-1}.

Assuming that $\mathbf{a}(t)$ is available, either via the adiabatic approximation or via an "exact" numerical computation, the probabilities to observe $|E_1\rangle$ and $|E_2\rangle$ as a function of time are given by

$$
\begin{aligned}
P_1 &\equiv |b_1|^2 = |a_1(t)\cos\alpha(t) + a_2(t)\sin\alpha(t)|^2, \\
P_2 &\equiv |b_2|^2 = |-a_1(t)\sin\alpha(t) + a_2(t)\cos\alpha(t)|^2,
\end{aligned}
\tag{11.20}
$$

and the overall dissociation probability P_d is

$$
P_d \equiv 1 - (P_1 + P_2).
\tag{11.21}
$$

To examine the accuracy of the adiabatic approximation, we compare, in Figure 11.1, the zeroth-order adiabatic solution for the probabilities with the exact numerical integration and with an iterative method [325] of solving Eq. (11.2). For this relatively slowly varying pulse and large detuning, the adiabatic solution is essentially exact. However, when going to shorter pulses, the adiabatic approximation fails to properly partition the probability, at long time, between levels $|E_1\rangle$ and $|E_2\rangle$. Further, it fails to display the oscillations at intermediate times, which are a result of the interference between the two adiabatic channels. This type of failure is expected because the shortening of the pulse induces rapid variation in the diagonalizing transformation, resulting in the breakdown of the adiabatic condition.

A different type of failure of the adiabatic approximation, unique to the case where the final state is in the continuum, is demonstrated in Figure 11.2. In this case the detuning Δ_1 is decreased with respect to the intermediate level to 1 cm^{-1}, and the pulse duration is kept at 100 ps. Clearly, the adiabatic approximation fails to display the oscillations at intermediate times that result from the interference between two adiabatic channels. A reduction in detuning of the intermediate level does not affect the bound state adiabatic condition, $\Omega^0\Delta\tau \gg 1$, which appears to hold (as suggested by the large number of oscillations during pulse time). However, the more exact statement of the adiabatic condition [Eq. (9.40)] shows that the breakdown of the adiabatic approximation is due to the near divergence of the complex mixing angle α of Eq. (11.12) at small Δ_1, the complex nature of which is a result of the presence of the final continuum.

11.2 PULSED INCOHERENT INTERFERENCE CONTROL

An extension of the three-level scheme discussed in the last section, through the addition of a third laser frequency, results in the pulsed version of incoherent interference control, a scenario introduced in Section 5.4.1. In that scenario two cw sources coupled two bound states to the continuum, resulting in control over the branching between final fragment channels, that is, control over the selectivity of the process. The approach was particularly interesting because it was shown to be insensitive to the relative phase between the two light sources and hence did not

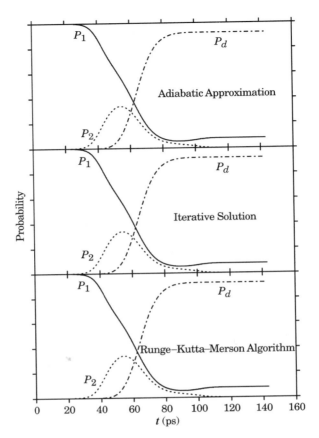

Figure 11.1 Populations of initial (P_1), intermediate (P_2), and continuum (P_d) states vs. time, in the zero-order (adiabatic) approximation (upper), and the exact solution, obtained by iterative corrections to the adiabatic approximation [325] (middle), and by direct numerical intergration (lower). Pulses, lasting 100 ps, were coincident and detuned by $\Delta_1 = 10$ cm^{-1}. (Taken from Fig. 2, Ref. [325].)

require coherent lasers. Studies of cw-induced controlled photodissociation are, however, relatively limited in scope. The fact that the laser is always on prevents an analysis of the effects of laser parameters, such as the relative pulse orderings and the pulse shapes, which characterize pulsed sources, on the yield. In this section we extend the theory of incoherent interference control to the case of laser pulses and consider both selectivity of different final photodissociation channels, and conditions necessary to achieve high photodissociation yield.

Note that the approach introduced below relies exclusively on the computation of material matrix elements, as in the weak-field domain. As a result, one need only compute these matrix elements once in order to obtain dissociation rates and probabilities for a variety of pulse configurations and field strengths.

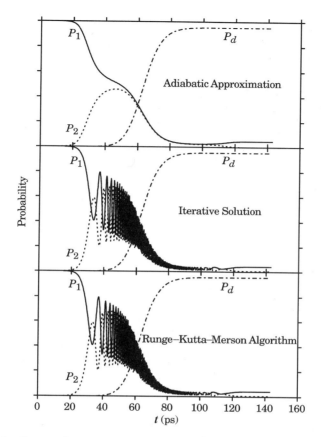

Figure 11.2   Same as in Fig. 11.1 for $\Delta_1 = 1$ cm^{-1}. (Taken from Fig. 5, Ref. [325].)

As shown in Figure 11.3, we consider a molecule (in this case Na$_2$), initially in state $|E_0\rangle$, which is excited by the combined action of two laser pulses, with central frequencies ω_0 and ω_1, to a continuum of states associated with two or more different product channels at energy E. The frequency ω_0 is assumed to be in near resonance with the transition to an intermediate bound state $|E_1\rangle$, and ω_1 is in near resonance with the transition frequency between $|E_1\rangle$ and the continuum. Thus ω_0 carries the system from $|E_0\rangle$ to $|E_1\rangle$, and ω_1 carries the system from $|E_1\rangle$ to the continuum. To avoid confusion, note that the energy levels are labeled somewhat differently from that in the previous section.

In addition, as shown in Figure 11.3, the continuum is coupled to a third bound state $|E_2\rangle$ by a laser of central frequency ω_2. Basically, the three-level two-laser scheme described in Section 11.1 is being extended here to a four-level three-laser control scheme.

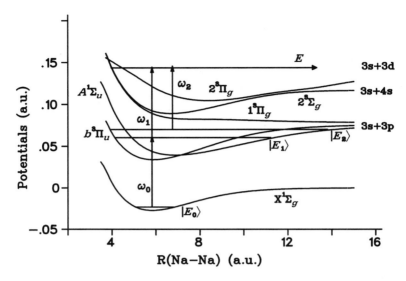

Figure 11.3   Incoherent interference control (IIC) scheme and potential energy curves for $Na_2 \rightarrow Na + Na(3d)$, Na(4s), Na(3p). In this scheme an $\omega_0 + \omega_1$ photon excitation to the continuum interferes with an ω_2 photon from an initially unpopulated state. Two-photon absorption proceeds from an initial state, $|E_0\rangle$ (in Na_2 it is taken to be the $v = 5, J = 37$ state), via the $|E_1\rangle$ ($v = 35$, $J = 36, 38$) intermediate resonance belonging to the interacting $A^1\Sigma_u/^3\Pi_u$ electronic states. The ω_2 photon couples the continuum to the (initially unpopulated) $|E_2\rangle$ ($v = 93$, $J = 36$ or $v = 93$, $J = 38$) level of the $A^1\Sigma_u/^3\Pi_u$ electronic states. (Taken from Fig. 1, Ref. [201].)

The total Hamiltonian of the system is now given by

$$H_{\text{tot}} = H_M - 2 \, \text{Re}\{\mathbf{d}_0 \cdot \hat{\varepsilon}_0 \mathcal{E}_0(t) \exp(i\omega_0 t) + \mathbf{d}_1 \cdot \hat{\varepsilon}_1 \mathcal{E}_1(t) \exp(i\omega_1 t)$$
$$+ \mathbf{d}_2 \cdot \hat{\varepsilon}_2 \mathcal{E}_2(t) \exp(i\omega_2 t)\}, \tag{11.22}$$

where H_M is the molecular Hamiltonian. The solution to the time-dependent Schrödinger equation is obtained by expanding the total wave function $\Psi(t)$ in a basis set, this time composed of three bound states and a set of continuum states, $|E, \mathbf{n}^-\rangle$, as

$$|\Psi(t)\rangle = b_0|E_0\rangle \exp(-iE_0t/\hbar) + b_1|E_1\rangle \exp(-iE_1t/\hbar) + b_2|E_2\rangle \exp(-iE_2t/\hbar)$$
$$+ \sum_{\mathbf{n}} \int dE \, b_{E,\mathbf{n}}(t)|E, \mathbf{n}^-\rangle \exp\left(-\frac{iEt}{\hbar}\right). \tag{11.23}$$

Here $\mathbf{n} = \{\mathbf{m}, q\}$ where $q = 1, 2 \ldots$ denotes the product arrangement channel and $\mathbf{m}$ denotes the remaining quantum numbers other than the energy.

Substituting Eq. (11.23) into the time-dependent Schrödinger equation and using the orthogonality of the basis functions results in a set of first-order differential equations for the expansion coefficients, which, in the rotating-wave approximation, is given by

$$i\hbar \frac{db_0}{dt} = -\mathcal{E}_0(t) d_{0,1} \exp(-i\Delta_1 t) b_1(t), \tag{11.24}$$

$$i\hbar \frac{db_1}{dt} = -\mathcal{E}_0^*(t) d_{1,0} \exp(i\Delta_1 t) b_0(t)$$
$$- \mathcal{E}_1(t) \sum_{\mathbf{n}} \int dE\, d(1|E, \mathbf{n}) \exp(-i\Delta_{E,1} t) b_{E,\mathbf{n}}(t), \tag{11.25}$$

$$i\hbar \frac{db_2}{dt} = -\mathcal{E}_2(t) \sum_{\mathbf{n}} \int dE\, d(2|E, \mathbf{n}) \exp(-i\Delta_{E,2} t) b_{E,\mathbf{n}}(t), \tag{11.26}$$

$$i\hbar \frac{db_{E,\mathbf{n}}}{dt} = -\mathcal{E}_1^*(t) d(E, \mathbf{n}|1) \exp(i\Delta_{E,1} t) b_1(t)$$
$$- \mathcal{E}_2^*(t) d(E, \mathbf{n}|2) \exp(i\Delta_{E,2} t) b_2(t), \tag{11.27}$$

where

$$
\begin{array}{llll}
d_{1,0} \equiv \langle E_1 | \mathbf{d}_0 \cdot \hat{\varepsilon}_0 | E_0 \rangle, & d(E, \mathbf{n}|i) \equiv \langle E, \mathbf{n}^- | \mathbf{d}_i \cdot \hat{\varepsilon}_i | E_i \rangle, & i = 1, 2, \\
\Delta_1 \equiv (E_1 - E_0)/\hbar - \omega_0, & \Delta_{E,i} \equiv (E - E_i)/\hbar - \omega_i, & i = 1, 2.
\end{array}
\tag{11.28}
$$

Equation (11.27) shows that the contribution to the continuum amplitude $b_{E,\mathbf{n}}$ derives from excitations from levels $|E_1\rangle$ and $|E_2\rangle$. Below we assume $b_1(0) = b_2(0) = 0$. Hence state $|E_2\rangle$ is populated by the ω_2 pulse via the continuum state that was itself populated by the ω_1 pulse via $|E_1\rangle$. The two lowest order routes to dissociation are $|E_0\rangle \to |E_1\rangle \to |E, \mathbf{n}^-\rangle$ and $|E_0\rangle \to |E_1\rangle \to |E, \mathbf{n}^-\rangle \to |E_2\rangle \to |E, \mathbf{n}'^-\rangle$. Contributions from these, and higher order terms, provide multiple pathways to the product in a given channel at energy E, which interfere either constructively or destructively with one another, and affords the opportunity for quantum control.

To consider the nature of solutions to Eqs. (11.25) to (11.27), we substitute the formal solution of Eq. (11.27):

$$b_{E,\mathbf{n}}(t) = \frac{-1}{i\hbar} \left\{ d(E, \mathbf{n}|1) \int_0^t dt'\, \mathcal{E}_1^*(t') \exp(i\Delta_{E,1} t') b_1(t') \right.$$
$$\left. + d(E, \mathbf{n}|2) \int_0^t dt'\, \mathcal{E}_2^*(t') \exp(i\Delta_{E,2} t') b_2(t') \right\}, \tag{11.29}$$

into Eqs. (11.26) and (11.27) to obtain

$$
i\hbar \frac{db_1}{dt} = -d_{1,0}\mathcal{E}_0^*(t)\exp(i\Delta_1 t)b_0(t)
$$
$$
+ \frac{\mathcal{E}_1(t)}{i\hbar}\sum_{\mathbf{n}}\int dE |d(E,\mathbf{n}|1)|^2 \exp(-i_{E,1}t)\int_0^t dt'\, \mathcal{E}_1^*(t')\exp(i\Delta_{E,1}t')b_1(t')
$$
$$
+ \frac{\mathcal{E}_1(t)}{i\hbar}\sum_{\mathbf{n}}\int dE\, d(1|E,\mathbf{n})d(E,\mathbf{n}|2)\exp(-i\Delta_{E,1}t)
$$
$$
\times \int_0^t dt'\, \mathcal{E}_2^*(t')\exp(i\Delta_{E,2}t')b_2(t'), \tag{11.30}
$$

$$
i\hbar \frac{db_2}{dt} = \frac{\mathcal{E}_2(t)}{i\hbar}\sum_{\mathbf{n}}\int dE\, d(2|E,\mathbf{n})d(E,\mathbf{n}|1)\exp(-i\Delta_{E,2}t)\int_0^t dt'\, \mathcal{E}_1^*(t')\exp(i\Delta_{E,1}t')b_1(t')
$$
$$
+ \frac{\mathcal{E}_2(t)}{i\hbar}\sum_{\mathbf{n}}\int dE |d(E,\mathbf{n}|2)|^2 \exp(-i\Delta_{E,2}t)\int_0^t dt'\, \mathcal{E}_2^*(t')\exp(i\Delta_{E,2}t')b_2(t'). \tag{11.31}
$$

Invoking the slowly varying continuum approximation (SVCA) [Eq. (10.13)],

$$
\sum_{\mathbf{n}} d(i|E,\mathbf{n})d(E,\mathbf{n}|j) \approx \sum_{\mathbf{n}} d(i|E_i + \hbar\omega_i, \mathbf{n})d(E_j + \hbar\omega_j, \mathbf{n}|j) \equiv \frac{\mathcal{I}_{i,j}}{\pi}, \qquad i,j = 1,2, \tag{11.32}
$$

we have that

$$
\int dE \sum_{\mathbf{n}} d(i|E,\mathbf{n})d(E,\mathbf{n}|j)\exp[-i\Delta_{E,i}(t-t')] = 2\hbar\mathcal{I}_{i,j}\delta(t-t'). \tag{11.33}
$$

Utilizing the SVCA greatly simplies Eqs. (11.29) and (11.30), giving

$$
i\hbar \frac{db_1}{dt} = -d_{1,0}\mathcal{E}_0^*(t)\exp(i\Delta_1 t)b_0(t) - i|\mathcal{E}_1(t)|^2 \mathcal{I}_{1,1}b_1(t)
$$
$$
- i\mathcal{E}_1(t)\mathcal{E}_2^*(t)\mathcal{I}_{1,2}\exp(-i\Delta t)b_2(t), \tag{11.34}
$$
$$
i\hbar \frac{db_2}{dt} = -i\mathcal{E}_2(t)\mathcal{E}_1^*(t)\mathcal{I}_{2,1}\exp(i\Delta t)b_1(t) - i|\mathcal{E}_2(t)|^2 \mathcal{I}_{2,2}b_2(t), \tag{11.35}
$$

where $\Delta \equiv \Delta_{E,1} - \Delta_{E,2} = (E_2 - E_1)/\hbar + \omega_2 - \omega_1$. Note that in obtaining these results the factor of two from Eq. (11.33) has been canceled by a factor of one-half, which arises from evaluating the integrals of $\delta(t-t')$ at the integration end point $t' = t$.

The resultant coupled equations are still not in a convenient form to solve numerically, or to examine from the point of view of the adiabatic approximation. To this end we define a modified coefficient three-vector:

$$\mathbf{c} \equiv (c_0, c_1, c_2)^{\mathbf{T}} \equiv [\exp(i\Delta_1 t)b_0, b_1, b_2 \exp(-i\Delta t)]^{\mathbf{T}}, \tag{11.36}$$

where the superscript $\mathbf{T}$ denotes the transpose. Equations (11.35) and (11.25) can then be rewritten as the matrix equation:

$$\frac{d\mathbf{c}(t)}{dt} = \frac{i}{\hbar}\underline{\underline{f}}(t) \cdot \mathbf{c}(t), \tag{11.37}$$

where

$$\underline{\underline{f}}(t) = \begin{pmatrix} \hbar\Delta_1 & d_{0,1}\mathcal{E}_0 & 0 \\ d_{1,0}\mathcal{E}_0^* & i|\mathcal{E}_1|^2\mathcal{I}_{1,1} & i\mathcal{E}_1\mathcal{E}_2^*\mathcal{I}_{1,2} \\ 0 & i\mathcal{E}_1^*\mathcal{E}_2\mathcal{I}_{2,1} & i|\mathcal{E}_2|^2\mathcal{I}_{2,2} - \hbar\Delta \end{pmatrix}, \tag{11.38}$$

and the explicit time dependence of the field envelopes has been omitted for brevity. By writing

$$d_{0,1} = |d_{0,1}|\exp(i\phi_{0,1}), \qquad \mathcal{I}_{1,2} = |\mathcal{I}_{1,2}|\exp(i\phi_{1,2}),$$
$$\mathcal{E}_i(t) = |\mathcal{E}_i(t)|\exp[i\theta_i(t)], \qquad i = 0, 1, 2, \tag{11.39}$$

we can factorize the $\underline{\underline{f}}(t)$ matrix as

$$\underline{\underline{f}}(t) = \underline{\underline{\hat{e}}}^{-1}(t) \cdot \underline{\underline{g}}(t) \cdot \underline{\underline{\hat{e}}}(t), \tag{11.40}$$

where $\underline{\underline{\hat{e}}}(t)$ is a diagonal matrix containing the phase factors:

$$\underline{\underline{\hat{e}}}(t) = \begin{pmatrix} 1 & 0 & 0 \\ 0 & \exp[i(\phi_{0,1} + \theta_0)] & 0 \\ 0 & 0 & \exp[i(\phi_{1,2} + \phi_{0,1} + \theta_0 + \theta_1 - \theta_2)] \end{pmatrix}, \tag{11.41}$$

and

$$\underline{\underline{g}}(t) = \begin{pmatrix} \hbar\Delta_1 & |d_{0,1}\mathcal{E}_0| & 0 \\ |d_{0,1}\mathcal{E}_0| & i|\mathcal{E}_1|^2\mathcal{I}_{1,1} & i|\mathcal{E}_1\mathcal{E}_2\mathcal{I}_{1,2}| \\ 0 & i|\mathcal{E}_1\mathcal{E}_2\mathcal{I}_{1,2}| & i|\mathcal{E}_1|^2\mathcal{I}_{2,2} - \hbar\Delta \end{pmatrix}. \tag{11.42}$$

The $\underline{\underline{\hat{e}}}$ matrix contains all the information about the laser phases.

We now introduce a new three-component vector defined as

$$\mathbf{a}(t) \equiv (a_0, a_1, a_2)^{\mathrm{T}} = \hat{\underline{\mathbf{e}}}(t) \cdot \mathbf{c}(t), \tag{11.43}$$

and combine Eq. (11.43) with Eq. (11.37) to obtain that

$$\frac{d\mathbf{a}(t)}{dt} = \frac{i}{\hbar} \{\underline{\mathbf{g}} + \underline{\dot{\theta}}\} \cdot \mathbf{a}(t), \tag{11.44}$$

where $\underline{\dot{\theta}}$ is a diagonal time-derivative phase matrix

$$\underline{\dot{\theta}} = \hbar \begin{pmatrix} 0 & 0 & 0 \\ 0 & \dot{\theta}_0 & 0 \\ 0 & 0 & \dot{\theta}_0 + \dot{\theta}_1 - \dot{\theta}_2 \end{pmatrix}, \tag{11.45}$$

where $\dot{\theta} \equiv (d/dt)\theta$. Note that were it not for the variation of the laser phases with time, embodied in the $\underline{\dot{\theta}}$ matrix, $\mathbf{a}$ would be independent of the laser phases.

Using Eqs. (11.36) and (11.43), we can express the full expansion coefficients b_1 and b_2 in terms of a_1 and a_2 as

$$b_1 = c_1 = a_1 \exp[-i(\phi_{0,1} + \theta_0)], \tag{11.46}$$
$$b_2 = c_2 \exp(i\Delta t) = a_2 \exp[-i(\phi_{0,1} + \theta_0 + \phi_{1,2} + \theta_1 - \theta_2 - \Delta t)]. \tag{11.47}$$

After substituting these expressions for b_1 and b_2 into Eq. (11.27), we obtain an explicit expression for the continuum coefficient:

$$i\hbar \frac{db_{E,\mathbf{n}}}{dt} = -[\mathrm{d}(E, \mathbf{n}|1)|\mathcal{E}_1(t)|a_1(t) - \mathrm{d}(E, \mathbf{n}|2)|\mathcal{E}_2(t)|e^{i\phi_{1,2}} a_2(t)]e^{i\Delta_{E,1}t - i(\phi_{0,1} + \theta_0 + \theta_1)} \tag{11.48}$$

The total product probability P_q of producing product in channel q is then given as

$$P_q = \sum_{\mathbf{m}} \int dE \, P_{\mathbf{m},q}(E) = \sum_{\mathbf{m}} \int dE |b_{E,\mathbf{m},q}(t \to \infty)|^2. \tag{11.49}$$

The desired dissociation probability can therefore be found by first solving Eqs. (11.44) and (11.48) and inserting the result into Eq. (11.49). If we assume that $\dot{\theta}$ is time independent, then the $\underline{\dot{\theta}}$ term drops out of Eq. (11.44) and it follows from Eq. (11.48) that the $b_{E,\mathbf{n}}$ coefficients become independent of the ω_2 laser phase θ_2. Qualitatively this is so because, as discussed in Section 5.4.1, in the $|E_0\rangle \to |E_1\rangle \to |E, \mathbf{n}^-\rangle \to |E_2\rangle \to |E, \mathbf{n}'^-\rangle$ excitation pathway, the stimulated emission of an ω_2 photon in the $|E, \mathbf{n}^-\rangle \to |E_2\rangle$ step contributes, in accord with Section 5.4.1, a factor of $\exp[-i\theta_2]$ to the molecular wave function that exactly cancels the $\exp[i\theta_2]$ phase factor that accompanies the absorption of the same frequency photon in the $|E_2\rangle \to |E, \mathbf{n}'^-\rangle$ step.

In addition to the loss of the θ_2 phase dependence, when the laser-phase matrix is time independent, the dependence on the phases of the other lasers is also lost as well. This is because the time-independent $\exp[-i(\phi_{0,1} + \theta_0 + \theta_1)]$ term factors out of the expression for $b_{E,\mathbf{n}}$ in Eq. (11.48), leaving unity when $|b_{E,\mathbf{n}}|^2$ is computed. As noted above, when the laser-phase matrix is time independent, $\mathbf{a}(t)$ of Eq. (11.44) becomes independent of all laser phases. Hence, using Eq. (11.49), it follows that in that case P_q is totally independent of the phases of all three lasers. As a consequence, for time-independent laser phases, we can control branching reactions in this scenario by using rather *incoherent* laser sources, a result similar to the situation with cw sources (see [309] and Section 5.4.2), hence the name "incoherent interference control" (IIC). However, as a related consequence, the laser phase is no longer a useful control variable. Rather, the control knob in IIC is the $\omega_2 - \omega_1$ difference, and to a smaller extent the laser intensities, pulse widths, and pulse orderings.

When the laser phases are time dependent, we can gain insight into the phase dependence of the control by investigating the adiabatic states. We follow to some extent the analysis of Kobrak and Rice [310] who have considered a similar, though not identical, system in which population transfer between two bound states proceeds via an intermediate continuum, dominated by a single resonance. Rather than diagonalizing the entire $\underline{\underline{g}} + \dot{\underline{\underline{\theta}}}$ matrix of Eq. (11.44) (as Kobrak and Rice have done in their 3×3 case), which would yield highly complicated expressions, we only diagonalize the lower-right 2×2 block of the $\underline{\underline{g}} + \dot{\underline{\underline{\theta}}}$ matrix.

We transform the $\underline{\underline{g}} + \dot{\underline{\underline{\theta}}}$ matrix of Eq. (11.42) as

$$\underline{\underline{h}}(t) = \underline{\underline{U}}^{\mathbf{T}}(t) \cdot \{\underline{\underline{g}} + \dot{\underline{\underline{\theta}}}\} \cdot \underline{\underline{U}}(t), \tag{11.50}$$

where

$$\underline{\underline{U}} = \begin{pmatrix} 1 & 0 & 0 \\ 0 & & \\ 0 & & \underline{\underline{u}} \end{pmatrix}, \tag{11.51}$$

with $\underline{\underline{u}}$ being a 2×2 complex-orthogonal matrix, parametrized as in Eq. (11.11) by a complex mixing angle α. Here $\underline{\underline{u}}$ is the diagonalizing transformation of the 2×2 lower block of the $\underline{\underline{g}} + \dot{\underline{\underline{\theta}}}$ matrix, which we denote $\underline{\underline{r}}(t)$:

$$\underline{\underline{r}}(t) = \begin{pmatrix} i|\mathcal{E}_1|^2 \mathcal{I}_{1,1} + \hbar\dot{\theta}_0 & i|\mathcal{E}_1 \mathcal{E}_2 \mathcal{I}_{1,2}| \\ i|\mathcal{E}_1 \mathcal{E}_2 \mathcal{I}_{1,2}| & i|\mathcal{E}_2|^2 \mathcal{I}_{2,2} - \hbar(\Delta - \dot{\theta}_0 - \dot{\theta}_1 + \dot{\theta}_2) \end{pmatrix}. \tag{11.52}$$

We have that

$$\underline{\underline{u}}^{\mathbf{T}} \cdot \underline{\underline{r}} \cdot \underline{\underline{u}} = \underline{\underline{\lambda}}, \tag{11.53}$$

with $\underline{\lambda}$ being the 2×2 eigenvalue matrix with diagonal elements

$$
\begin{aligned}
\lambda_{1,2} = \frac{1}{2} \{ & i|\mathcal{E}_1|^2 \mathcal{I}_{1,1} + i|\mathcal{E}_2|^2 \mathcal{I}_{2,2} + \hbar[2\dot\theta_0 + \dot\theta_1 - \dot\theta_2 - \Delta] \\
& \pm [(i|\mathcal{E}_1|^2 \mathcal{I}_{1,1} - i|\mathcal{E}_2|^2 \mathcal{I}_{2,2} - \hbar[\dot\theta_1 - \dot\theta_2 - \Delta])^2 - 4|\mathcal{E}_1 \mathcal{E}_2 \mathcal{I}_{1,2}|^2]^{1/2} \}.
\end{aligned}
\tag{11.54}
$$

With the above definitions of $\underline{U}$ and $\underline{u}$, the transformation of $\underline{g} + \underline{\dot\theta}$ [Eq. (11.50)] results in the following matrix:

$$
\underline{h}(t) = \begin{pmatrix}
\hbar\Delta_1 & \cos\alpha|d_{0,1}\mathcal{E}_0| & \sin\alpha|d_{0,1}\mathcal{E}_0| \\
\cos\alpha|d_{0,1}\mathcal{E}_0| & \lambda_1 & 0 \\
\sin\alpha|d_{0,1}\mathcal{E}_0| & 0 & \lambda_2
\end{pmatrix}.
\tag{11.55}
$$

Allowing the $\underline{U}$ matrix to operate on the expansion coefficients vectors, $\mathbf{a}' \equiv \underline{\underline{U}} \cdot \mathbf{a}$, has the effect of transforming Eqs. (9.8), which describes the time evolution of the system, to:

$$
\frac{d}{dt}\mathbf{a}'(t) = \frac{i}{\hbar} \{\underline{\underline{h}}(t) + \underline{\underline{A}}\} \cdot \mathbf{a}'(t),
\tag{11.56}
$$

where the nonadiabatic coupling matrix $\underline{A}$ is defined in Eq. (9.13). Assuming that the $\mathcal{E}_0$ pulse (whose sole purpose is to excite $|E_0\rangle$ to $|E_1\rangle$, and in no way to control the process) is weak, and assuming the adiabatic approximation (in which we neglect $\underline{A}$), we can easily solve Eq. (11.56) to obtain that

$$
a_0'(t) = \exp(i\Delta_1 t)a_0'(0),
\tag{11.57a}
$$

$$
a_1'(t) = \frac{i}{\hbar}\cos\alpha|d_{0,1}\mathcal{E}_0|a_0'(0) \int_0^t dt' \exp[i\Delta_1 t' + i\int_0^{t'} dt'' \; \lambda_1(t'')],
\tag{11.57b}
$$

$$
a_2'(t) = \frac{i}{\hbar}\sin\alpha|d_{0,1}\mathcal{E}_0|a_0'(0) \int_0^t dt' \exp[i\Delta_1 t' + i\int_0^{t'} dt'' \lambda_2(t'')].
\tag{11.57c}
$$

When we use the back transformation from $\mathbf{a}'$ to $\mathbf{a}$, $\mathbf{a} = \underline{\underline{U}}^T \cdot \mathbf{a}'$, we obtain the desired solution:

$$a_1(t) = \frac{i}{\hbar} |d_{0,1} \mathcal{E}_0| \int_0^t dt' \exp[i\Delta_1 t'] \left\{ \cos^2 \alpha \exp \left[i \int_0^{t'} dt'' \, \lambda_1(t'') \right] \right.$$
$$\left. - \sin^2 \alpha \exp \left[i \int_0^{t'} dt'' \, \lambda_2(t'') \right] \right\}, \tag{11.58a}$$

$$a_2(t) = \frac{i}{\hbar} |d_{0,1} \mathcal{E}_0| \int_0^t dt' \exp[i\Delta_1 t'] \cos \alpha \sin \alpha \left\{ \exp \left[i \int_0^{t'} dt'' \, \lambda_1(t'') \right] \right.$$
$$\left. + \exp \left[i \int_0^{t'} dt'' \, \lambda_2(t'') \right] \right\}, \tag{11.58b}$$

where we have used the fact that $a_0(t) = c_0(t)$, and that at $t = 0$ all the population resides in state $|E_0\rangle$, i.e., that $c_0(0) = 1$.

To obtain the probabilities and branching ratios to various channels, given by Eq. (11.49), we can substitute the expressions for a_1 and a_2 from Eq. (11.58) into Eq. (11.48). Because of the dependence of $\lambda_{1,2}$ on $\dot{\theta}_{0,1,2}$ [see Eq. (11.54)], some dependence of P_q on the laser phases remains for time-dependent phases. A numerical solution [337] of Eq. (11.37) shows, however, that this dependence is tiny: the a_1 and a_2 coefficients obtained numerically were found to be unaffected (changes of less than 1% were computed), even when rapidly varying time-dependent phases were introduced.

As an example of pulsed IIC, we consider the case shown in Figure 11.3 of the two-photon dissociation of Na_2 to yield the $Na(3p) + Na(3s)$, $Na(4s) + Na(3s)$, and $Na(3d) + Na(3s)$ products. We denote the probabilities of their formation as $P(3p)$, $P(4s)$, and $P(3d)$, and choose $|E_0\rangle$ as the $v = 0$, $J = 33$ level of the Na_2 ground $^1\Sigma_g$ state (the high J state is chosen to mimic thermal conditions in a high-temperature heat pipe [201]); $|E_1\rangle$ as the mixed state formed by spin-orbit coupling the $v = 33$, $J = 31, 33$ level of the $A^1\Sigma_u$ and the $v = 33$, $J = 32$ level of the $b^3\Pi_u$ electronic state [336]; and $|E_2\rangle$ as the $v = 93$, $J = 32$ or $v = 93$, $J = 34$ state of the $b^3\Pi_u$ state. The laser frequency $\omega_0 = 17,844$ cm^{-1}, that is, in close resonance with the $|E_0\rangle$ to $|E_1\rangle$ transition (a detuning of -0.21 cm^{-1}); the $\omega_1 = 17,712$ cm^{-1}. The frequency ω_2 of the control laser is tuned over the range of 12,929 to 13,017 cm^{-1}, which corresponds to varying the detuning Δ from -43.8 to $+43.8$ cm^{-1}. The laser pulse profiles in Eq. (11.22) are taken to be Gaussians of the following form:

$$\mathcal{E}_i(t) = \bar{\mathcal{E}}_i \exp[-(t - \tau_i)^2 / 2\alpha_i^2], \qquad (i = 0, 1, 2), \tag{11.59}$$

where $\bar{\mathcal{E}}_i$ is the peak amplitude, and τ_i, α_i are the temporal center position and width of the ith pulse.

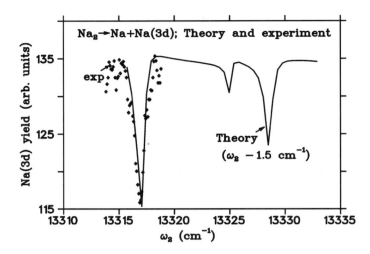

Figure 11.4   Comparison of experimental and theoretical $Na_2 \rightarrow Na + Na(3d)$ yields as a function of ω_2 for fixed ω_1. In calculation, an intermediate $v = 33$, $J = 31, 33$ resonance is used and ω_1 is fixed at $17,720.7 \, cm^{-1}$. Intensities of two laser fields are $I(\omega_1) = 1.72 \times 10^8 \, W/cm^2$ and $I(\omega_2) = 2.84 \times 10^8 \, W/cm^2$. The ω_2 frequency axis of the calculated results was shifted by $-1.5 \, cm^{-1}$ in order to better compare predicted and measured line shapes. (Taken from Fig. 3, Ref. [201].)

The numerical solutions to the original equations [Eqs. (11.30) and (11.31)] are plotted in Figures 11.4 and 11.5 and contrasted with the experimental results of Ref. [201]. These experiments are made easier by the fact that the IIC control scenario does not require laser coherence, hence, generally available, nontransform-limited

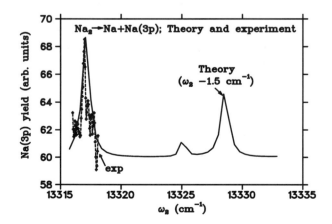

Figure 11.5   Comparison of experimental and theoretical $Na_2 \rightarrow Na + Na(3p)$ yields as a function of ω_2, with parameters as in Fig. 11.4. (Taken from Fig. 4, Ref. [201].)

nanosecond dye lasers can be used. In the experiment [201], the IIC scenario was studied for just two channels of the Na_2 two-photon dissociation process:

$$Na(3s) + Na(3p) \leftarrow Na_2 \rightarrow Na(3s) + Na(3d).$$

Two dye lasers pumped by a frequency-doubled Nd–Yag laser were used. One dye laser, of frequency ω_2 tuned between 13,312 and 13,328 cm^{-1}, was used to couple the continuum to the $|E_2\rangle$ vib-rotational state of the $A^1\Sigma_u/^3\Pi_u$ mixed electronic state [202, 309] of Na_2. The other dye laser, of frequency $\omega_1 = 17{,}474.12$ cm^{-1}, was used to induce a two-photon dissociation of the $|E_0\rangle$ $(v = 5, J = 37$ ground state of Na_2, through intermediate resonances $|E_1\rangle$ $(v = 35, J = 38$ and $v = 35, J = 36)$ of the $A^1\Sigma_u/^3\Pi_u$ mixed state. The ω_1 and ω_2 pulses, both of $\sim$5-ns duration with the stronger among them (ω_2) having an energy of $\sim$3.5 mJ, were made to overlap in a heat pipe containing Na vapor at 370 to 410°C. Spontaneous emission from the excited Na atoms [Na(3d) $\rightarrow$ Na(3p) and Na(3p) $\rightarrow$ Na(3s)] resulting from the Na_2 photodissociation was detected and dispersed in a spectrometer and a detector with a narrow bandpass filter.

Figure 11.4 shows the experimental Na(3d) emission as a function of ω_2 at a fixed ω_1. Each point represents an average over a few hundred laser shots, each chosen to have an ω_2 pulse energy that deviates by less than 5% from 3.5 mJ (intensity of $\sim$10^7 W/cm^2). In Figure 11.5 we plot the experimental Na(3p) yield obtained in the same experiment.

Also shown in Figures 11.4 and 11.5 are the theoretical calculations [201] described above. We see that the Na(3d) signal yield dips, and the Na(3p) signal peaks, as a function of the $\omega_2 - \omega_1$ frequency difference, yielding a $\sim$30% modulation in the Na(3p)/Na(3d) branching ratio. Considering the uncertainties in the theoretical potentials used [97, 201], the agreement between theory and experiment [especially in the Na(3d) signal] is remarkable. Additional computations [200, 329] suggest that the observed experimental substructures may be due to the excitation of numerous additional, as yet unassigned, thermally populated vib-rotational Na_2 energy levels.

11.3 RESONANTLY ENHANCED TWO-PHOTON ASSOCIATION

The set of equations used in Section 11.1 for two-photon dissociation problems can be used to address another significant problem—that of resonantly enhanced two-photon *association*, depicted schematically in Figure 10.1c.

The significance of resonantly enhanced two-photon association stems from the possibility of using it to form ultracold molecules, a topic of considerable interest. Laser cooling schemes that work for atoms [338–340] tend to fail for molecules, mainly due to the presence of many near-resonance lines and the presence of other degrees of freedom, in addition to translation (rotations, vibrations, etc.), that must be cooled. Thus far all suggestions on methods to cool warm molecules using light [341, 342] have not been realized because of a variety of difficulties.

Rather than attempting to cool warm molecules one can try to *synthesize* cold molecules by associating cold atoms. The molecules thus formed are expected to maintain the translational temperature of the recombining atoms because the center-of-mass motion remains unchanged in the association process (save for the little momentum imparted by the photon). This idea was first proposed by Julienne and co-workers [343, 344] who envisioned a multistep association, first involving the continuum-to-bound excitation of translational continuum states of cold trapped atoms to an excited vibrational level in an excited electronic molecular state. This step was followed by bound–bound spontaneous emission to the ground electronic state.

An undesirable feature of this scheme is that the spontaneous nature of the second step allows the molecules to end up in a large range of vibrational levels. As a consequence, the use of *stimulated* emission [308, 345–351], discussed below, is preferable insofar as it allows population transfer to a particular final molecular state of interest.

11.3.1 Theory of Photoassociation of a Coherent Wave Packet

In photoassociation the initial state is the scattering state, and the goal is to transfer the population to the final *bound* state $|E_1\rangle$. We therefore consider a pair of colliding atoms described by scattering states $|E, \mathbf{n}^+\rangle$, [defined in Eq. (2.52)] with $\mathbf{n}$ incorporating the quantum indices specifying the electronic states of the separated atoms and E being the total collision energy. As explained in conjunction with Eq. (2.52), the plus notation signifies, in contrast with the minus states that were previously used to describe dissociation processes, that the *initial* state of the fragments is known.

Following Ref. [345], we focus attention on a Λ-type system, shown in Figure 10.1c, subjected to the combined action of two laser pulses of central frequencies ω_1 and ω_2. Here ω_2 is in near resonance with the transition from the $|E, \mathbf{n}^+\rangle$ continuum to an intermediate bound state $|E_2\rangle$ and ω_1 is in near resonance with the transition from $|E_2\rangle$ to $|E_1\rangle$.

With the total Hamiltonian of the system given by Eq. (9.2) and the material wave function of the system expanded as in Eq. (11.1), we obtain a set of first-order differential equations for the expansion coefficients that is essentially identical to that of Eq. (11.2), except that the bound-continuum dipole matrix elements are now of the form $\langle E_2 | \mathbf{d}_2 \cdot \boldsymbol{\epsilon}_2 | E, \mathbf{n}^+\rangle$, involving the $|E, \mathbf{n}^+\rangle$, rather than the $|E, \mathbf{n}^-\rangle$, states.

In the photoassociation case, contrary to the dissociation cases discussed above, the continuum is initially populated, that is $b_{E,\mathbf{n}}(0) \neq 0$. Hence the formal solution of Eq. (11.2c) is now of the form:

$$b_{E,\mathbf{n}}(t) = b_{E,\mathbf{n}}(t = 0) + i \int_0^t dt' \, \Omega_{2,E,\mathbf{n}}^*(t') \exp(i\Delta_E t') b_2(t'). \tag{11.60}$$

Substituting this solution into Eq. (11.2b) gives

$$
\begin{aligned}
\frac{db_2}{dt} &= i\Omega_1(t)\exp(i\Delta_1 t)b_1(t) + i\sum_{\mathbf{n}} \int dE\, \Omega_{2,E,\mathbf{n}}(t)\exp(-i\Delta_E t)b_{E,\mathbf{n}}(t=0) \\
&\quad - \sum_{\mathbf{n}} \int dE \int_0^t dt'\, \Omega_{2,E,\mathbf{n}}(t)\Omega_{2,E,\mathbf{n}}^*(t')\exp[-i\Delta_E(t-t')]b_2(t').
\end{aligned}
\tag{11.61}
$$

If the molecular continuum is unstructured, we can invoke the SVCA and replace the energy-dependent bound-continuum dipole matrix elements by their value at the pulse center, given (in the Λ configuration of Fig. 10.1c) as $E_L = E_2 - \hbar\omega_2$. This is the case, for example, for Na_2 at threshold energies, where the bound-continuum dipole matrix elements vary with energy by less than 1% over a typical nanosecond-pulse bandwidth. Within the SVCA, Eq. (11.61) becomes

$$
\frac{db_2}{dt} = i\Omega_1(t)\exp(i\Delta_1 t)b_1(t) - \Omega_2^I(t)b_2(t) + iF(t),
\tag{11.62}
$$

where

$$
F(t) \equiv \mathcal{E}_2(t)\bar{\mathrm{d}}_2(t)/\hbar,
\tag{11.63}
$$

with

$$
\bar{\mathrm{d}}_2(t) \equiv \sum_{\mathbf{n}} \int dE\, \langle E_2 | \mathbf{d}_2 \cdot \hat{\varepsilon}_2 | E, \mathbf{n}^+ \rangle \exp(-i\Delta_E t)b_{E,\mathbf{n}}(t=0),
\tag{11.64}
$$

and where $\Omega_2^I(t)$ is defined as in Eq. (11.7). Equations (11.62) and (11.5) can be expressed in matrix notation as

$$
\frac{d}{dt}\mathbf{b} = i\{\underline{\underline{H}} \cdot \mathbf{b}(t) + \mathbf{f}\},
\tag{11.65}
$$

where

$$
\mathbf{f}(t) = (0, F(t))^{\mathrm{T}},
\tag{11.66}
$$

with $\mathbf{b}$ as defined in Eq. (11.8) and $\underline{\underline{H}}$ by Eq. (11.9).

The *net association rate* $R(t)$ is the rate of population change in the bound manifold, given by $d/(dt)(|b_1|^2 + |b_2|^2)$. It can be written, using Eq. (11.65) and its complex conjugate, as

$$R(t) = \frac{d}{dt}(|b_1|^2 + |b_2|^2) = \frac{d}{dt}|\mathbf{b}|^2 = \mathbf{b}^\dagger \cdot \left(\frac{d}{dt}\mathbf{b}\right) + \left(\frac{d}{dt}\mathbf{b\dagger}\right) \cdot \mathbf{b}$$

$$= i\{\mathbf{b}^\dagger \cdot (\underline{\underline{H}} - \underline{\underline{H}}^\dagger) \cdot \mathbf{b} + \mathbf{b}^\dagger \cdot \underline{f} - \underline{f}^\dagger \cdot \mathbf{b}\} = 2I_m[F^*(t)b_2(t)] - 2\Omega_2^I(t)|b_2(t)|^2.$$

$$(11.67)$$

The first term in Eq. (11.67) represents the association rate,

$$R_{\text{rec}}(t) \equiv 2\,\text{Im}[F^*(t)b_2(t)], \tag{11.68}$$

and the second term is the back-dissociation rate,

$$R_{\text{diss}}(t) \equiv 2\Omega_2^I(t)|b_2(t)|^2. \tag{11.69}$$

As expected, the net association rate [Eq. (11.67)] is the difference between the association rate and the back-dissociation rates.

As in Eq. (9.10), we can solve Eq. (11.65) adiabatically by diagonalizing the $\underline{\underline{H}}$ matrix. Operating with $\underline{U}(t)$ on Eq. (11.65), with $\mathbf{a}(t)$ defined as in Eq. (11.13), we obtain that

$$\frac{d}{dt}\mathbf{a} = \{i\underline{\lambda}(t) + \underline{A}\} \cdot \mathbf{a} + i\mathbf{g}, \tag{11.70}$$

where the source vector $\mathbf{g}$ is given as

$$\mathbf{g}(t) = \begin{pmatrix} F(t)U_{1,2}(t) \\ F(t)U_{2,2}(t) \end{pmatrix} = \begin{pmatrix} F(t)\sin\theta(t) \\ F(t)\cos\theta(t) \end{pmatrix}. \tag{11.71}$$

Invoking the adiabatic approximation, we obtain from Eq. (11.70) that

$$\frac{d}{dt}\mathbf{a} = i\underline{\lambda}(t) \cdot \mathbf{a}(t) + i\mathbf{g}(t). \tag{11.72}$$

In the association process the initial conditions are such that

$$\mathbf{a}(t = 0) = 0, \tag{11.73}$$

so that the adiabatic solutions are

$$\mathbf{a}(t) = \underline{v}(t) \cdot \mathbf{q}(t), \tag{11.74}$$

where

$$\underline{v}(t) = \exp\left\{ i \int_0^t \underline{\underline{\lambda}}(t') \, dt' \right\},$$

(11.75)

and

$$\mathbf{q}(t) = i \int_0^t \underline{v}^{-1}(t') \cdot \mathbf{g}(t') \, dt',$$

(11.76)

with $\lambda_{1,2}$ given by Eq. (11.17).

Using Eqs. (11.13) and (11.8), we obtain for $b_1(t)$ and $b_2(t)$ in the adiabatic approximation:

$$b_1(t) = i \left\{ \cos \theta(t) \int_0^t \exp\left[i \int_{t'}^t \lambda_1(t'') \, dt'' \right] F(t') \sin \theta(t') \, dt' \right.$$

$$\left. - \sin \theta(t) \int_0^t \exp\left[i \int_{t'}^t \lambda_2(t'') \, dt'' \right] F(t') \cos \theta(t') \, dt' \right\} \exp(i\Delta_1 t),$$

$$b_2(t) = i \left\{ \sin \theta(t) \int_0^t \exp\left[i \int_{t'}^t \lambda_1(t'') \, dt'' \right] F(t') \sin \theta(t') \, dt' \right.$$

$$\left. + \cos \theta(t) \int_0^t \exp\left[i \int_{t'}^t \lambda_2(t'') \, dt'' \right] F(t') \cos \theta(t') \, dt' \right\}.$$

(11.77)

Given $b_2(t)$, the continuum coefficients $b_{E,n}(t)$ are obtained directly via Eq. (11.60).

It is instructive to study the adiabatic solution when there is insignificant temporal overlap between the two laser pulses. Assuming in that case that the ω_2 pulse precedes the ω_1 pulse, we have during the ω_2 pulse that $\mathcal{E}_2 \gg \mathcal{E}_1$; hence by Eq. (11.71) $\lambda_1 \approx \Delta_1$, $\lambda_2 \approx i\Omega_2^I$ and $\theta(t) = 0$. Substituting these values into Eq. (11.77) gives that during the ω_2 pulse (when the ω_1 pulse is off):

$$b_1(t) = 0; \qquad b_2(t) = i \int_0^t \exp\left\{ - \int_{t'}^t \Omega_2^I(t'') \, dt'' \right\} F(t') \, dt'.$$

(11.78)

From Eq. (11.63) it is clear that the source term $F(t)$ is linearly proportional to the pulse amplitude. On the other hand, since $\Omega_2 > 0$ and $t' < t$, the $\exp\{- \int_{t'}^t \Omega_2^I(t'') \, dt''\}$ factor (describing dissociation back to the continuum) decays exponentially with increasing intensity. Thus, merely increasing the laser power does not necessarily increase the association yield. There exists some optimal intensity beyond which the association probability decreases. Below we display some pulse configurations for a realistic photoassociation case.

As an example of this formulation we consider pulsed photoassociation of a coherent wave packet of cold Na atoms [345]. The colliding atoms are described by an (energetically narrow) normalized Gaussian packet of $J = 0$ radial waves:

$$|\Psi(t = 0)\rangle = \int dE \, b_E(t = 0)|E, 3s + 3s\rangle, \tag{11.79}$$

where $|E, 3s + 3s\rangle$ are the translational Na–Na s waves with the atoms in the 3s state, and b_E at time zero is taken as

$$b_E(t = 0) = (\delta_E^2 \pi)^{-1/4} \exp\{-(E - E_{col})^2/2\delta_E^2 + i\Delta_E t_0\}. \tag{11.80}$$

Here, t_0 denotes the instant of maximum overlap of the Na + Na wave packet with the $|E_2\rangle$ state. In the simulations, E_{col}, the mean collision energy, varies between $E_{col} = 0.00695$–0.0695 cm$^{-1} \approx 0.01$ K $- 0.1$ K and the wave packet widths, δ_E, vary over the range $\delta_E = 10^{-4}$–10^{-3} cm^{-1}. State $\langle E_1|$ is chosen as the $(X^1\Sigma_g^+, v = 0, J = 0)$ state and $|E_2\rangle$ as the $(A^1\Sigma_u^+, v' = 34, J = 1)$ state, as shown in Figure 11.6. Thus, the combined effect of the two laser pulses is the transfer of population

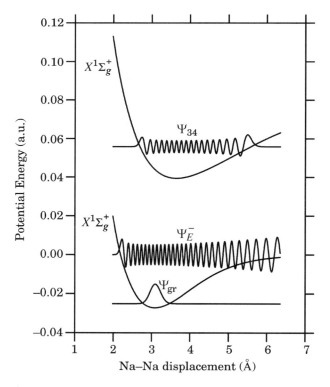

Figure 11.6 Potentials and vibrational wave functions used in the simulation of the Na + Na two-photon association. (Taken from Fig. 2, Ref. [345].)

from the continuum to the ground vib-rotational state $(X^1\Sigma_g^+, v = 0, J = 0)$, with the bound $(A^1\Sigma_u^+, v' = 34, J = 1)$ state acting as an intermediate state. To minimize spontaneous emission losses we concentrate on the "counterintuitive" [245] scheme where the "dump" pulse $\mathcal{E}_1(t)$ is applied *before* the $\mathcal{E}_2(t)$ "pump" pulse.

When either the final or initial state is in the continuum the Rabi frequency is imaginary, which changes the range of validity of the adiabatic approximation. For example, the adiabatic approximation does not necessarily hold even for "large area" $\int \Omega \, dt$ pulses [191, 325]. For example, as shown in Figure 11.2, in the presence of a continuum the adiabatic approximation tends to break down for small detunings. Despite this fact, we show below that with the proper choice of pulse parameters, it is possible to transfer the entire population contained in the continuum wave packet to the ground state, while keeping the intermediate state population low at all times. Moreover, for such pulse parameters the adiabatic solutions [Eqs. (11.77)] are in excellent agreement with exact-numerical solutions [345].

A typical population evolution is shown in Figure 11.7, obtained with pulse intensities of order 10^8 W/cm^2 and pulse durations of several nanoseconds. Such pulse intensities are sufficiently small to avoid unwanted photoionization, photodissociation, and other strong-field parasitic processes. However, working with pulses requires that the atoms be sufficiently close to one another during the laser pulse that

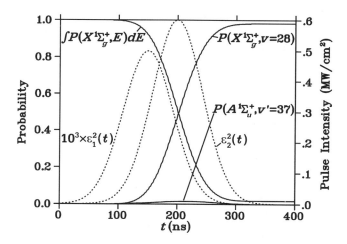

Figure 11.7 Results of the counterintuitive pulse sequence. Shown, as function of time, are integrated population of the wave packet of initial continuum states, population of $v = 34$, $J = 1$ intermediate state, and population of $v = 0, J = 0$ final ground state. Dashed lines are intensity profiles of two Gaussian pulses whose central frequencies are $\omega_1 = 18,143.775$ cm^{-1} and $\omega_2 = 12,277.042$ cm^{-1} (i.e., $\Delta_1 = \Delta_{E_{col}} = 0$). The maximum intensity of the dump pulse is 1.6×10^8 W/cm^2 and that of the pump pulse is 3.1×10^9 W/cm^2. Both pulses last 8.5 ns. Pump pulse peaks at $t_0 = 20$ ns, the peak time of Na + Na wave packet. Dump pulse peaks 5 ns before that time. Initial kinetic energy of Na atoms is 0.0695 cm^{-1} (or 0.1 K). (Taken from Fig. 4, Ref. [345].)

they can be recombined. In other words, the initial wave packet of continuum states considered here must be synchronized in time and duration with the recombining pulses. It is therefore of interest to see whether it is possible to employ longer pulses (of lower intensity) in order to increase the absolute number of recombining atoms and the overall duty cycle of the process.

Use of pulses of different intensity and different durations is illustrated for the counterintuitive scheme in Figures 11.8a and 11.8b, where the rates of association [R_{rec} of Eq. (11.68)], back-dissociation [R_{diss} of Eq. (11.69)], and net association [$R(t)$ of Eq. (11.67)] are plotted as a function of time. A short-pulse case is shown in Figure 11.8a and a long-pulse case, with a wider matter wave packet, is shown in Figure 11.8b. Both figures appear identical, though in Figure 11.8b the abscissa is scaled up by a factor of 10 and the ordinate is scaled down by a factor of 10.

The scaling behavior seen in Figure 11.8 is due to the existence of exact scaling relations in Eq. (11.65). This scaling is obtained when the initial wave packet width and the pulse intensities are scaled down as

$$\delta_E \to \frac{\delta_E}{s}, \qquad \mathcal{E}_1^0 \to \frac{\mathcal{E}_1^0}{s}, \qquad \mathcal{E}_2^0 \to \frac{\mathcal{E}_2^0}{\sqrt{s}}, \tag{11.81}$$

and the pulse durations are scaled up as

$$\Delta t_{1,2} \to \Delta t_{1,2} s. \tag{11.82}$$

It follows from Eq. (11.7) that under these transformations

$$F(t) \to \bar{F}_2(t) = \frac{F(t/s)}{s}, \frac{\Omega_{1,2}(t/s)}{s};$$

and Eq. (11.65) becomes

$$\frac{d}{dt/s}\mathbf{b}' = i\left\{\underline{\underline{H}}\left(\frac{t}{s}\right) \cdot \mathbf{b}' + \underline{f}\left(\frac{t}{s}\right)\right\}, \tag{11.83}$$

where $\mathbf{b}'$ denotes the vector of solutions of the scaled equations. Thus, the scaled coefficients at time t are identical to the unscaled coefficients at times t/s.

One of the results of the above scaling relations is that the pulses' durations can be made longer and their intensities concomitantly smaller, without changing the final population transfer yields. As noted above, lengthening of the pulses is beneficial because it causes more atoms to recombine within a given pulse.

There is a range of pulse parameters (such as the pulse area, $\Omega_{2,E_L}\Delta t_2$) that maximizes the association yield for a *fixed* initial wave packet. For both the intuitive and the counterintuitive schemes there is a clear maximum at a specific pulse area; merely increasing the pulse intensity does not lead to an improved association yield. We can attribute this to the fact that the association rate [R_{rec} of Eq. (11.68)] increases linearly with increasing pulse intensity, whereas the dissociation rate

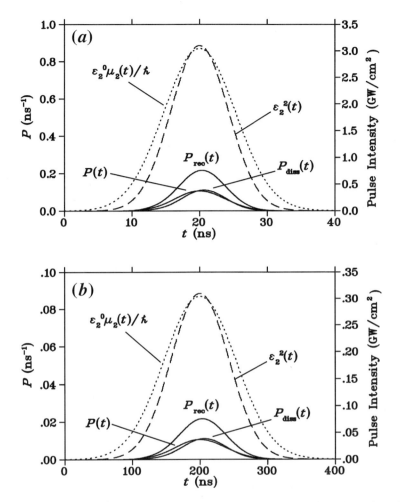

Figure 11.8 Rates of association (P_{rec}), back-dissociatioin (P_{diss}), and total molecule formation (P) vs. t in the counterintuitive scheme. Dashed lines are pulse intensity profile, dotted lines denote effective Rabi frequency $\mathcal{E}_2^0 \bar{d}_2(t)/\hbar$, where $\mathcal{E}_2^0$ is peak pulse intensity and $\bar{d}_2(t) \equiv \mu_2(t)$. (a) Initial wave packet width of $\delta_E = 10^{-3}$ cm^{-1} and other pulse parameters as in Fig. 11.7. (b) Dynamics scaled by $s = 10$ [Eq. (19.81)]: Initial wave packet width of $\delta_E = 10^{-4}$ cm^{-1}; both pulses lasting 85 ns; pump pulse peaking at $t_0 = 200$ ns and dump pulse peaking at 50 ns before that time. Peak intensity of dump pulse is 1.6×10^6 W/cm^2 and of the pump pulse is 3.1×10^8 W/cm^2. (Taken from Fig. 5, Ref. [345].)

[R_{diss} of Eq. (11.69)] increases exponentially with the intensity. Hence, as long as the energetic width of the initial wave packet stays fixed, the association yield peaks with increasing pulse area. The turn-over point is different for the two pulse schemes: in the counterintuitive case it occurs at a much higher intensity (area).

The existence of a window of intensities for efficient association explains why it is not possible to increase the pulse durations *ad infinitum*, that is, to work with cw light. As $\Delta t_{1,2}$ increases, it follows from Eq. (11.81) that $|\mathcal{E}_2/\mathcal{E}_1|^2$ must also increase. Since $|\mathcal{E}_1|^2$ cannot vanish, $|\mathcal{E}_2|^2$ must diverge if one is to stay within the windows of intensities for efficient association in the cw limit. Hence radiative association as described in this section cannot take place in the cw regime.

Experimental confirmations of photoassociation via two-photon transition as discussed above have been obtained [352–358]. Evidence that counterintuitive pulse ordering might result in large photoassociation cross sections as suggested above has also been presented [359].

11.4 LASER CATALYSIS

Over the past two decades a number of strong-field scenarios for laser acceleration and suppression of dissociation processes and chemical reactions have been proposed [30, 360–372]. The theme common to many of these schemes is quite different from that of coherent control in that the lasers are so strong as to affect chemical reactivity by modifying the molecular potential to produce so-called dressed potentials (see Chapter 12). Success is attained by designing dressed potentials that promote a given reaction. The main difference between this type of laser enhancement of chemical reactions and ordinary photochemistry is that the former involves no net absorption of laser photons. The concept of *laser catalysis* [308, 372–374], that is, a process in which a laser field returns to its exact initial state after altering a reaction, is a refinement of such scenarios and is the subject of this section.

Consider first some of the factors affecting the design of such laser catalysis schemes. Ground electronic state based laser enhancement schemes [216, 364, 366] rely on the induction of nuclear dipole moments to aid in promoting a desired reaction [30, 367]. For example, the use of infrared (IR) radiation has been proposed to overcome reaction barriers on the ground electronic state [30, 367]. However, this proposal requires powers on the order of terawatts per centimeter squared (TW/cm^2). At these powers nonresonant multiphoton absorption, which invariably leads to ionization and/or dissociation, becomes dominant, drastically reducing the yield of the reaction of interest.

Continuum–continuum transitions involving *excited* electronic states [368] might be thought useful insofar as they ought to require less power than those occurring on the ground state because in this case the laser can couple to strong electronic transition dipoles. However, in this case the continuum–continuum nuclear factors lead to smaller transition dipole matrix elements, and moreover, once the system is deposited on an unbounded excited electronic surface, it is impossible to prevent reaction on that surface and the resultant retention of the absorbed photon. Such a chain of events resembles that of conventional (weak-field) photochemistry where the laser is used to impart energy to the reaction, rather than to catalyze it.

Scenarios [308, 372–374] employing transitions between scattering states on the ground electronic surface and bound excited electronic states may reduce the above power requirements, primarily because of the involvement of the strong bound-continuum nuclear factors. For an excited surface possessing reaction well(s), such schemes give rise to laser catalysis [308, 372–374] described below because the reagents, once excited, remain in the transition state region and shuttle freely between the reactants' side and the products' side of the ground state barrier (see Fig. 10.1*d*). If the energy available to the nuclei on the excited state is insufficient to break any bond, the system, not being able to escape the transition state region, eventually relaxes (radiatively or nonradiatively) back to the ground state. In doing so, it has *a priori* similar probabilities of landing on the products' side as on the reactants' side of the barrier. However, if the laser is strong enough, the stimulated radiative relaxation route, yielding back the same photon absorbed, overcomes the nonradiative channels, resulting in true laser catalysis.

We show below that the methodology developed in this book can account for many of the phenomena described above. The only required modification of the models discussed is to consider the coupling of a bound manifold to *two* continua (that of the reactants and that of the products) via laser pulses.

11.4.1 Coupling of a Bound State to Two Continua by a Laser Pulse

As a model, we consider the A + BC→AB + C exchange reaction described by the smooth one-dimensional potential barrier along a reaction coordinate (Fig. 11.9). The eigenstates of the system form a continuum of "outgoing" scattering states $|E, 1^+\rangle$ and $|E, 2^+\rangle$. In accordance with our general notation, the 1^+ and 2^+ indices are reminders that the reaction has originated in either arrangement channel 1, the A + BC channel, or arrangement channel 2, the AB + C channel. The situation is depicted in Figure 10.1*d*.

In accord with standard scattering theory [375], the asymptotic behavior of the $|E, 1^+\rangle$ and $|E, 2^+\rangle$ states is given by

$$\lim_{x \to -\infty} \langle x|E, 1^+\rangle = \sqrt{\frac{m}{k_1 h}} \exp(ik_1 x) + R_1(E) \exp(-ik_1 x), \qquad (11.84)$$

$$\lim_{x \to +\infty} \langle x|E, 1^+\rangle = T_1(E) \exp(ik_2 x), \qquad (11.85)$$

and

$$\lim_{x \to \infty} \langle x|E, 2^+\rangle = \sqrt{\frac{m}{k_2 h}} \exp(-ik_2 x) + R_2(E) \exp(ik_2 x), \qquad (11.86)$$

$$\lim_{x \to -\infty} \langle x|E, 2^+\rangle = T_2(E) \exp(-ik_1 x), \qquad (11.87)$$

where $k_{1,2} = \sqrt{2m[E - V(\mp\infty)]}/\hbar$. Here $R_i(E)$ denotes the amplitude for reflection, and $T_i(E)$ the amplitude for reaction when the system starts in channel i.

The laser catalysis scenario is shown in Figure 10.1*d*: Under the action of a laser pulse of central frequency ω, assumed to be in near resonance with the transition

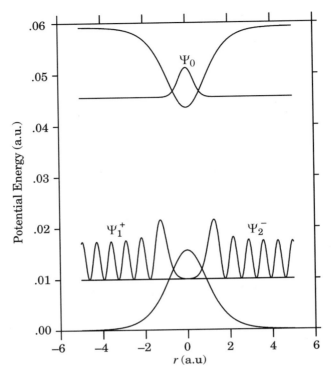

Figure 11.9 Eckart potentials and wave functions used in the simulation of laser catalysis. Potential parameters were $A = 0$ a.u., $B = 6.247$ a.u., $l = 4.0$ a.u., and $m = 1060.83$ a.u. Ψ_1^+ denotes $|E, 1^+\rangle$ state of text, Ψ_2^- denotes $|E, 2^-\rangle$ state of text, and Ψ_0 denotes $|E_0\rangle$ state of text. (Taken from Fig. 2, Ref. [308].)

from the continuum $|E, i^+\rangle$ to an intermediate bound state $|E_0\rangle$, population is transferred from states $|E, 1^+\rangle$ to a set of "incoming" scattering states $|E, 2^-\rangle$, with the asymptotic behavior

$$\lim_{x \to \infty} \langle x|E, 2^-\rangle = \sqrt{\frac{m}{k_2 h}} \exp(ik_2 x) + R_2^*(E) \exp(-ik_2 x), \qquad (11.88)$$

$$\lim_{x \to -\infty} \langle x|E, 2^-\rangle = T_2^*(E) \exp(ik_1 x). \qquad (11.89)$$

To address the laser catalysis problem, with the total Hamiltonian of the system given by Eq. (9.2), we expand the material wave function of the system as

$$|\Psi(t)\rangle = b_0|E_0\rangle \exp\left(-\frac{iE_0 t}{\hbar}\right) + \int dE \, [b_{E,1}(t)|E, 1^+\rangle + b_{E,2}(t)|E, 2^+\rangle] \exp\left(-\frac{iEt}{\hbar}\right), \qquad (11.90)$$

where $|E_0\rangle$ and $|E, n^+\rangle$ satisfy the material Schrödinger equations

$$[E_0 - H_M]|E_0\rangle = [E - H_M]|E, n^+\rangle = 0, \qquad n = 1, 2. \qquad (11.91)$$

Substituting the expansion [Eq. (11.90)] into the time-dependent Schrödinger equation, and using the orthogonality of the $|E_0\rangle$, $|E, 1^+\rangle$, and $|E, 2^+\rangle$ basis states, results in a set of first-order differential equations similar to Eq. (11.2):

$$\frac{db_0}{dt} = i \sum_{n=1,2} \int dE\, \Omega_{0,E,n}(t) \exp(i\Delta_E t) b_{E,n}(t), \qquad (11.92a)$$

$$\frac{db_{E,m}}{dt} = i\Omega^*_{0,E,m}(t) \exp(-\Delta_E t) b_0(t), \qquad m = 1, 2, \qquad (11.92b)$$

where

$$\Omega_{0,E,\mathbf{n}}(t) \equiv \langle E_0|\hat{\varepsilon} \cdot \mathbf{d}|E, n^+\rangle \mathcal{E}(t)/\hbar, \qquad n = 1, 2,$$
$$\Delta_E \equiv (E_0 - E)/\hbar - \omega. \qquad (11.93)$$

Substituting the formal solution of Eq. (11.92b), where t_0 denotes the initial time,

$$b_{E,n}(t) = b_{E,n}(t_0) + i \int_{t_0}^{t} dt'\, \Omega^*_{0,E,n}(t') \exp(-i\Delta_E t') b_0(t'), \qquad (11.94)$$

into Eq. [11.92(a)], we obtain

$$\frac{db_0}{dt} = i \sum_{n=1,2} \int dE\, \Omega_{0,E,n}(t) \exp(i\Delta_E t) b_{E,n}(t_0)$$
$$- \sum_{n=1,2} \int dE\, \Omega_{0,E,n}(t) \int_{t_0}^{t} dt'\, \Omega^*_{0,E,n}(t') \exp[(i\Delta_E(t - t')] b_0(t'). \qquad (11.95)$$

We now invoke the SVCA, which in the context of the two-photon Λ configuration of Figure 10.1d, reads

$$\sum_n |\langle E, n^-|\hat{\varepsilon} \cdot \mathbf{d}|E_0\rangle|^2 \approx \sum_n |\langle E_L, n^+|\hat{\varepsilon} \cdot \mathbf{d}|E_0\rangle|^2, \qquad (11.96)$$

where $E_L = E_0 - \hbar\omega$. Upon substituting Eqs. (11.3) and (11.96) into Eq. (11.95), we obtain

$$\frac{db_0}{dt} = iF(t) - \sum_{n=1,2} \Omega^I_n(t) b_0(t), \qquad (11.97)$$

where

$$\Omega_n^I(t) \equiv \pi |\langle E_L, n^+|\hat{\varepsilon} \cdot \mathbf{d}|E_0\rangle \mathcal{E}(t)|^2 /\hbar, \qquad n = 1, 2. \tag{11.98}$$

and where the source term $F(t)$ is given by

$$F(t) = \sum_{n=1,2} F_n(t) = \mathcal{E}(t) \sum_{n=1,2} \frac{\bar{\mathsf{d}}_n(t)}{\hbar}, \tag{11.99}$$

where

$$\bar{\mathsf{d}}_n(t) = \int dE \ \langle 0|\hat{\varepsilon} \cdot \mathbf{d}|E, n^+\rangle \exp(i\Delta_E t) b_{E,n}(t_0), \qquad n = 1, 2. \tag{11.100}$$

As in the photoassociation case, we can obtain analytical solutions of Eq. (11.97):

$$b_0(t) = v(t)q(t) + b_0(t_0)v(t), \tag{11.101}$$

where

$$v(t) = \exp\left\{-\int_{t_0}^t \left[\Omega_1^I(t') + \Omega_2^I(t')\right] dt'\right\} \tag{11.102}$$

and

$$q(t) = i \int_{t_0}^t \frac{F(t')}{v(t')} dt'. \tag{11.103}$$

In the laser catalysis process, the initial conditions are such that $b_0(t_0) = 0$ and $b_{E,2}(t_0) = 0$ for all E. Therefore, we obtain

$$b_0(t) = i \int_{t_0}^t F_1(t') \exp\left\{-\int_{t'}^t \left[\Omega_1^I(t'') + \Omega_2^I(t'')\right] dt''\right\} dt'. \tag{11.104}$$

Given $b_0(t)$, the continuum population distributions $b_{E,1}(t)$ and $b_{E,2}(t)$ are obtained directly via Eq. (11.94). Typical potentials and eigenfunctions used to simulate one-photon laser catalysis are plotted in Figure 11.9.

As an illustration, consider laser catalysis with an Eckart potential [376, 377] for the ground state:

$$V_{\text{ground}}(x) = V[\xi(x)] = -\frac{A\xi}{1 - \xi} - \frac{B\xi}{(1 - \xi)^2}; \qquad \xi = -\exp\left(\frac{2\pi x}{l}\right), \tag{11.105}$$

with $A = 0$, $B = 6.247 \times 10^{-2}$ a.u., $l = 40$ a.u., and an inverted Eckart potential,

$$V_{\text{excited}}(x) = \mathcal{E} - V[\xi(x)] \tag{11.106}$$

having a well, for the excited state. The particle's mass m was chosen as that of the $H + H_2 \rightarrow H_2 + H$ reaction, that is $m = 1060.83$ a.u. The intermediate state was the $v = 0$ level of the inverse Eckart potential given in Eq. (11.106), with the same parameters as above [378, 379]. The resulting potential curves, similar to the $H + H_2$ reaction path, are plotted in Figure 11.9. Given these parameters, eigenfunctions and eigenenergies were obtained using the formulae of Ref. [308].

The initial state of the system is described by a normalized Gaussian wave packet of outgoing scattering states:

$$|\Psi(t = 0)\rangle = \int dE\, b_{E,1}(t = 0)|E, 1^+\rangle, \tag{11.107}$$

where $b_{E,1}(t = 0)$ is given by Eq. (11.80). Simulations were made for initial collision energies of $E_{\text{col}} = 0.005 - 0.03$ a.u. and wave packet widths $\delta_E = 10^{-4} - 10^{-3}$ cm^{-1}.

The time dependence of the expansion coefficients is shown in Figure 11.10. Initial collision energy for this calculation was 0.01 a.u. At this energy the non-radiative reaction probability is negligible. Here the effect of the laser pulse is to induce a near-complete ($> 99\%$) population transfer from the wave packet of

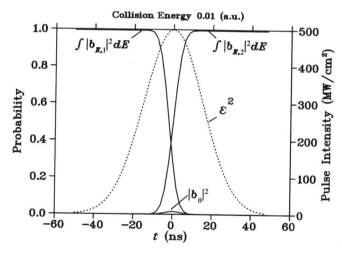

Figure 11.10 Integrated populations of incoming and outgoing continuum states and the population of the $v = 0$ intermediate state vs. time. Dashed line is intensity profile of Gaussian pulse whose maximum intensity is 5×10^8 W/cm^2. The FWHM of the pulse is 30 ns and its central frequency was chosen so that $\Delta_{E_{\text{col}}} = 0$. Initial reactant collision energy is 0.01 a.u. and initial wave packet width is $\delta_E = 10^{-3}$ cm^{-1}. (From Fig. 5, Ref. [308].)

$|E, 1^+\rangle$ states (localized to the left of the potential barrier) to a wave packet of $|E, 2^-\rangle$ states (localized to the right of the barrier), while keeping the population of the $|E_0\rangle$ states to a bare minimum. The latter serves to minimize spontaneous emission losses.

Due to the invariance of Eq. (11.97) upon the rescaling of Eq. (11.81), it is possible to freely vary the pulse durations and intensity, as long as the integrated pulse power $|\mathcal{E}^0|^2 \times \Delta t$ is kept fixed. Thus the dynamics remains the same if the time evolution of the system is scaled up by a factor of 10 and the pulse intensity is scaled down by the same factor. The advantage of long pulses is that only reactants that collide during the laser pulse will react. One anticipates that longer pulses would increase the number of product molecules formed within a single pulse duration. The disadvantage of longer pulses is that the power requirements become increasingly more difficult to fulfill because the peak power must go down exactly as $1/\Delta t$, whereas in most practical devices the power goes down much faster with increasing pulse durations.

As mentioned above, by keeping the population of the intermediate resonance low (as is the case in Fig. 11.10), the spontaneous emission losses are effectively eliminated. Figure 11.11 shows the intermediate level population as a function of t at four different pulse intensities. The reaction probability for all plotted intensities is near unity. However, it is evident that the intermediate state population throughout the process decreases with increasing pulse intensity. Thus, to avoid spontaneous emission losses, high pulse intensities should be used.

Calculated reactive line shapes (i.e., the reaction probability as a function of the pulse center frequency) at three pulse intensities are shown in Figure 11.12. The initial collision energy is 0.014 a.u., that is, slightly closer to the barrier maximum

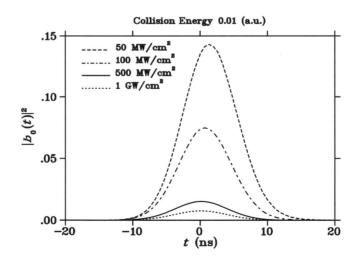

Figure 11.11   Intermediate state population vs. time at pulse intensities of 50 MW/cm², 100 MW/cm², 500 MW/cm², and 1 GW/cm². (Taken from Fig. 7, Ref. [308].)

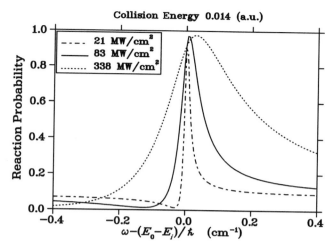

Figure 11.12 Calculated reactive line shapes at 21 MW/cm², 83 MW/cm², and 338 MW/cm². The FWHM of pulse is 20 ns. Reactants collision energy is 0.014 a.u. and initial wave packet width is $\delta_E = 10^{-3}$ cm⁻¹. (Taken from Fig. 8, Ref. [308].)

than before. At this energy the non-radiative reaction probability is ∼9%. This causes the line shapes to assume an asymmetric form due to the interference between the nonradiative tunneling pathway and the laser-catalyzed pathway. We see that the reaction probability is enhanced for a positive (blue) detuning and suppressed for a negative (red) detuning.

The fact that there is a point where the reaction probability assumes the value of unity is best understood by adopting the "(photon) dressed states" picture, discussed extensively in Section 12.3. Here it suffices to say that the only difference between the ordinary potential matrix and the dressed-potential matrix is that in our (2 × 2) case, the (1, 1) diagonal matrix element of the dressed-potential matrix is a sum of the material potential and the energy of the laser photon. The (2, 2) potential matrix element is just the excited state potential, and the off-diagonal matrix elements are the field-dipole coupling terms. When this matrix is diagonalized, one obtains the two "field-matter" eigenvalues shown in Figure 11.13. As demonstrated in this figure, the ground field-matter eigenvalue assumes the shape of a double-barrier potential, and the excited eigenvalue assumes the shape of a double-well potential. The separation between these eigenvalues increases as the coupling field strength is increased.

In the adiabatic approximation, particles starting out in the remote past in the ground state remain on the lowest eigenvalue at all times. These particles experience resonance scattering by a double-barrier potential. It is known that in this situation there is one energy point with unity tunneling probability, irrespective of the details of the potential [380–385]. This point occurs when the incident energy is near a bound state of the well contained within the barriers. Similar phenomena have been

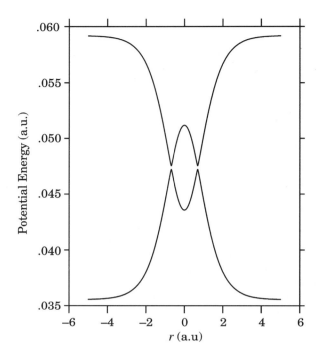

Figure 11.13   Dressed state potentials for laser catalysis process at maximum pulse intensity. Initial kinetic energy is 0.01 a.u.

noted for semiconductor devices [380–382], in the context of the Ramsauer–Townsend effect [8] and for Fabry–Perot interferometers [386]. Related observations, linking the phenomenon of field-induced transparency with the emergence of a field-dressed double-barrier potential, were made by Vorobeichik et al. [387].

The point where the tunneling is suppressed (i.e., when the tunneling probability of Fig. 11.12 is zero) appears, in the dressed states picture, as a result of the breakdown of the adiabatic approximation: At the energy of tunneling suppression, the flux leaking to the excited double-well eigenvalue interferes destructively with the flux remaining on the low double-barrier potential.

CHAPTER 12

STRONG-FIELD COHERENT CONTROL

12.1 QUANTIZATION OF THE ELECTROMAGNETIC FIELD

Photodissociation and control by strong electromagnetic fields is best treated by introducing states of the matter-plus-radiation field, the so-called dressed states. Since this is most naturally done within the framework of a *quantized* electromagnetic field, we begin this chapter by introducing the theory of quantized fields. For a full treatment of this topic see Ref. [3].

In complete analogy with the procedure for quantizing the coordinates and momenta of particles (Chapter 1), we first express H_R, the classical Hamiltonian of the electromagnetic field, in terms of canonical coordinates, Q_k, and momenta P_k. To do so we introduce the relevant canonical variables Q_k, P_k that are related to the field-mode amplitudes, A_k and A_k^* [Eq. (1.24)] by the definitions:

$$ Q_{\mathbf{k}} \equiv (\epsilon_0 V/c^2)^{1/2}(A_{\mathbf{k}} + A_{\mathbf{k}}^*), \qquad P_{\mathbf{k}} \equiv i\omega_k(\epsilon_0 V/c^2)^{1/2}(A_{\mathbf{k}} - A_{\mathbf{k}}^*), \qquad (12.1) $$

where the cavity volume V is assumed, in this chapter, to be finite. As a consequence, the number of field modes is discrete. It can be easily verified that both Q_k and P_k satisfy Eq. (1.21), that is, the dynamical equations for a set of harmonic oscillators.

Using Eq. (12.1) to express A_k and A_k^* in terms of Q_k and P_k gives

$$ A_{\mathbf{k}} = c(4\epsilon_0 V\omega_k^2)^{-1/2}(\omega_k Q_{\mathbf{k}} - iP_{\mathbf{k}}), \qquad A_{\mathbf{k}}^* = c(4\epsilon_0 V\omega_k^2)^{-1/2}(\omega_k Q_{\mathbf{k}} + iP_{\mathbf{k}}), $$

$$ (12.2) $$

and the cycle-averaged energy [Eq. (1.34)] can therefore be written as

$$\bar{H}_R = \frac{1}{2}\sum_{\mathbf{k}}(\omega_k^2 Q_{\mathbf{k}}^2 + P_{\mathbf{k}}^2).$$

(12.3)

Since the energy of the electromagnetic field fluctuates during one cycle, it is the cycle-averaged energy $\bar{H}_R$ that represents the energy of the field after many cycles. Hence, this is the quantity that we now proceed to quantize. We do so, in the coordinate representation, by replacing the classical momenta with the operators

$$P_{\mathbf{k}} \rightarrow -i\hbar\frac{\partial}{\partial Q_{\mathbf{k}}}.$$

(12.4)

As a result, the radiative Hamiltonian $\bar{H}_R$ [Eq. (12.3)] assumes the quantized form:

$$H_R = \frac{1}{2}\sum_{\mathbf{k}}\left(\omega_k^2 Q_{\mathbf{k}}^2 - \hbar^2\frac{\partial^2}{\partial^2 Q_{\mathbf{k}}}\right).$$

(12.5)

Using the annihilation and creation operators, $\hat{a}_{\mathbf{k}}$ and $\hat{a}_{\mathbf{k}}^\dagger$, defined as

$$\hat{a}_{\mathbf{k}} = (2\hbar\omega_k)^{-1/2}(\omega_k Q_{\mathbf{k}} + iP_{\mathbf{k}}) = (2\hbar\omega_k)^{-1/2}\left(\omega_k Q_{\mathbf{k}} + \hbar\frac{\partial}{\partial Q_{\mathbf{k}}}\right),$$

$$\hat{a}_{\mathbf{k}}^\dagger = (2\hbar\omega_k)^{-1/2}(\omega_k Q_{\mathbf{k}} - iP_{\mathbf{k}}) = (2\hbar\omega_k)^{-1/2}\left(\omega_k Q_{\mathbf{k}} - \hbar\frac{\partial}{\partial Q_{\mathbf{k}}}\right),$$

(12.6)

allows us to rewrite the radiative Hamiltonian as

$$H_R = \sum_{\mathbf{k}}\hbar\omega_k\left(\hat{a}_{\mathbf{k}}^\dagger\hat{a}_{\mathbf{k}} + \frac{1}{2}\right),$$

(12.7)

where we have used the commutations relations,

$$[\hat{a}_{\mathbf{k}}, \hat{a}_{\mathbf{k}}^\dagger] \equiv \hat{a}_{\mathbf{k}}\hat{a}_{\mathbf{k}}^\dagger - \hat{a}_{\mathbf{k}}^\dagger\hat{a}_{\mathbf{k}} = 1.$$

(12.8)

Similarly, comparing Eqs. (12.2) and (12.6) we obtain the operators $\hat{A}_{\mathbf{k}}$ and $\hat{A}_{\mathbf{k}}^\dagger$ as quantum analogs of $A_{\mathbf{k}}$ and $A_{\mathbf{k}}^*$:

$$\hat{A}_{\mathbf{k}} = c\left(\frac{\hbar}{2\epsilon_0 V\omega_k}\right)^{1/2}\hat{a}_{\mathbf{k}}, \qquad \hat{A}_{\mathbf{k}}^\dagger = c\left(\frac{\hbar}{2\epsilon_0 V\omega_k}\right)^{1/2}\hat{a}_{\mathbf{k}}^\dagger.$$

(12.9)

Because Eq. (12.7) is separable, the eigenstates of H_R are products of eigenstates $|N_\mathbf{k}\rangle$ of the different harmonic mode oscillators. That is,

$$H_R|\mathbf{N}\rangle = \mathbf{N}|\mathbf{N}\rangle, \qquad (12.10)$$

$$|\mathbf{N}\rangle \equiv \Pi_\mathbf{k}|N_\mathbf{k}\rangle, \qquad (12.11)$$

where $|N_\mathbf{k}\rangle$ are eigenstates of the operator $\hat{N}_\mathbf{k} \equiv \hat{a}_\mathbf{k}^\dagger \hat{a}_\mathbf{k}$,

$$\hat{N}_\mathbf{k}|N_\mathbf{k}\rangle = \hat{a}_\mathbf{k}^\dagger \hat{a}_\mathbf{k}|N_\mathbf{k}\rangle = N_\mathbf{k}|N_\mathbf{k}\rangle. \qquad (12.12)$$

It follows from the properties of harmonic oscillators that the eigenvalues $N_\mathbf{k}$ are non-negative integers ($N_\mathbf{k} = 0, 1, 2, \ldots$). Hence, the $|N_\mathbf{k}\rangle$ states are called *number states* and the $\hat{N}_\mathbf{k}$ operator is called the *number operator*. We say that the $N_\mathbf{k}$th level of a given field mode, labeled by $\mathbf{k}$, has $N_\mathbf{k}$ *photons* in that mode. Note also that we can define $e^{i\hat{\theta}_\mathbf{k}}$, the exponential-phase operator of the radiation field, through the relationship $\hat{a}_\mathbf{k} = (\hat{N}_\mathbf{k} + 1)^{1/2} e^{i\hat{\theta}_\mathbf{k}}$. It is clear from the commutation relations for $\hat{a}_\mathbf{k}$ and $\hat{a}_\mathbf{k}^\dagger$ that $\hat{N}_\mathbf{k}$ and $e^{i\hat{\theta}_\mathbf{k}}$ do not commute. As a result, a state with a well-defined number of photons has an ill-defined phase.

Using Eq. (12.7), the time-independent Schrödinger equation for free radiation field assumes the form:

$$H_R|\mathbf{N}\rangle = \left[\sum_\mathbf{k} \hbar\omega_k \left(N_\mathbf{k} + \frac{1}{2}\right)\right]|\mathbf{N}\rangle. \qquad (12.13)$$

Notice that we have adopted the *Schrödinger representation* here insofar as all operators, such as $\hat{a}_\mathbf{k}$ and $\hat{a}_\mathbf{k}^\dagger$, have no time dependence; rather, the time dependence is contained in the wave functions. It is, of course, possible to define another representation (the Heisenberg representation) where the operators do vary with time and where the wave functions are time independent. It is also possible to define a mixed *interaction* representation where the operators assume the time dependence of the free field. It is in the latter representation that the quantized electric field most resembles the classical form.

12.2 LIGHT–MATTER INTERACTION

To obtain the Schrödinger equation for the interaction of a molecule with the quantized radiation field, that is, the Schrödinger equation for the (matter + radiation) system, we need the quantum analog of H_{MR}, the matter–radiation interaction. In the dipole approximation H_{MR} depends, according to Eq. (1.51), on the transverse electric field. The required quantized electric field is obtained by substituting Eq. (12.9) into Eq. (1.27) to give

$$\hat{\mathbf{E}}(\mathbf{r}) = i \sum_{\mathbf{k}} \left(\frac{\hbar \omega_k}{2\epsilon_0 V}\right)^{1/2} \hat{\varepsilon}_{\mathbf{k}} [\hat{a}_{\mathbf{k}} e^{i\mathbf{k}\cdot\mathbf{r}+i\phi_{\mathbf{k}}} - \hat{a}_{\mathbf{k}}^\dagger e^{-i\mathbf{k}\cdot\mathbf{r}-i\phi_{\mathbf{k}}}]. \tag{12.14}$$

Here, in accord with Eq. (2.9), we have added an extra phase, $\phi_{\mathbf{k}}$, to each plane wave field mode in Eq. (12.15), representing the phase shifts accumulated by the light in the $\mathbf{k}$ mode as it travels from the source to the sample.

By substituting Eq. (12.14) into Eq. (1.51) we obtain the quantum analog of the radiation–matter interaction as

$$H_{\mathrm{MR}} = -i \sum_{\mathbf{k}} \left(\frac{\hbar \omega_k}{2\varepsilon_0 V}\right)^{1/2} \hat{\varepsilon}_{\mathbf{k}} \cdot \mathbf{d} [\hat{a}_{\mathbf{k}} e^{i\mathbf{k}\cdot\mathbf{r}+i\phi_{\mathbf{k}}} - \hat{a}_{\mathbf{k}}^\dagger e^{-i\mathbf{k}\cdot\mathbf{r}-i\phi_{\mathbf{k}}}], \tag{12.15}$$

where $\mathbf{d}$, as before, is the dipole operator.

The quantum analog of the total (matter + radiation) Hamiltonian now assumes the form

$$H = H_M + H_R + H_{\mathrm{MR}} \equiv H_f + H_{\mathrm{MR}}, \tag{12.16}$$

where

$$H_f = H_M + H_R \tag{12.17}$$

is called the *radiatively decoupled* Hamiltonian, it being the sum of the independent material and radiative parts. By contrast, H is the total Hamiltonian; its eigenstates are called *fully interacting states*.

12.3 STRONG-FIELD PHOTODISSOCIATION WITH CONTINUOUS-WAVE QUANTIZED FIELDS

In a photodissociation process induced by continuous wave (cw) radiation, we envision the molecule as existing initially ($t = -\infty$) in a state that does not interact with the field, that is, in an eigenstate $|E_i, N_i\rangle$ of H_f:

$$(E_i + N_i \hbar \omega_i - H_f)|E_i, N_i\rangle = 0. \tag{12.18}$$

Here $|E_i, N_i\rangle = |E_i\rangle|N_i\rangle$, where $|E_i\rangle$ is a bound state of H_M [defined in Eq. (2.1)], and $|N_i\rangle$ denotes a free radiation state with N_i photons in the $\mathbf{k}_i$ mode, whose frequency is ω_i. The molecule then interacts with the field, and we are interested in determining the photodissociation probability, defined as the probability of eventually (i.e., as $t \to \infty$) populating an eigenstate of H_f, where the molecule is dissociated. [Strictly speaking, as discussed in Chapter 2, the photodissociation probability is the probability of populating a (radiatively decoupled *and* materially decoupled) continuum eigenstate of $H_0 + H_R$, where H_0 is the free Hamiltonian of Eq. (2.39). However, as shown in Chapter 2, the nature of the incoming eigenstates of H_M, $|E, \mathbf{n}^-\rangle$, is such that the probability of populating a specific $|E, \mathbf{n}^-\rangle$ state at asymptotic times is identical to the probability of populating the $|E, \mathbf{n}; 0\rangle$ eigenstate of H_0.]

Following our strategy in the weak-field domain (see Chapter 3), we do not obtain the photodissociation probability by actually following the dynamics for long times. Rather, we calculate, at any given time, the transition probability to the particular fully interacting state that is guaranteed to evolve to the radiatively decoupled state of interest as $t \to \infty$.

The fully interacting incoming Hamiltonian eigenstates are denoted $|E, \mathbf{n}^-, \mathbf{N}^-\rangle$, where $\mathbf{N}$ is a vector of photon occupation numbers, $\mathbf{N} = (N_{\mathbf{k}_1}, N_{\mathbf{k}_2}, \ldots, N_{\mathbf{k}_m}, \ldots)$. These states are the strong-field analogs of the material incoming states, $|E, \mathbf{n}^-\rangle$, defined in Eq. (2.52). In order to find such fully interacting incoming states, we consider a particular radiatively decoupled state of interest in the distant future and evolve it backward in time to the present. This back-evolution is done as the field–matter interaction H_{MR} is switched on. To do so it is convenient to introduce the following notation:

$$|E, \mathbf{n}^-, \mathbf{N}^-, t\rangle \equiv \exp(-iHt/\hbar)|E, \mathbf{n}^-, \mathbf{N}^-\rangle = \exp(-iE_T t/\hbar)|E, \mathbf{n}^-, \mathbf{N}^-\rangle, \quad (12.19)$$

where $E_T = E + \sum_{\mathbf{k}} N_{\mathbf{k}} \hbar \omega_{\mathbf{k}}$ is the total (matter + radiation) energy of the state $|E, \mathbf{n}^-, \mathbf{N}^-\rangle$, and

$$|E, \mathbf{n}^-, \mathbf{N}, t\rangle \equiv |E, \mathbf{n}^-, t\rangle |\mathbf{N}\rangle \equiv |E, \mathbf{n}^-\rangle |\mathbf{N}\rangle \exp(-iEt/\hbar). \quad (12.20)$$

Note that the absence of the minus superscript in the state $|E, \mathbf{n}^-, \mathbf{N}, t\rangle$ implies that the molecule and radiation field are no longer coupled.

The desired fully interacting state is then

$$|E, \mathbf{n}^-, \mathbf{N}^- t\rangle = \lim_{t_1 \to \infty} \exp\left[-\frac{iH(t - t_1)}{\hbar}\right]|E, \mathbf{n}^-, \mathbf{N}^-, t_1\rangle$$

$$= \lim_{t_1 \to \infty} \exp\left[-\frac{iH(t - t_1)}{\hbar}\right]|E, \mathbf{n}^-, t_1\rangle |\mathbf{N}\rangle. \quad (12.21)$$

Equation (12.21) uses the fact that we can prove, in complete analogy to the proof given in the context of the material incoming states [see Eq. (2.57)], that as $t \to \infty$ the fully interacting states go over to the states wherein the matter and radiation are decoupled, but where the matter itself is interacting. That is,

$$|E, \mathbf{n}^-, \mathbf{N}^-, t\rangle \overset{t \to \infty}{\to} |E, \mathbf{n}^-, t\rangle |\mathbf{N}\rangle. \quad (12.22)$$

In turn, as the molecular fragments separate in the photodissociation process, these states go over to the noninteracting radiatively decoupled states,

$$|E, \mathbf{n}^-, t\rangle |\mathbf{N}\rangle \overset{t \to \infty}{\to} |E, \mathbf{n}; 0\rangle |\mathbf{N}\rangle \exp\left(-\frac{iEt}{\hbar}\right). \tag{12.23}$$

As in the case of the incoming states for the pure material part [Eq. (2.52)], the $|E, \mathbf{n}^-, \mathbf{N}^-\rangle$ states [Eq. (12.19)] satisfy a modified Schrödinger equation:

$$\lim_{\epsilon \to +0}\left(E - i\epsilon + \sum_{\mathbf{k}} N_{\mathbf{k}} \hbar \omega_k - H\right)|E, \mathbf{n}^-, \mathbf{N}^-\rangle = 0. \tag{12.24}$$

Note that the total energy of the system is $E_T = E + \sum_{\mathbf{k}} N_{\mathbf{k}} \hbar \omega_k$, but neither E nor $\mathbf{N}$ are good quantum numbers at other than asymptotic times. That is, the molecule and radiation field exchange energy. As in the case of $\mathbf{n}^-$ for the pure material case [Eq. (2.52)], the $\mathbf{N}^-$ notation simply serves as a reminder that the incoming fully interacting states correlate at long times with the radiatively decoupled state, $|E, \mathbf{n}^-\rangle |\mathbf{N}\rangle$.

The proper limiting behavior is, as in the pure material case, achieved by adding $-i\epsilon$ to the energy [see Eq. (2.52)]. This is equivalent to multiplying the radiation–matter interaction by a slowly decaying function to produce a time-dependent $H_{\mathrm{MR}}(t)$, where

$$H_{\mathrm{MR}}(t) \equiv H_{\mathrm{MR}} \exp\left(-\frac{\epsilon t}{\hbar}\right) \overset{t \to \infty}{\to} 0. \tag{12.25}$$

Since the state $|E, \mathbf{n}^-, \mathbf{N}^-, t\rangle$ contains the effect of the full Hamiltonian at time t, then the photodissociation amplitude $A(E, \mathbf{n}, \mathbf{N}, t|i, N_i)$ into the final state with energy E, internal quantum numbers $\mathbf{n}$ and radiation field described by $\mathbf{N}$, starting in the initial state $|E_i, N_i\rangle$ is simply the overlap between the radiatively decoupled initial state and the incoming fully interacting state. That is,

$$A(E, \mathbf{n}, \mathbf{N}, t|i, N_i) = \langle E, \mathbf{n}^-, \mathbf{N}^-, t|E_i, N_i\rangle = \langle E, \mathbf{n}^-, \mathbf{N}^-| \exp(iHt/\hbar)|E_i, N_i\rangle, \tag{12.26}$$

where the second equality in Eq. (12.26) arises from Eq. (12.19).

This expression is most readily evaluated for the slow-turn-on interaction of Eq. (12.25) through the formula:

$$e^{iHt/\hbar} = \lim_{\epsilon \to +0, t_1 \to \infty}\left[1 - \left(\frac{i}{\hbar}\right)\int_t^{t_1} dt' \, e^{iHt'/\hbar} H_{\mathrm{MR}}(t') e^{-iH_f t'/\hbar}\right] e^{iH_f t/\hbar}. \tag{12.27}$$

To obtain Eq. (12.27) we use the following equality:

$$-i\hbar \frac{d}{dt}\left[\exp\left(\frac{iHt}{\hbar}\right)\exp\left(-\frac{iH_f t}{\hbar}\right)\right] = \exp\left(\frac{iHt}{\hbar}\right)H_{MR}(t)\exp\left(-\frac{iH_f t}{\hbar}\right)$$

$$+ \exp\left(\frac{iHt}{\hbar}\right)t\frac{dH_{MR}(t)}{dt}\exp\left(-\frac{iH_f t}{\hbar}\right).$$

Integrating both sides of this equation, and making the time dependence of the Hamiltonians explicit, gives

$$\exp\left[\frac{iH(t)t}{\hbar}\right]\exp\left(-\frac{iH_f t}{\hbar}\right) = \exp\left[\frac{iH(t_1)t_1}{\hbar}\right]\exp\left(-\frac{iH_f t_1}{\hbar}\right)$$

$$-\frac{i}{\hbar}\left\{\int_t^{t_1}\exp\left[\frac{iH(t')t'}{\hbar}\right]H_{MR}(t')\exp\left(-\frac{iH_f t'}{\hbar}\right)dt'\right.$$

$$\left.+\int_t^{t_1}\exp\left[\frac{iH(t')t'}{\hbar}\right]t'\frac{dH_{MR}(t')}{dt'}\exp\left(-\frac{iH_f t'}{\hbar}\right)dt'\right\}.$$

$$(12.28)$$

Given the behavior of $H_{MR}(t)$ [Eq. (12.25)], the second term in the curly brackets vanishes in the $\epsilon \to 0$ limit. Rearranging terms, we obtain Eq. (12.27) in the $\epsilon \to 0$ and $t_1 \to \infty$ limits.

Substituting Eq. (12.27) in Eq. (12.26) and using Eq. (12.24) gives

$$A(E, \mathbf{n}, \mathbf{N}, t|i, N_i) = \left(-\frac{i}{\hbar}\right)\lim_{\epsilon \to +0}\lim_{t_1 \to \infty},$$

$$\langle E, \mathbf{n}^-, \mathbf{N}^-|\int_t^{t_1} dt'\, e^{i[(E+i\epsilon)/\hbar + \sum_\mathbf{k} N_\mathbf{k}\omega_k]t'}H_{MR}e^{-i(E_i/\hbar + N_i\omega_i)(t'-t)}|E_i, N_i\rangle$$

$$= \lim_{\epsilon \to +0}\frac{\langle E, \mathbf{n}^-, \mathbf{N}^-|H_{MR}|E_i, N_i\rangle \exp\{i[(E+i\epsilon)/\hbar + \sum_\mathbf{k} N_\mathbf{k}\omega_k]t\}}{E + i\epsilon - E_i + \sum_\mathbf{k} N_\mathbf{k}\hbar\omega_k - N_i\hbar\omega_i}, \quad (12.29)$$

where we have used the orthogonality between the bound and the continuum eigenstates of the material Hamiltonian, that is, $\langle E_i|E, \mathbf{n}^-\rangle = 0$.

The cw rate of transition $R(E, \mathbf{n}, \mathbf{N}|i, N_i)$, into the radiatively decoupled state $|E, \mathbf{n}^-, \mathbf{N}\rangle$, can be obtained as the rate of change of the photodissociation probability as the interaction is being slowly switched on:

$$R(E, \mathbf{n}, \mathbf{N}|i, N_i) = -\frac{d}{dt}|A(E, \mathbf{n}, \mathbf{N}, t|i, N_i)|^2. \quad (12.30)$$

The minus sign is introduced in Eq. (12.30) because in our expressions for $|A(E, \mathbf{n}, \mathbf{N}, t|i, N_i)|^2$ the interaction is being switched off, rather than switched on. Using Eq. (12.30) we obtain

$$R(E, \mathbf{n}, \mathbf{N}|i, N_i) = \lim_{\epsilon \to +0} \frac{2\epsilon}{\hbar} \frac{\exp(-2\epsilon t/\hbar)|\langle E, \mathbf{n}, \mathbf{N}^-|H_{\mathrm{MR}}|E_i, N_i\rangle|^2}{\left(E - E_i + \sum_{\mathbf{k}} N_{\mathbf{k}}\hbar\omega_k - N_i\hbar\omega_i\right)^2 + \epsilon^2}$$

$$= \frac{2\pi}{\hbar}|\langle E, \mathbf{n}^-, \mathbf{N}^-|H_{\mathrm{MR}}|E_i, N_i\rangle|^2 \delta\left(E - E_i + \sum_{\mathbf{k}} N_{\mathbf{k}}\hbar\omega_k - N_i\hbar\omega_i\right).$$

(12.31)

The adiabaticity of the switch-off (i.e., the $\epsilon \to +0$ limit) has yielded a cw rate that is independent of time. Notice that the δ function that appears in the transition rate expression guarantees that the transition rate is zero if the total energy is not conserved. By contrast, $A(E, \mathbf{n}, \mathbf{N}, t|i, N_i)$, the instantaneous transition amplitude, which is the overlap between a radiatively decoupled and a fully interacting state, permits a spread in the final energies observed.

Given the explicit form of H_{MR} [Eq. (12.15)], the expression for the transition rate assumes the form:

$$R(E, \mathbf{n}, \mathbf{N}|i, N_i) = \delta\left(E - E_i \sum_{\mathbf{k}} N_{\mathbf{k}}\hbar\omega_k - N_i\hbar\omega_i\right) \sum_{\mathbf{k}} \frac{\pi\omega_k}{\varepsilon_0 V} |\langle E, \mathbf{n}^-, \mathbf{N}^-|\hat{\varepsilon}_k \cdot \mathbf{d}$$

$$\times \left[\hat{a}_{\mathbf{k}} \exp(i\mathbf{k} \cdot \mathbf{r} + i\phi_{\mathbf{k}}) - \hat{a}_{\mathbf{k}}^\dagger \exp(-i\mathbf{k} \cdot \mathbf{r} - i\phi_{\mathbf{k}})\right]|E_i, N_i\rangle|^2.$$

(12.32)

Since we are dealing with strong fields, we can assume that N_i, the number of photons in the incident beam, is very large compared to unity. Using the following properties of the creation and annihilation operators,

$$\hat{a}_{\mathbf{k}}^\dagger|N_i\rangle = \begin{cases} (N_i + 1)^{1/2}|N_i + 1\rangle, & \mathbf{k} = \mathbf{k}_i, \\ 2^{1/2}|1_{\mathbf{k}}\rangle|N_i\rangle, & \mathbf{k} \neq \mathbf{k}_i, \end{cases}$$

(12.33)

$$\hat{a}_{\mathbf{k}}|N_i\rangle = \begin{cases} N_i^{1/2}|N_i - 1\rangle, & \mathbf{k} = \mathbf{k}_i, \\ 0, & \mathbf{k} \neq \mathbf{k}_i, \end{cases}$$

(12.34)

we obtain when $N_i \gg 1$, that is, when $N_i + 1 \approx N_i$, that

$$
R(E, \mathbf{n}, \mathbf{N} | i, N_i) = \frac{\pi \omega_i N_i}{\varepsilon_0 V} \delta \left(E - E_i + \sum_{\mathbf{k}} N_{\mathbf{k}} \hbar \omega_k - N_i \hbar \omega_i \right)
$$

$$
\times \left| \langle E, \mathbf{n}^-, \mathbf{N}^- | \hat{\varepsilon}_i \cdot \mathbf{d} \left[|N_i - 1\rangle \exp \left(\frac{i \omega_i z}{c} + i \phi_i \right) \right. \right. \tag{12.35}
$$

$$
\left. \left. - |N_i + 1\rangle \exp \left(-\frac{i \omega_i z}{c} - i \phi_i \right) \right] | E_i \rangle \right|^2,
$$

where ϕ_i denotes $\phi(\omega_i)$ and z is the direction of propagation of the incident light beam. Here the contributions from the states $|1_{\mathbf{k}}\rangle$ with $\mathbf{k} \neq \mathbf{k}_i$ are neglected since $N_i \gg 1$. Equation (12.35) applies to transitions to any combination of final photon number states (i.e., to any multiphoton process) and is correct to all orders of the radiation strength.

In the weak-field limit, and when only one photon is absorbed, we can approximate $|E, \mathbf{n}^-, \mathbf{N}^-\rangle$ by $|E, \mathbf{n}^-\rangle |N_i - 1\rangle$. In addition, for visible light $\exp(i \omega_i z / c)$ is essentially constant over atomic and molecular dimensions (the dipole approximation). Under these conditions, with the initial state $|N_i\rangle$, we obtain from Eq. (12.32) the result, here called $R^{(1)}$:

$$
R^{(1)}(E, \mathbf{n}, N_i - 1 | i, N_i) = \frac{\pi \omega_i N_i}{\varepsilon_0 V} |\langle E, \mathbf{n}^- | \hat{\varepsilon}_i \cdot \mathbf{d} | E_i \rangle|^2 \delta(E - \hbar \omega_i - E_i). \tag{12.36}
$$

When we make the connection between $N_i^{1/2}$ and the incident radiation field envelope, we have

$$
\mathcal{E}_i = i \left(\frac{\hbar \omega_i N_i}{2 \varepsilon_0 V} \right)^{1/2} \exp \left(-\frac{i \omega_i z}{c} - i \phi_i \right), \tag{12.37}
$$

and we recover the weak-field expression of Eq. (2.78),

$$
R^{(1)}(E, \mathbf{n}, N_i - 1 | i, N_i) = \frac{2\pi}{\hbar} |\mathcal{E}_i \langle E, \mathbf{n}^- | \hat{\varepsilon}_i \cdot \mathbf{d} | E_i \rangle|^2. \tag{12.38}
$$

12.3.1 The Coupled-Channels Expansion

In order to compute $R(E, \mathbf{n}, \mathbf{N} | i, N_i)$ in the strong-field regime we must be able to evaluate the fully interacting wave functions $|E, \mathbf{n}^-, \mathbf{N}^-\rangle$. The "multichannel" aspect of the problem becomes quite involved for both dissociation of molecules [25, 64, 65, 388–390] and the ionization of atoms [391–395] in the strong pulse regime. General numerical methods for solving for the eigenfunctions of the radiatively coupled time-independent Schrödinger were developed by a number of research groups [24, 25, 388–390, 396].

A powerful way of achieving this goal uses the *coupled-channels* expansion, a method widely used in calculations of scattering cross sections [6]. In the context of quantized matter–radiation problems, the coupled-channels method amounts to expanding $|E, \mathbf{n}^-, \mathbf{N}^-\rangle$ in number states. Concentrating on the expansion in the ith mode, we write $|E, \mathbf{n}^-, \mathbf{N}^-\rangle$ as

$$|E, \mathbf{n}^-, \mathbf{N}^-\rangle = \sum_{N=N_i-m}^{N_i+m} |N\rangle \langle N|E, \mathbf{n}^-, \mathbf{N}^-\rangle \qquad (12.39)$$

where $|\mathbf{N}\rangle$ is the number of photons in the ith mode. Photons in the other modes are not explicitly indicated. Note that $\langle N|E, \mathbf{n}^-, \mathbf{N}^-\rangle$ are quantum states in the space of the material subsystem.

Using the orthogonality of the number states $\{|\mathbf{N}\rangle\}$ along with Eqs. (12.15), (12.33), and (12.34), we transform the Schrödinger equation [Eq. (12.24)] into a set of coupled differential equations, the so-called *coupled-channels equations*:

$$[E_i + (N_i - N)\hbar\omega_i - H_M]\langle N|E, \mathbf{n}^-, \mathbf{N}^-\rangle = -i\left(\frac{\hbar\omega}{2\varepsilon_0 V}\right)^{1/2}\hat{\varepsilon}_i \cdot \mathbf{d}$$

$$\times \left[(N+1)^{1/2}\exp(-i\omega_i z/c - i\phi_i)\langle N+1|E, \mathbf{n}^-, \mathbf{N}^-\rangle\right.$$

$$\left. - N^{1/2}\exp(i\omega_i z/c + i\phi_i)\langle N-1|E, \mathbf{n}^-, \mathbf{N}^-\rangle\right],$$

$$N = N_i - m, \ldots, N_i + m. \qquad (12.40)$$

For a dissociation process in which only one net photon is absorbed, the final photon occupation number state is $|N_i - 1\rangle \equiv |0, 0, \ldots, N_i - 1, 0, \ldots\rangle$, that is, $E = E_i + \hbar\omega_i$. If $N \gg 1$, we can equate $(N+1)^{1/2} \approx N^{1/2} \approx N_i^{1/2}$ and using Eq. (12.37), we obtain that

$$[E_i + (N_i - N)\hbar\omega_i - H_M]\langle N|E, \mathbf{n}^-, N_i - 1^-\rangle$$

$$= -\hat{\varepsilon}_i \cdot \mathbf{d} \cdot [\mathcal{E}_i \langle N+1|E, \mathbf{n}^-, N_i - 1^-\rangle + \mathcal{E}_i^*\langle N-1|E, \mathbf{n}^-, N_i - 1^-\rangle],$$

$$N = N_i - m, \ldots, N_i + m. \qquad (12.41)$$

This, or a similar, set of equations has been applied to a large number of problems associated with molecular dissociation in intense laser fields [360, 397].

We now focus on the case of just two electronic states, $|e\rangle$ and $|e'\rangle$ defined as two eigenstates of the electronic Hamiltonian H_{el} (see Section 2.3.2):

$$[H_{\text{el}}(\mathbf{y}|\mathbf{X}) - W_e(\mathbf{X})]\langle \mathbf{y}|e\rangle = 0,$$
$$[H_{\text{el}}(\mathbf{y}|\mathbf{X}) - W_{e'}(\mathbf{X})]\langle \mathbf{y}|e'\rangle = 0. \qquad (12.42)$$

Here, $\mathbf{y}$ denotes the electronic coordinates, and $\mathbf{X}$ denotes all the nuclear coordinates $\{\mathbf{R}_\alpha\}$, $\alpha = 1, \ldots, n$, where n is the number of nuclei in the problem. The $W_j(\mathbf{X})$ eigenvalue is the potential experienced by the nuclei in the jth electronic state.

We expand each $\langle N|E, \mathbf{n}^-, N_i - 1^-\rangle$ component in the two electronic states:

$$\langle N|E, \mathbf{n}^-, N_i - 1^-\rangle = (|e\rangle\langle e| + |e'\rangle\langle e'|)\langle N|E, \mathbf{n}^-, N_i - 1^-\rangle$$
$$= |e\rangle\langle N, e|E, \mathbf{n}^-, N_i - 1^-\rangle + |e'\rangle\langle N, e'|E, \mathbf{n}^-, N_i - 1^-\rangle;$$

note that $\langle N, e|E, \mathbf{n}^-, N_i - 1^-\rangle$ and $\langle N, e'|E, \mathbf{n}^-, N_i - 1^-\rangle$ are quantum states in the space of the nuclear degrees of freedom.

Substituting in Eq. (12.41), we obtain, using Eq. (12.42), that

$$\left[E_i + (N_i - N)\hbar\omega_i - \sum_\alpha K_\alpha - W_e(\mathbf{X})\right]\langle N, e|E, \mathbf{n}^-, N_i - 1^-\rangle$$
$$= -\hat{\varepsilon}_i \cdot \mathbf{d}_{e,e'}(\mathbf{X})[\mathcal{E}_i\langle N + 1, e'|E, \mathbf{n}^-, N_i - 1^-\rangle + \mathcal{E}_i^*\langle N - 1, e'|E, \mathbf{n}^-, N_i - 1^-\rangle],$$

$$(12.43a)$$

$$\left[E_i + (N_i - N)\hbar\omega_i - \sum_\alpha K_\alpha - W_{e'}(\mathbf{X})\right]\langle N, e'|E, \mathbf{n}^-, N_i - 1^-\rangle$$
$$= -\hat{\varepsilon}_i \cdot \mathbf{d}_{e',e}(\mathbf{X})[\mathcal{E}_i\langle N + 1, e|E, \mathbf{n}^-, N_i - 1^-\rangle + \mathcal{E}_i^*\langle N - 1, e|E, \mathbf{n}^-, N_i - 1^-\rangle],$$

$$N = N_i - m, \ldots, N_i + m, \quad (12.43b)$$

where K_α is the kinetic energy operator of the α nucleus, and $\hat{\varepsilon}_i \cdot \mathbf{d}_{e,e'}(\mathbf{X}) \equiv \langle e|\hat{\varepsilon}_i \cdot \mathbf{d}|e'\rangle$. We implicitly assume that the system has no permanent dipole moment, $\mathbf{d}_{e,e} = \mathbf{d}_{e',e'} = 0$.

The coupled channels expansion can be further simplified by introducing the (number state) rotating-wave approximation (RWA), valid only when the field is of moderate intensity and the system is near resonance. As pointed out above, given an initial photon number state $|N_i\rangle$, the components of $|E, \mathbf{n}^-, N_i - 1^-\rangle$ of greatest interest for a one-photon transition are $\langle N_i|E, \mathbf{n}^-, N_i - 1^-\rangle$ and $\langle N_i \pm 1|E, \mathbf{n}^-, N_i - 1^-\rangle$. If $|E_i\rangle$ is the ground material state, then the $\langle N_i + m, e|E, \mathbf{n}^-, N_i - 1^-\rangle$, $m > 0$ components, and the $\langle N_i + m, e'|E, \mathbf{n}^-, N_i - 1^-\rangle$, $m \geq 0$ components are closed since the total energy $E_i + N_i\hbar\omega_i$ is smaller than $E_j + (N_i + m)\hbar\omega_i$ for all E_j. This means that these components, though not necessarily zero at finite times, must vanish when the radiation and matter decouple at long times. For ω_i in the ultraviolet (UV) or visible range, the closed components are so far removed in energy from the initial state that for moderate field strengths they hardly affect $|E, \mathbf{n}^-, N_i - 1^-\rangle$. The same may be said about the $\langle e|\langle N_i - m|E, \mathbf{n}^-, N_i - 1^-\rangle$ open components. It is therefore reasonable in this case to only include the terms $\langle N_i, e | E, \mathbf{n}^-, N_i - 1^- \rangle$ and $\langle N_i - 1, e' | E, \mathbf{n}^-, N_i - 1^- \rangle$ in Eq. (12.43). Doing so, we obtain

$$\left[E_i - \sum_\alpha K_\alpha - W_e(\mathbf{X}) \right] \langle N_i, e | E, \mathbf{n}^-, N_i - 1^- \rangle$$
$$= -\mathcal{E}_i^* \hat{\varepsilon}_i \cdot \mathbf{d}_{e,e'}(\mathbf{X}) \langle N_i - 1, e' | E, \mathbf{n}^-, N_i - 1^- \rangle, \quad (12.44)$$

$$\left[E_i + \hbar \omega_i - \sum_\alpha K_\alpha - W_{e'}(\mathbf{X}) \right] \langle N_i - 1, e' | E, \mathbf{n}^-, N_i - 1^- \rangle$$
$$= -\mathcal{E}_i \hat{\varepsilon}_i \cdot \mathbf{d}_{e',e}(\mathbf{X}) \langle N_i, e | E, \mathbf{n}^-, N_i - 1^- \rangle. \quad (12.45)$$

Equations (12.44) and (12.45) constitute the number state RWA for moderate field photodissociation. They are nonperturbative within the basis set adopted but are approximate in that they only incorporate a small number of number states and they neglect the contribution of all modes other than that of the incident beam.

An example for one of the first applications of the coupled-channel equations with quantized fields for photodissociation problems is shown in Figure 12.1, where the dissociation of the IBr molecule by a two-photon (visible + IR) process was studied [388]. The results of the calculations, shown in Figure 12.2, demonstrate how the strong IR photon broadens the transition ("power broadening") allowing the system to be dissociated even if the first photon is tuned substantially away from resonance. This illustrates how multiphoton transitions induced by strong fields [392] are less restricted insofar as they need not be very close to an intermediate resonance, the situation described in Section 3.3.

Finally, we make a few additional remarks. First, note that a pure number state is a state whose phase $\theta_\mathbf{k}$ is evenly distributed between 0 and 2π. This is a consequence of the commutation relation [3] between $\hat{N}_\mathbf{k}$ and $e^{i\theta_\mathbf{k}}$. Nevertheless, dipole matrix elements calculated between number states are (as all quantum mechanical amplitudes) well-defined complex numbers, and as such they have well-defined phases. Thus, the phases of the dipole matrix elements in conjunction with the mode phase $\phi_\mathbf{k}$ [Eq. (12.15)] yield well-defined matter + radiation phases that determine the outcome of the photodissociation process. As in the weak-field domain, if only one incident radiation mode exists then the phase cancels out in the rate expression [Eq. (12.35)], provided that the RWA [Eqs. (12.44) and (12.45)] is adopted. However, in complete analogy with the treatment of weak-field control, if we irradiate the material system with two or more radiation modes then the *relative* phase between them may have a pronounced effect on the fully interacting state, so that phase control is possible.

Second, in the weak-field limit it is possible to obtain the transition rates of Eq. (12.36) directly by solving Eq. (12.45) in first-order perturbation theory. This is done by assuming that $\langle N_i, e | E, \mathbf{n}^-, N_i - 1^- \rangle$ always remains larger than $\langle N_i - 1, e' | E, \mathbf{n}^-, N_i - 1^- \rangle$. It is therefore possible to neglect the right-hand side of Eq. (12.44). Under these circumstances, Eq. (12.44) becomes identical to Eq.

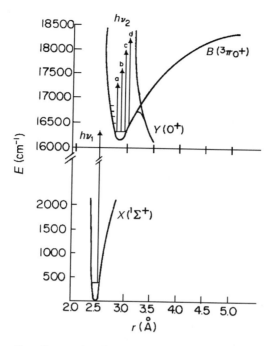

Figure 12.1   Energetics of two-color dissociation of IBr. Ground $X^1\Sigma^+$ potential curve and excited $Y(0^+)$ and $B^3\Pi_{0^+}$ states are shown. The intermediate level, at $16{,}333.03\ \mathrm{cm}^{-1}$ above the ground vibrational level of the molecule, is accessed by the weak $h\nu_1$ photon. Four $h\nu_2$ transitions were studied and are marked as (a) $960\ \mathrm{cm}^{-1}$, (b) $1282\ \mathrm{cm}^{-1}$, (c) $1652\ \mathrm{cm}^{-1}$, and (d) $1880\ \mathrm{cm}^{-1}$. (Taken from Fig. 1, Ref. [388].)

(2.1), whose solution is $|E_i\rangle$. This solution can be substituted in Eq. (12.45), which now becomes an inhomogeneous differential equation with a *known* source:

$$\left[E_i + \hbar\omega_i - \sum_\alpha K_\alpha - W_{e'}(\mathbf{X})\right]\langle e'|E, \mathbf{n}^-\rangle = -\mathcal{E}_i\hat{\boldsymbol{\epsilon}}_i \cdot \mathbf{d}_{e',e}(\mathbf{X})\langle e|E_i\rangle. \qquad (12.46)$$

The continuum component $\langle e'|E, \mathbf{n}^-\rangle$ is therefore "driven" by the bound state times the radiative coupling term. Driven equations of the type written here are implicitly used in all practical computational schemes developed for weak-field molecular photodissociation problems [14, 26].

12.4   QUANTIZED FIELDS: PULSES

Thus far we have considered the radiation field in the number representation. This corresponds to cw excitation. By contrast, using quantized fields in *pulsed* domain

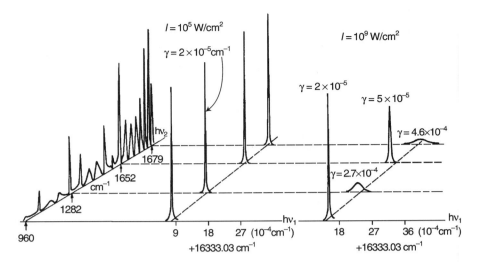

Figure 12.2   Effect of a strong laser field on the line shape for dissociation of an intermediate level at four hv_2 IR frequencies and at two intensities of the IR laser. The spectrum on the hv_2 axis (left-hand side) is computed IBr absorption spectrum in weak-field limit, starting from $960\,\mathrm{cm}^{-1}$ above the intermediate level (which is $16{,}333.03\,\mathrm{cm}^{-1}$ above ground vibrational level). We see that broadening of the $16{,}333.03\,\mathrm{cm}^{-1}$ line occurs for $I = 10^9\,\mathrm{W/cm^2}$ whenever the hv_2 photon is in near resonance with a strong predissociating line. (Taken from Fig. 2, Ref. [388].)

causes computation difficulties. This is due to the fact that to represent a pulse we must use *multi-mode* number states. Because of the presence of a large (or even a continuous) number of modes, coupled-channels expansions, such as the one introduced in the cw domain, cannot be used.

Over the years a number of computational methods have been developed to solve the problem of the interaction of a pulse of radiation with molecular and atomic systems [398–410]. To keep the treatment as simple as possible, we only consider solving the problem in the adiabatic approximation, introduced in Chapter 9 for classical fields.

Replacing the electric field amplitude $\mathcal{E}_i$ by a time-dependent pulse envelope $\varepsilon_i(t)$, we can write the time-dependent Schrödinger equation, in the approximations that led to Eqs. (12.44) and (12.45), as two coupled equations of the form

$$\left[i\hbar \frac{d}{dt} - \sum_\alpha K_\alpha - W_e(\mathbf{X}) \right] \langle N_i, e | E, \mathbf{n}^-, N_i - 1^-, t \rangle$$
$$= -\varepsilon_i^*(t)\hat{\varepsilon}_i \cdot \mathbf{d}_{e,e'}(\mathbf{X}) \langle N_i - 1, e' | E, \mathbf{n}^-, N_i - 1^-, t \rangle, \quad (12.47)$$

$$\left[i\hbar \frac{d}{dt} + \hbar\omega_i - \sum_\alpha K_\alpha - W_{e'}(\mathbf{X}) \right] \langle N_i - 1, e' | E, \mathbf{n}^-, N_i - 1^-, t \rangle$$
$$= -\varepsilon_i(t)\hat{\varepsilon}_i \cdot \mathbf{d}_{e',e}(\mathbf{X}) \langle N_i, e | E, \mathbf{n}^-, N_i - 1^-, t \rangle. \quad (12.48)$$

Defining a solution vector

$$\boldsymbol{\psi}(t) \equiv \begin{pmatrix} \langle N_i, e|E, \mathbf{n}^-, N_i - 1^-, t \rangle \\ \langle N_i - 1, e'|E, \mathbf{n}^-, N_i - 1^-, t \rangle \end{pmatrix},$$ (12.49)

we can write Eqs. (12.47) and (12.48) in matrix notation as

$$\left[\left(i\hbar \frac{d}{dt} - \sum_\alpha K_\alpha \right) \underline{\hat{\mathrm{I}}} - \underline{\underline{\mathrm{W}}}(t) \right] \cdot \boldsymbol{\psi}(t) = 0,$$ (12.50)

where $\underline{\mathrm{I}}$ is 2×2 unity matrix and $\underline{\underline{\mathrm{W}}}(t)$ is defined as

$$\underline{\underline{\mathrm{W}}}(\mathbf{X}, t) \equiv \begin{pmatrix} W_e(\mathbf{X}) & -\varepsilon_i^*(t)\hat{\varepsilon}_i \cdot \mathbf{d}_{e,e'}(\mathbf{X}) \\ -\varepsilon_i(t)\hat{\varepsilon}_i \cdot \mathbf{d}_{e',e}(\mathbf{X}) & W_{e'}(\mathbf{X}) - \hbar\omega_i \end{pmatrix}.$$ (12.51)

By doing this, the problem has been transformed to a scattering problem on two potential energy surfaces, one being the ground surface $q = e$, and the other being a "field-dressed" surface in which the excited state ($q = e'$) energy $W_{e'}$ has been lowered by the incident photon energy of $\hbar\omega_i$ to energy $W_{e'}(\mathbf{X}) - \hbar\omega_i$. The two surfaces are coupled by the transition dipole-moment matrix element times the field.

12.4.1 Light-Induced Potentials

Following George et al. [364] we can gain insight into Eqs. (12.50) and (12.51) by diagonalizing the $\underline{\underline{\mathrm{W}}}$ matrix for every space–time $(\mathbf{X}, t)$ point according to

$$\underline{\underline{\mathrm{U}}}^\dagger(\mathbf{X}, t) \cdot \underline{\underline{\mathrm{W}}}(\mathbf{X}, t) \cdot \underline{\underline{\mathrm{U}}}(\mathbf{X}, t) = \underline{\hat{\lambda}}(\mathbf{X}, t),$$ (12.52)

where $\underline{\hat{\lambda}}(\mathbf{X}, t)$ is a 2×2 diagonal matrix of field-dressed adiabatic surfaces. These diagonal elements are the so-called *light-induced potentials* (LIPs) [364, 397, 411–416], which are given by

$$\lambda_{1,2}(\mathbf{X}, t) = \tfrac{1}{2} \Big\{ W_e(\mathbf{X}) + W_{e'}(\mathbf{X}) - \hbar\omega_i \pm \big[(W_e(\mathbf{X}) - W_{e'}(\mathbf{X}) + \hbar\omega_i)^2 $$
$$+ 4|\varepsilon_i(t)\hat{\varepsilon}_i \cdot \mathbf{d}_{e',e}(\mathbf{X})|^2 \big]^{1/2} \Big\}.$$ (12.53)

If, as in the classical field case [Eq. (9.16)], we neglect the nonadiabatic coupling terms (in this case, the transformation to LIP entails nonadiabatic coupling terms arising from derivatives with respect to $\mathbf{X}$, embodied in the kinetic energy K_α operators, *and* the time derivative), the problem decouples to two Schrödinger equations:

$$i\hbar \frac{d}{dt} \psi_i'(\mathbf{X}, t) = \left[\sum_\alpha K_\alpha + \lambda_i(\mathbf{X}, t) \right] \psi_i'(\mathbf{X}, t), \qquad i = 1, 2.$$ (12.54)

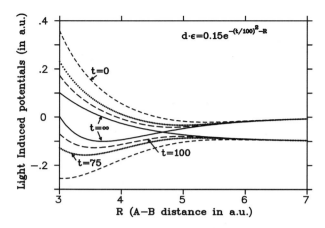

Figure 12.3 Opening of the gap between adiabatic states with rise of the coupling pulse whose form is shown in upper-right corner.

We see that the potentials have now been replaced by the $\lambda_i(\mathbf{X}, t)$ LIPs. The solution to Eq. (12.50) is then obtained by back-transforming the solution of Eq. (12.54), that is, $\mathbf{\psi} = \underline{\underline{U}} \cdot \mathbf{\psi}'$.

As depicted in Figure 12.3, the situation (which is analogous to a radiation-free curve-crossing problem [417]) is that of two ($\mathbf{X}$ and t dependent) eigenvalues that repel one another in the near crossing region. The stronger the electric field, the larger the repulsion between curves. Thus, in the pulsed case, as the field amplitude rises, the "gap" between the two eigenvalues opens up, and then closes as the pulse falls. For weak fields, or for strong fields during the initial rise time of the pulse, if the system has been deposited (by an excitation pulse) on the excited potential surface, it will stay on this surface and will dissociate to products linked to the original excited surface. If, however, the system reaches the crossing region at the maximum of the pulse, and the pulse is strong enough, the gap that opens up between the eigenvalues will force the system to remain on the higher eigenvalue and to dissociate to the products that are linked to the original ground potential surface.

This type of *strong-field control* was demonstrated theoretically by Giusti-Suzor et al. [411, 418] and Bandrauk et al. [397] in their study of the strong-field photodissociation of H_2^+, and experimentally by various researchers [419–422] in their study of the photodissociation of H_2 in intense laser fields. In these studies, another phenomenon, that of *bond softening*, was demonstrated. Here [360, 418, 422, 423], due to the opening of the gap, the lower eigenvalue is pushed down, thereby reducing the binding energy felt by the molecule. Moderate amounts of excitation can then be enough to give the molecule enough energy to cross the gap and be dissociated, provided it reaches the crossing point at the instant the gap opens. We see that an important element of strong-field control is, in addition to having a sufficiently strong field, the exact timings of the wave packet motion. This timing can be

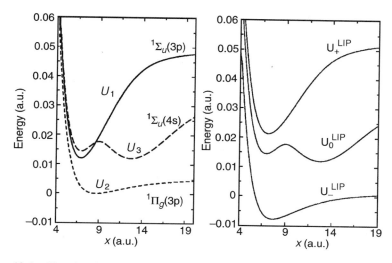

Figure 12.4 "Bond softening" experienced by the $\lambda_i(\mathbf{X}, t)$ LIP in the three-surface case. Radiation free curves are shown on the left, and the LIP are shown on the right. (Taken from Fig. 2, Ref. [416].)

controlled by varying the time delay between the excitation pulse which "lifts" the wave packet to the excited state, and the "control" pulse, responsible for creating the gap. This bond softening is observed at energies below the minimum of U_0^{LIP}. However, *bond hardening* or *vibrational trapping* occur in U_+^{LIP} where the molecule has an increased lifetime [424–426].

The extension of Eq. (12.50) to the three-surface case [413–416], shown in Figure 12.4, allows for adiabatic population transfer of entire *wave packets* (comprising many vibrational states) between electronic states, in complete analogy to the adiabatic population transfer in the three-state Λ system, treated in Section 9.1. As shown in Figure 12.4, the "bond softening" experienced by the $\lambda_i(\mathbf{X}, t)$ LIP in the three-surface case results in a reduction in the barrier separating reactants and products of the nondiagonalized dressed potentials, that is, $W_e(\mathbf{X})$ and $W_{e'}(\mathbf{X}) - \hbar\omega_i$. In complete analogy with the three-level STIRAP problem, in the three-surfaces case, a counterintuitive pulse sequence would be capable of inducing a complete population transfer between *wave packets*, each composed of many states.

12.5 CONTROLLED FOCUSING, DEPOSITION, AND ALIGNMENT OF MOLECULES

12.5.1 Focusing and Deposition

The light-induced potentials (LIPs) defined above are valid for both on-resonance cases, when $W_{e'}(\mathbf{X}) - W_e(\mathbf{X}) - \hbar\omega_i \approx 0$, and off-resonance processes, where this condition is not satisfied. Many multiphoton excitation experiments involve irradiating a molecule at a frequency that is far from any molecular transition or at energies far from that of the difference between potential energy surfaces. In this case, if the field is not too strong, we can assume that $[W_e(\mathbf{X}) + \hbar\omega_i - W_{e'}(\mathbf{X})]^2 \gg 4|\varepsilon_i(t)\hat{\varepsilon}_i \cdot \mathbf{d}_{e',e}(\mathbf{X})|^2$ and we can expand the square root in Eq. (12.53) to obtain the following form for the two LIPs:

$$\lambda_1(\mathbf{X}, t) = W_e(\mathbf{X}) + \frac{|\varepsilon_i(t)\hat{\varepsilon}_i \cdot \mathbf{d}_{e',e}(\mathbf{X})|^2}{W_e(\mathbf{X}) + \hbar\omega_i - W_{e'}(\mathbf{X})},$$

$$\lambda_2(\mathbf{X}, t) = W_{e'}(\mathbf{X}) - \hbar\omega_i - \frac{|\varepsilon_i(t)\hat{\varepsilon}_i \cdot \mathbf{d}_{e',e}(\mathbf{X})|^2}{W_e(\mathbf{X}) + \hbar\omega_i - W_{e'}(\mathbf{X})}. \quad (12.55)$$

The second term in Eqs. (12.55) can be rewritten as the product of a *dynamic polarizability tensor* of the molecule, given as

$$\underline{\underline{\alpha}}(\omega_i, \mathbf{X}) \equiv \frac{\mathbf{d}_{e,e'}(\mathbf{X}) \otimes \mathbf{d}_{e',e}(\mathbf{X})}{W_{e'}(\mathbf{X}) - W_e(\mathbf{X}) - \hbar\omega_i}, \quad (12.56)$$

times the polarization directions, $\hat{\varepsilon}_i$, times the square of the field. (The $\otimes$ symbol denotes the *outer product* of two vectors. For example, $\mathbf{a} \otimes \mathbf{b}$, where $\mathbf{a}$ and $\mathbf{b}$ each have x, y, and z components, is a 3×3 matrix whose elements are $a_x b_x, a_x b_y$, etc.) That is, the field-dependent component $\Delta W(\mathbf{X}, t)$ of the LIP felt by the system is given by

$$\Delta W(\mathbf{X}, t) = -\hat{\varepsilon}_i \cdot \underline{\underline{\alpha}}(\omega_i, \mathbf{X}) \cdot \hat{\varepsilon}_i |\varepsilon_i(t)|^2. \quad (12.57)$$

This contribution may be viewed classically as being due to the interaction of an induced dipole, $\mathbf{d}^{\text{ind}}(t) = \underline{\underline{\alpha}}(\omega_i) \cdot \hat{\varepsilon}_i \varepsilon_i(t)$ created by the field, with the field that created it, with the energy of interaction being given by $\Delta W(\mathbf{X}, t) = -\varepsilon_i(t)\hat{\varepsilon}_i \cdot \mathbf{d}^{\text{ind}}$.

The only "nonclassical" aspect of the far off-resonance limit result, Eq. (12.57), is the dependence of the dynamic polarizability [Eq. (12.56)] on the nuclear-coordinates $\mathbf{X}$. This dependence is in accord with qualitative classical expectations since we expect different molecular configurations to be more easily polarizable than others. However, it is only quantum mechanics, via Eq. (12.56), that advises how to compute this shape-dependent polarizability.

This off-resonance LIP [Eq. (12.57)] gives rise to the so-called dipole force. This force has been used in the focusing [427, 428] and trapping [429] of atoms passing through strong (usually standing-wave) electromagnetic fields, and for nanodeposition (i.e., deposition on a nanometer size scale) of atoms on surfaces. In these experiments, the atoms are first cooled (e.g., by "laser cooling" [430]). They then pass through the optical standing waves, before impinging on a surface, where one observes the formation of periodic submicron atomic patterns. There are far fewer results for molecules (for which cooling is difficult), the most noteworthy being experiments [431, 432] and theory [433, 434] on the focusing of molecules using intense laser fields.

To better understand these effects, we consider an atom interacting with a standing-wave field of the form

$$\mathbf{E}_i(x, y, z, t) = 4\hat{\varepsilon}\mathcal{E}(z) \cos(ky) \cos(\omega_i t), \tag{12.58}$$

where the polarization direction $\hat{\varepsilon}$ is assumed to lie along the z axis. For nano-deposition cases, y is chosen to be parallel to the plane of the surface on which the atoms are to be deposited. It follows directly from Eq. (12.57) that an atom transversing this field will experience a periodic potential of the form,

$$\Delta W = -\alpha(\omega_i)\mathcal{E}^2(z) \cos^2(ky), \tag{12.59}$$

where, as appropriate for atoms, we have suppressed the $\mathbf{X}$ (nuclear coordinate) dependence of ΔW and $\underline{\alpha}(\omega_i)$. Further, the $\underline{\alpha}$ tensor was replaced by an α scalar because, for atoms, the induced dipole is always parallel to the external field. The situation in molecules will be examined later.

The force acting on the atom in the direction parallel to the surface is given by

$$F_y = -\frac{\partial \Delta W}{\partial y} = -k\alpha(\omega_i)\mathcal{E}^2(z) \sin(2ky). \tag{12.60}$$

For positive $\alpha(\omega_i)$ (i.e., when $W_{e'} - W_e - \hbar\omega_i > 0$), this force deflects the atoms towards the $y = n\pi/k$, $n = 0, 1, \ldots$ points, where the field intensity is maximal, and away from the $y = (n + 1/2)\pi/k$, $n = 0, 1, \ldots$ points, where the field intensity is zero. Close to one of the $y = n\pi/k$ stable points, we can expand the force to yield

$$F_y \approx \frac{\partial F_y}{\partial y}\left(y - \frac{n\pi}{k}\right) = -2k^2\alpha(\omega_i)\mathcal{E}^2(z)\left(y - \frac{n\pi}{k}\right). \tag{12.61}$$

Thus, when the atom encounters the standing waves close to one of the $y \approx n\pi/k$ points, it experiences a harmonic restoring force proportional to $y - n\pi/k$. Assuming, for simplicity, a "flat-top" field profile in the z direction $\mathcal{E}(z) = \mathcal{E}$ for $0 < z < a$ and $\mathcal{E}(z) = 0$ for $z \leq 0$ or $z \geq a$, the frequency of the harmonic motion near one of the $y \approx n\pi/k$ points is given according to Eq. (12.61) as

$$\omega = k\mathcal{E}[2\alpha(\omega_i)/M]^{1/2}, \tag{12.62}$$

where M is the atomic mass. Since for harmonic potentials the period is independent of the degree of stretching of the oscillator relative to its equilibrium point, and provided that $|\, y - n\pi/k|$ is small enough for Eq. (12.61) to hold, it follows that all the atoms entering the field region will reach the $y = n\pi/k$ points after a quarter of a period, that is, at a time equal to

$$\tau = \frac{\pi}{2\omega} = \frac{\pi}{2k\mathcal{E}} \left[\frac{M}{2\alpha(\omega_i)} \right]^{1/2}. \tag{12.63}$$

Therefore, if the incident atomic velocity v_z is normal to the surface, the best position to place the surface on which the atoms are to be deposited is at a distance of

$$\delta z = v_z \tau = \frac{v_z \pi}{2k\mathcal{E}} \left[\frac{M}{2\alpha(\omega_i)} \right]^{1/2}, \tag{12.64}$$

below the $z = 0$ line where the field starts. This will result in the best focusing of atoms at the $y = n\pi/k, n = 0, 1, \ldots$ points. This result, and a set of computed atomic trajectories leading to it, is demonstrated in Figure 12.5. Note, in particular, the focusing that occurs near 0.07 μs.

The focusing noted above will deteriorate due to: (a) the failure of the harmonic approximation for atoms hitting y points for which Eq. (12.61) does not hold, (b) a nonuniform v_z distribution, and (c) non-zero v_y transverse-velocity values, that is, deviations of the atomic directions from the normal. The last two effects can be minimized by cooling the atomic beam and selecting its directionality by using a sequence of slits.

The first effect cannot, however, be overcome in this manner. To estimate the fraction of atoms that will not focus well on the surface, we note that the range of harmonicity, which is the range of linearity of the $\sin 2ky$ function, is approximately $-\frac{1}{4} < 2ky < \frac{1}{4}$. This means that in each $[-\pi/2, \pi/2]$ interval, only a fraction of $\approx 1/2\pi \approx 16\%$ of the atoms will be tightly focused on the surface. This is so even if the surface is placed at the optimal distance of δz of Eq. (12.64) below the onset of the light field.

It is possible to overcome this lack of tight focusing by scattering away from the surface those atoms whose y impact parameter lies outside the $-1/8k < y < -1/8k$ tight-focusing range. One way of doing that is by changing the sign of $\alpha(\omega_i)$, thereby making the potential repulsive, for just these impact parameters. This can be accomplished [436], as explained below, by a straightforward application of the bichromatic control scenario of Section 3.1.1.

In essence, bichromatic control can be used to change the sign of $\alpha(\omega_i)$ as a function of y. That is, we prepare the molecular (or atomic) beam in a superposition of two energy eigenstates,

$$|\psi(t)\rangle = c_1 |E_1\rangle \exp(-iE_1 t/\hbar) + c_2 |E_2\rangle \exp(-iE_2 t/\hbar). \tag{12.65}$$

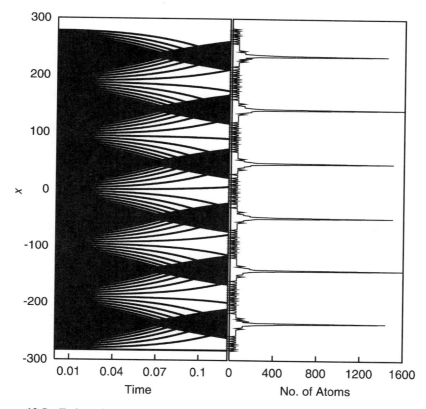

Figure 12.5 Trajectories and associated deposition for Rb atoms in the 16^2S Rydberg state passing through a field of intensity $I = 1.9 \times 10^7$ W/cm^2 and wavelength 188,495.6 nm. Ordinate shows position along y axis, parallel to the surface. Abscissa of the left one of the two figures shows time in microseconds associated with the paths of trajectories, that are themselves shown as lines, incident on surface. Note the visible "waist" associated with the focusing effect. Right panel shows the intensity of the atoms incident on the surface at the focal waist. (From Ref. [435].)

Instead of using the single-field Eq. (12.58), we pass the molecules in this superposition state through a bichromatic standing-wave field of the form:

$$\mathbf{E}(y, t) = 4\varepsilon[\mathcal{E}_1^{(0)} \cos(k_1 y) \cos(\omega_1 t) + \mathcal{E}_2^{(0)} \cos(k_2\, y + \theta_F) \cos(\omega_2 t)] \equiv \mathbf{E}(\omega_1) + \mathbf{E}(\omega_2).$$
$$(12.66)$$

As in the usual bichromatic control scenario, we use two laser frequencies that satisfy the $\omega_2 - \omega_1 = (E_1 - E_2)/\hbar$ relation; that is, we create two-pathway interference at the energy $E = E_1 + \hbar\omega_1 = E_2 + \hbar\omega_2$. Notice that as in the refractive index control discussed in Section 6.2, this energy is far off resonance from any molecular energy level.

The bichromatic off-resonance LIP obeys the same general relation to the field as does the monochromatic LIP, namely, $\Delta W(y) = -\mathbf{d}^{\text{ind}} \cdot \mathbf{E}(y, t)$. All that one need do is calculate the dipole induced in the material superposition state by the bichromatic field. Following our discussion of the control of refractive indices in Section 6.2, the induced dipole is given by

$$
\begin{aligned}
\mathbf{d}^{\text{ind}} &= [\underline{\alpha}^{\text{in}}(\omega_1) + \underline{\alpha}^n(\omega_1)] \cdot \mathbf{E}(\omega_1) + [\underline{\alpha}^{\text{in}}(\omega_2) + \underline{\alpha}^n(\omega_2)] \cdot \mathbf{E}(\omega_2) \\
&+ \underline{\alpha}^{\text{in}}(\omega_{2,1} + \omega_1) \cdot \mathbf{E}(\omega_{2,1} + \omega_1) + \underline{\alpha}^{\text{in}}(\omega_{2,1} - \omega_2) \cdot \mathbf{E}(\omega_{2,1} - \omega_2),
\end{aligned}
\tag{12.67}
$$

where

$$
\mathbf{E}(\omega_{2,1} + \omega_1) = 4\hat{\varepsilon}\mathcal{E}_1^{(0)} \cos(k_1 y) \cos[(\omega_{2,1} + \omega_1)t]
$$

and

$$
\mathbf{E}(\omega_{2,1} - \omega_2) = 4\hat{\varepsilon}\mathcal{E}_2^{(0)} \cos(k_2 y + \theta_F) \cos[(\omega_{2,1} - \omega_2)t].
$$

As explained in Section 6.2, $\underline{\alpha}^{\text{in}}(\omega)$ is that part of the polarizability tensor that results from the interference between the two paths associated with the field-dressed energy E. It is a function of the $c_1^* c_2$ coherence between the two states that make up the superposition state in Eq. (12.65). By contrast, $\underline{\alpha}^n(\omega)$ denotes the ordinary polarizability tensor, resulting from noninterfering terms. It is only a function of the populations, $|c_1|^2$ and $|c_2|^2$, of the $|E_1\rangle$ and $|E_2\rangle$ states.

The total polarizability at ω_i, $\underline{\alpha}^{\text{in}}(\omega_i) + \underline{\alpha}^n(\omega_i)$ is then a function of both θ_F and of θ_M, where $\theta_M = \arg(c_1/c_2)$. For example, Figure 12.6 shows the interference contribution to the $\alpha_{zz}(\omega)$ for rubidium at parameters indicated in the figure caption. The range of $\alpha^{\text{in}}(\omega)$ is seen to be large and the $\alpha^n(\omega)$ contribution (not shown) is insignificant on the scale of Figure 12.6.

Assuming for simplicity that (as in atoms) the induced dipole moment is pointing in the direction of the field [i.e., only $\alpha_{zz}(\omega)$ is operative], we obtain the bichromatic LIP as

$$
\Delta W(y) = -\mathbf{d}^{\text{ind}} \cdot \mathbf{E}(y, t) = V^n(y) + V^{\text{in}}(y),
\tag{12.68}
$$

where the noninterfering contribution is given by

$$
-V^n(y) = 2[4\mathcal{E}_1^{(0)^2} \cos^2(k_1 y)\alpha^n(\omega_1) + 4\mathcal{E}_2^{(0)^2} \cos^2(k_2 y + \theta_F)\alpha^n(\omega_2)],
\tag{12.69}
$$

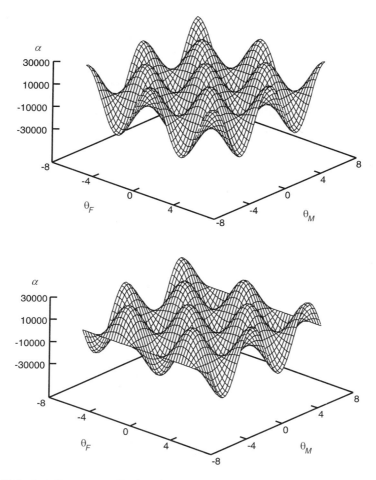

Figure 12.6 Interference contribution α^{in} to the polarizability (in a.u.) of Rb plotted against θ_M and θ_F for a superposition of the $|16s\rangle$ and $|16d\rangle$ atomic states. Two fields used are of intensity 19.1 W/cm^2 and 1.912×10^5 W/cm^2 at wavelengths of 1782.53 and 1832.31 nm. Upper plot is the real component of α^{in} and lower plot is the imaginary component. (From Ref. [435].)

and the interfering part is given by

$$-V^{\text{in}}(y) = 2[4\mathcal{E}_1^{(0)^2} \cos(k_1 y) \cos(k_2 y + \theta_F)\alpha_r^{\text{in}}(\omega_1) + 4\mathcal{E}_2^{(0)^2} \cos(k_1 y) \cos(k_2 y + \theta_F)\alpha_r^{\text{in}}(\omega_2)], \qquad (12.70)$$

where $\alpha_r^{\text{in}}(\omega_1)$ is the real part of the interference polarizability term [Eq. (6.21)]. In deriving these expressions, we have neglected the time-dependent parts of the LIP, emanating from the $\mathbf{E}(\omega_{2,1} + \omega_1)$ and $\mathbf{E}(\omega_{2,1} - \omega_2)$ terms because these terms oscillate with high frequency and were found not to affect the trajectory of the atoms as they pass through the field.

Depending on the value of the $c_1^* c_2$ coherence, where c_1 and c_2 are the coefficients of the superposition state of Eq. (12.65), the $\alpha_r^{in}(\omega_1)$ term can either be positive or negative. Thus, we can control the magnitude of the negative polarizability, and hence the strength of the repulsive LIP that is added to the usual attractive LIP experienced in the monochromatic case when $\hbar\omega_i \ll W_{e'} - W_e$. As explained above, the controlled addition of a repulsive potential allows us to repel atoms that hit certain ranges of the y impact parameters. In particular, we can make the potential that is outside the $-1/8k < y < -1/8k$ tight-focusing range repulsive, thereby rejecting the atoms impinging at these impact parameters, while leaving the potential inside this range essentially harmonic. Hence, this affords the possibility of achieving much better focusing.

This is in fact what happens. Figure 12.7 shows the results in the presence and absence of interference contributions for a superposition comprised of two vibra-

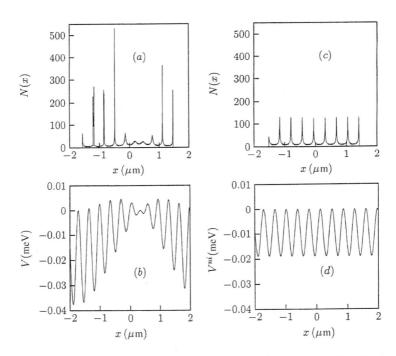

Figure 12.7 Molecular deposition of N_2 and associated optical potential for the initial superposition $0.2^{1/2}|0, 0, 0\rangle + 0.8^{1/2}|0, 2, 0\rangle$ due to $\Delta W(y) = V^{in}(y) + V^n(y)$ [(a) and (b)], and due to $\Delta W^{in}(y)$ only. (Here V^n is denoted V^{ni}). Here $|i, j, k\rangle$ denotes the state with vibrational quantum number i, rotational quantum number j, and projection k of total angular momentum along the z axis. System parameters are $\mathcal{E}_2^0/\mathcal{E}_1^0 = 1.0 \times 10^4$, $\mathcal{E}_1^0 = 1.0 \times 10^2$ V/cm, $\lambda_1 = 0.628$ μm, $\lambda_2 = 0.736$ μm, $\theta_F = -2.65$ radian, and $t_{int} = 0.625$ μs, where t_{int} is interaction time of molecules with field. (From Fig. 2, Ref. [436].)

tional states of the N_2 molecule. Figures 12.7a and 12.7b show the pattern of deposition, and the associated optical potential, for dynamics in the presence of $\Delta W(y) = V^{in}(y) + V^n(y)$. For comparison we show, in Figures 12.7c and 12.7d, the corresponding results assuming that there is no coherence between $|E_1\rangle$ and $|E_2\rangle$, i.e., neglecting $V^{in}(y)$. In the absence of molecular coherence the optical potential is seen to be (Fig. 12.7d) periodic, resulting in a series of short periodic deposition peaks (Fig. 12.7c). By contrast, the inclusion of interference contributions (Figs. 12.7a and 12.7b) result in significant enhancement and narrowing of peaks (full-width at half-maximum of less than 4 nm) as well as the appearance of an aperiodic potential and associated aperiodic deposition pattern.

Control can also be achieved by changing the relative phase θ_F between the two light fields, defined by Eq. (12.66). Figures 12.8a and 12.8b show significant differences in both the position and intensity of the peaks as a function of θ_F. By contrast, an analogous plot (not shown) where only V^n is included shows no variation in peak intensity as a function of θ_F. Similarly, Figures 12.8c and 12.8d show the strong

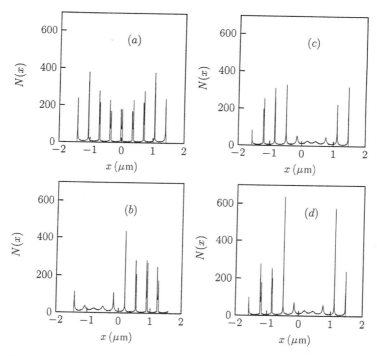

Figure 12.8 (a) and (b): Molecular deposition of N_2 associated with $0.8^{1/2}|0, 0, 0\rangle + 0.2^{1/2}|1, 2, 0\rangle$ for varying θ_F; i.e., (a) $\theta_F = 0$, and (b) $\theta_F = 2.0$ radians. (c) and (d): Sample variation of deposition with changes in $|c_1|$, $|c_2|$: (c) $[0.99^{1/2}]|0, 0, 0\rangle + 0.1|0, 2, 0\rangle$ and (d) $0.4^{1/2}|0, 0, 0\rangle + 0.6^{1/2}|0, 2, 0\rangle$. Other parameters and notation are as in Fig. 12.7. (From Fig. 3, Ref. [436].)

dependence of the deposition upon the magnitude of the coefficients of the created superposition.

12.5.2 Strong-Field Alignment

This discussion can be extended to consider the *alignment* effect that strong laser fields can have on molecules. That is, in addition to focusing, the off-resonance LIP tends to align molecules in space [68, 437–446]. The alignment effect was first observed in multiphoton ionization experiments of diatomic molecules by Normand et al. [437]. It was explained in terms of the off-resonance LIP by Friedrich and Herschbach [439], who showed, as described below, that rather than rotate freely, molecules in the presence of the field execute a "pendular"-like motion. Trapping and strong alignment of molecules [444–446] was subsequently observed, as well as the acceleration of rotational motion of laser-aligned molecules by the *optical centrifuge* effect [447–449], which is also discussed below.

The alignment effect is seen to emanate from Eq. (12.57) by noting that, in molecules, the induced dipole is not necessarily parallel to the field that induced it. In fact, the induced dipole can have three perpendicular components, d^X, d^Y, and d^Z, in the X, Y, Z molecular-fixed coordinate system. Given these components, we can express $\hat{\varepsilon}_i \cdot \mathbf{d}_{e'e}$, the projection of the transition dipole onto the laboratory-fixed z axis, appearing in Eq. (12.55) as [450]

$$\hat{\varepsilon}_i \cdot \mathbf{d} = d^Z \cos \Theta + \sin \Theta[-d^X \cos \xi + d^Y \sin \xi] = d^Z \cos \Theta - d^\perp \sin \Theta \cos(\xi + \beta),$$
$$(12.71)$$

where (see Fig. 12.9), Θ is the polar angle of orientation of the molecular Z axis relative to the laboratory z axis (along which the field polarization $\hat{\varepsilon}$ lies); ξ is an azimuthal angle describing rotation of the molecule about the molecular Z axis; and

$$d^\perp \equiv \sqrt{|d^X|^2 + |d^Y|^2} \quad \text{and} \quad \tan \beta \equiv \frac{d^Y}{d^X}. \qquad (12.72)$$

For brevity we have omitted the e', e subscripts.

Using Eq. (12.71) in Eq. (12.56), we have that

$$\hat{\varepsilon} \cdot \underline{\underline{\alpha}}(\omega_i) \cdot \hat{\varepsilon}_i = \frac{|d^Z \cos \Theta - d^\perp \sin \Theta \cos(\xi + \beta)|^2}{W_{e'} - W_e - \hbar\omega_i}. \qquad (12.73)$$

Therefore, according to Eq. (12.57), the molecule feels a LIP ΔW that depends on the orientation of the molecular Z axis relative to the field polarization and the azimuthal rotation angle of the molecular frame about the Z axis.

Consider then the case of diatomic molecules in Σ electronic states. Here the e and e' electronic states have cylindrical symmetry about the Z axis, and the transition dipole matrix element $\mathbf{d}_{e,e'}$ is parallel to the molecular (Z) axis. Hence only the d^Z

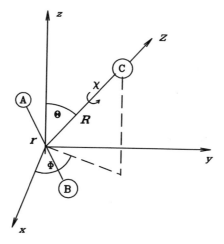

Figure 12.9 Body-fixed and space-fixed coordinates of a triatomic molecule.

component survives. In this case, the polarizability tensor [Eq. (12.56)] reduces to $\underline{\underline{\alpha}}^{\parallel}(\omega_i)$, which is

$$\underline{\underline{\alpha}}^{\parallel}(\omega_i) = \frac{\mathbf{d}^Z \otimes \mathbf{d}^Z}{W_{e'} - W_e - \hbar\omega_i}, \qquad (12.74)$$

and the LIP reduces to

$$\Delta W(t) = -\hat{\varepsilon}_i \cdot \underline{\underline{\alpha}}^{\parallel}(\omega_i) \cdot \hat{\varepsilon}_i |\varepsilon_i(t)|^2 = -\frac{|\mathbf{d}^Z \varepsilon_i(t)|^2 \cos^2 \Theta}{W_{e'} - W_e - \hbar\omega_i}. \qquad (12.75)$$

The resultant force tends to align the molecule along the laboratory z axis, that is, toward $\Theta = 0$.

The other diatomic extreme exists when the e electronic state is of Σ symmetry, and the e' electronic state is of Π symmetry (i.e., it has one unit of angular momentum about the molecular axis). In that case the dipole matrix element $\mathbf{d}_{e,e'}$ is perpendicular to the molecular axis, that is, only the $\mathbf{d}_{e,e'}^{\perp}$ vector survives. The polarizability (Eq. (12.56)] now reduces to $\underline{\underline{\alpha}}^{\perp}(\omega_i)$:

$$\underline{\underline{\alpha}}^{\perp}(\omega_i) = \frac{\mathbf{d}^{\perp} \otimes \mathbf{d}^{\perp}}{W_{e'} - W_e - \hbar\omega_i}, \qquad (12.76)$$

and the LIP is

$$\Delta W(t) = -\frac{|\mathbf{d}^\perp \varepsilon_i(t)|^2 \sin^2 \Theta \cos^2(\xi + \beta)}{W_{e'} - W_e - \hbar\omega_i}. \tag{12.77}$$

Averaging the $\cos^2(\xi + \beta)$ term of the rapidly revolving ξ angle yields $\frac{1}{2}$, and we obtain for the LIP for this case that

$$\Delta W(t) = -\frac{|\mathbf{d}^\perp \varepsilon_i(t)|^2 \sin^2 \Theta}{2(W_{e'} - W_e - \hbar\omega_i)}. \tag{12.78}$$

This potential tends to align the molecule *perpendicular* to the z axis, that is, toward $\Theta = \pi/2$.

In general, more than two electronic states contribute to the polarizability and to the LIP. To include more than one excited state we write $\underline{\underline{\alpha}}(\omega_i)$ as

$$\underline{\underline{\alpha}}(\omega_i) = \sum_{e'} \frac{\mathbf{d}_{e,e'} \otimes \mathbf{d}_{e',e}}{W_{e'} - W_e - \hbar\omega_i}. \tag{12.79}$$

For diatomic molecules some of the transition dipole matrix elements will be parallel to the molecular axis and some will be perpendicular to it. Hence

$$\underline{\underline{\alpha}}(\omega_i) = \sum_{e'\parallel} \frac{\mathbf{d}_{e,e'} \otimes \mathbf{d}_{e'e}}{W_{e'} - W_e - \hbar\omega_i} + \sum_{e'\perp} \frac{\mathbf{d}_{e,e'} \otimes \mathbf{d}_{e',e}}{W_{e'} - W_e - \hbar\omega_i}, \tag{12.80}$$

and the LIP can be written as

$$\Delta W(t) = -\varepsilon_i^2(t)[\alpha^\parallel(\omega_i)\cos^2\Theta + \tfrac{1}{2}\alpha^\perp(\omega_i)\sin^2\Theta], \tag{12.81}$$

where

$$\begin{aligned} \alpha^\parallel(\omega_1) &\equiv \sum_{e'\parallel} \frac{|\mathbf{d}_{e',e}^Z|^2}{W_{e'} - W_e - \hbar\omega_i}, \\ \alpha^\perp(\omega_i) &= \sum_{e'\perp} \frac{|\mathbf{d}_{e',e}^\perp|^2}{W_{e'} - W_e - \hbar\omega_i}. \end{aligned} \tag{12.82}$$

The LIP of Eq. (12.81) has two components, one attempting to align the molecule perpendicular to the field and one trying to align it parallel to the field. In diatomic molecules, the latter is usually much larger. Hence, the molecule attempts to align itself along the field direction. However, due to its initial rotational energy, the molecule cannot do so instantaneously. Since the dynamic polarizability is positive for $\hbar\omega_i \ll W_{e'} - W_e$, the LIP of Eq. (12.81) is usually negative, that is, it is purely attractive. Hence the initial effect of the LIP is to accelerate the rotational motion whenever Θ nears the potential minimum. However, in addition to the aligning force, the molecule feels the deflection force that draws it toward the high field region. As it gets drawn more and more into the high field region it feels a greater and greater aligning force that further accelerates its rotational motion. Since the LIP is a function of both y and Θ these motions are no longer separable and energy can flow from the rotational motion to the translational motion and vice versa. It is therefore entirely possible that the molecule will lose enough kinetic energy in the Θ coordinate in favor of the y coordinate to have insufficient energy to execute a full rotation. At that point the motion (which resembles that of a pendulum) is called *pendular*. Pendular motions are routinely observed in other situations, especially for molecules with a permanent dipole moment in a direct current (DC) electric field [451].

The reverse can also happen, namely that the molecule will accelerate its rotational motion at the expense of the y motion and become trapped due to the y dependence of the LIP well. The probability of either phenomenon occurring depends on the degree of cooling of both the rotational motion and the translational motion of the molecule prior to its entrance into the high field region [439, 452]. If the temperature is sufficiently low, the field sufficiently high, and relaxation mechanisms to other degrees of freedom possible, trapping [429] and alignment [445] of molecules by highly off-resonance light fields becomes feasible. Once a molecule is trapped and aligned, it can be manipulated to control a variety of processes, like enhancing reactivity [445, 453].

An interesting application of the optical alignment effect is the ability to accelerate the rotational motion and impart to the molecule hundreds of $\hbar$ units of angular momentum. In this case, one *rotates* the direction of polarization in space, causing a molecule to rotate with the field polarization, onto which it is aligned. By accelerating the rotational motion of the field polarization one accelerates the rotational motion of the molecule aligned with it. This phenomenon, which was suggested by Ivanov et al. [447, 449] and demonstrated experimentally by Corkum's group [448], is termed an *optical centrifuge*.

To best understand how the optical centrifuge works, it is convenient to consider the field as polarized in the x–y plane, rather than along the z axis as above. Here we reserve the z axis for the direction of quantization of the molecular angular momentum, J.

Considering now $\hat{\varepsilon}$ to lie in the x–y plane, and remembering that for diatomic molecules, ξ, the azimuthal angle describing rotation of the molecule about the molecular Z axis (see Fig. 12.9), can be chosen to be 0, the projections of $\mathbf{d}$ on x and y [instead of on z as was done in Eq. (12.71)], are given by [450]

$$\begin{aligned}
d^x &= d^Z \sin\Theta\cos\Phi + d^X\cos\Theta\cos\Phi - d^Y\cos\Theta\sin\Phi, \\
d^y &= d^Z \sin\Theta\sin\Phi + d^X\cos\Theta\sin\Phi + d^Y\cos\Theta\cos\Phi,
\end{aligned} \tag{12.83}$$

where (see Fig. 12.9) Φ and Θ are the azimuthal and polar angle of orientation of the molecular Z axis relative to the laboratory z axis (the axis of quantization).

Consider now a field whose linear polarization rotates in the x–y plane:

$$\mathbf{E}_i(t) = \varepsilon(t)\cos(\omega_i t)[\hat{\mathbf{x}}\cos\Phi_L(t) + \hat{\mathbf{y}}\sin\Phi_L(t)], \qquad (12.84)$$

where $\hat{\mathbf{x}}$ and $\hat{\mathbf{y}}$ are unit vectors in the x and y directions, respectively. We have that $\hat{\varepsilon}_i \cdot \mathbf{d}$ is now given as

$$
\begin{aligned}
\hat{\varepsilon}_i \cdot \mathbf{d} &= \hat{\varepsilon}_x \mathrm{d}^x + \hat{\varepsilon}_y \mathrm{d}^y = \mathrm{d}^Z \sin\Theta\cos[\Phi - \Phi_L(t)] + \mathrm{d}^X \cos\Theta\cos[\Phi - \Phi_L(t)] \\
&\quad - \mathrm{d}^Y \cos\Theta\sin[\Phi - \Phi_L(t)],
\end{aligned}
\qquad (12.85)
$$

or

$$\hat{\varepsilon}_i \cdot \mathbf{d} = \hat{\varepsilon}_x \mathrm{d}^x + \hat{\varepsilon}_y \mathrm{d}^y = \mathrm{d}^Z \sin\Theta\cos[\Phi - \Phi_L(t)] + \mathrm{d}^\perp \cos\Theta\cos[\Phi + \beta - \Phi_L(t)], \qquad (12.86)$$

where $\mathrm{d}^\perp$ and β are defined in Eq. (1.8). Substituting Eq. (12.86) in Eqs. (12.56) and (12.57) the LIP is obtained as

$$\Delta W(t) = \varepsilon_i^2(t)\frac{|\mathrm{d}^Z \sin\Theta\cos[\Phi - \Phi_L(t)] + \mathrm{d}^\perp \cos\Theta\cos[\Phi + \beta - \Phi_L(t)]|^2}{W_{e'} - W_e - \hbar\omega_i}. \qquad (12.87)$$

The first (parallel) term in the numerator creates an LIP that tends to align the molecule in the polarization (x–y) plane, whereas the second term tends to align the molecule along the z axis—perpendicular to the polarization plane. In both cases the molecule will execute a nutational motion in which it rotates with the frequency of $\dot{\Phi}_L(t)$ about the z axis and executes a pendular motion in Θ. For the parallel term of the LIP the equilibrium point for the pendular motion is $\Theta = \pi/2$, that is, in the x–y polarization plane. For the perpendicular part, the molecular pendulum moves about $\Theta = 0$, that is, perpendicular to the polarization plane.

In practice it is possible to rotate the direction of the polarization by building the linearly polarized field of Eq. (12.84) as a sum of two *circularly polarized* light fields whose phases vary (the phase variance with time is known as "chirp") in the opposite sense:

$$
\begin{aligned}
\mathbf{E}_i(t) &= \frac{\varepsilon(t)}{2}\{\hat{\mathbf{x}}\cos[\omega_i t + \Phi_L(t)] + \hat{\mathbf{y}}\sin(\omega_i t + \Phi_L(t))\}. \\
&\quad + \frac{\varepsilon(t)}{2}\{\hat{\mathbf{x}}\cos[\omega_i t - \Phi_L(t)] + \hat{\mathbf{y}}\sin[\omega_i t - \Phi_L(t)]\}.
\end{aligned}
\qquad (12.88)
$$

The time dependence of $\Phi_L(t)$ therefore determines the rate of rotation of the linearly polarized field. A molecule trapped in the minimum of the LIP would tend to follow faithfully the rotation of the linearly polarized field. As the rate of this rotation is accelerated so is the rate of the molecular rotation of the molecule.

The optical centrifuge is capable of imparting so much rotational energy to a diatomic molecule that the bond breaks. To see how this happens, note that the effective radial potential seen by a diatomic molecule is a sum of its "real" potential $V(R)$ and a centrifugal one associated with its angular motion. That is, the vibrational motion sees the effective potential

$$V_{\text{eff}}(R) = V(R) + \frac{\hbar^2 J(J+1)}{2\mu R^2},\qquad(12.89)$$

with a J-dependent equilibrium distance $R_{\text{eq}}(J)$, given by the implicit equation

$$R_{\text{eq}} = \left\{ \frac{\hbar^2 J(J+1)}{\mu(dV/dR|_{R_{\text{eq}}})} \right\}^{1/3}.$$

Figure 12.10 displays the effective potentials for the Cl_2 molecule [449] and shows how the binding energy of the molecule is reduced with increasing J, until, at $J = 400$, the effective potential is completely repulsive and the molecule dissociates.

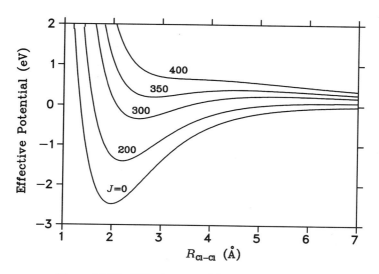

Figure 12.10   Effective potentials as a function of J.

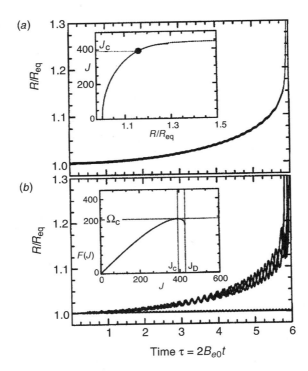

Figure 12.11 Sample trajectories for Cl_2. Pulse is 70 ps long (equal to one ground rotational period of Cl_2, corresponding to $\tau = 2\pi$ in dimensionless units), with 5.5 ps turn-on; intensity is 1.7×10^{13} W/cm^2; $\omega_{final} = 0.02$ eV. J_C indicates "critical" J at which the system breaks apart (which is implicit in Fig. 12.10). (a) Quiet trajectory; inset shows excellent agreement between numerical calculation (dashed line) and analytic approximation (solid line). (b) Four sample trajectories $R(t)$ for Cl_2. Initially, $J = 30$ and a random orientation was chosen. Three trajectories show dissociation, and the fourth was not trapped. Inset shows $F(J) \equiv J/\mu R_{eq}^2(J)$. (Taken from Fig. 1, Ref. [447].)

Sample trajectories of the Cl_2 molecule on its way to dissociation are shown in Figure 12.11. We see a "quiet" trajectory, which follows adiabatically the potential minimum as the system is rotated more and more by the LIP until it breaks apart, as well as "nonquiet" trajectories, which deviate from the position of the minimum.

CHAPTER 13

CASE STUDIES IN OPTIMAL CONTROL

Chapter 4 introduced the essential principles of optimal control. Here we describe a number of applications to the control of molecular processes.

13.1 CREATING EXCITED STATES

One of the earliest objectives of optimal control studies was to control the population of a specified vibrational state. Interest in this topic dates back to the early days of Infrared multiphoton dissociation (IRMPD) [454], when people sought methods to preferentially populate a given vibrational state before, as was commonly thought, the deleterious effects of intramolecular vibrational relaxation (IVR) set in. In particular, the ideas discussed in Section 3.5 were prevalent; that is, it was thought that one could "beat" out the rate of IVR by populating a given level while it remains "pure." However, as discussed in Section 3.5, the correct way to achieve this objective in the long-time limit is to focus on properties of energy eigenstates, and not on the time dependence of IVR.

The ability to populate specific energy levels using laser pulse sequences has been realized in a number of ways. An example is the STIRAP approach discussed in Section 9.1. A number of alternative approaches are discussed below.

Consider first the less general, though nontrivial, task of populating a target wave packet of states by a perturbative N-photon process. In particular we focus on the preparation of "bright" states, with the term "bright" to be defined precisely below.

Consider a molecule with Hamiltonian H_M, with energy eigenstates $|E_n\rangle$ of energy E_n, subjected to an optical pulse. Specifically, consider N-photon absorption by the molecule that is initially in state $|E_i\rangle$. We define two classes of states, *bright* states, $|\phi_s\rangle$, that are accessible by N-photon absorption from $|E_i\rangle$, and *dark* states $|\chi_m\rangle$ that are not accessible by this process. That is, the dark states satisfy

$$\langle E_i|T^{(N)}(\omega)|\chi_m\rangle \approx 0, \tag{13.1}$$

where $T^{(N)}(\omega)$ is the lowest order N-photon transition operator [whose three-photon analog is given in Eq. (3.44)], given by

$$T^{(N)}(\omega) = \prod_{k=1,2,\ldots,N-1} [\hat{\epsilon} \cdot \mathbf{d}(E_i - H_M + k\hbar\omega)^{-1}\hat{\epsilon} \cdot \mathbf{d}]. \tag{13.2}$$

For one-photon transitions $T^{(N)}(\omega) = \hat{\epsilon} \cdot \mathbf{d}$.

Whenever the lowest (Nth) order perturbation theory for the N-photon problem is valid, it is possible to generate a wave packet of bright states, assuming that such bright states exist. To see this, partition the excited state manifold into bright states $|\phi_s\rangle$ and dark states $|\chi_m\rangle$. The eigenstates $|E_n\rangle$ of the molecular Hamiltonian H_M can therefore be written as

$$|E_n\rangle = \sum_s |\phi_s\rangle\langle\phi_s|E_n\rangle + \sum_m b_{m,n}|\chi_m\rangle. \tag{13.3}$$

Consider then excitation with an optical pulse of the form

$$\varepsilon(t) = \int d\omega\, \epsilon(\omega)\{\exp(-i\omega\mathbf{t}) + \text{c.c.}\}, \tag{13.4}$$

where $\mathbf{t} \equiv t - z/c$, and z is the direction of propagation, incident on the molecule in state $|E_i\rangle$. By analogy with Section 3.3.1, at the end of the pulse, and in the (Nth-order) perturbative limit, the molecule will be in a superposition state of the form

$$\begin{aligned}
|\psi(t)\rangle &= \frac{2\pi i}{\hbar}\sum_n \bar{\epsilon}^N(\omega_n)|E_n\rangle\langle E_n|T^{(N)}(\omega_n)|E_i\rangle \exp\left(-\frac{iE_n t}{\hbar}\right)\\
&= \frac{2\pi i}{\hbar}\sum_{n,s}\langle\phi_s|T^{(N)}(\omega_n)|E_i\rangle\bar{\epsilon}^N(\omega_n)|E_n\rangle\langle E_n|\phi_s\rangle \exp\left(-\frac{iE_n t}{\hbar}\right),
\end{aligned} \tag{13.5}$$

where $\omega_n \equiv (E_n - E_i)/N\hbar$, $\bar{\epsilon}(\omega_n) \equiv \epsilon(\omega_n)\exp(i\omega_n z/c)$ and where we have used Eq. (13.1) in the last equality.

If the frequency width of the pulse is sufficiently large (i.e., the pulse is sufficiently short in time) and ω_n is far from an intermediate resonance, then we can invoke the SVCA [Eq. (10.13)]. Alternatively, the power broadening may be large enough to smooth out the intermediate resonances. (Note that *power broadening* is the tendency of spectral absorption lines to broaden when measured using intense fields [215].) In this case invoking the SVCA means assuming that $T^{(N)}(\omega_n)\bar{\epsilon}^N(\omega_n)$

varies more slowly with n than does $\langle E_n | \phi_s \rangle$. We can then take $T^{(N)}(\omega_n)\bar{\epsilon}^N(\omega_n)$ out of the sum in Eq. (13.5), approximate it by a constant $T^{(N)}(\bar{\omega})\bar{\epsilon}^N(\bar{\omega})$, and obtain,

$$
\begin{aligned}
|\psi(t)\rangle &= \frac{2\pi i}{\hbar} \sum_s \langle \phi_s | T^{(N)}(\bar{\omega}) | E_i \rangle \bar{\epsilon}^N(\bar{\omega}) \sum_n |E_n\rangle \exp\left(-\frac{iE_n t}{\hbar}\right) \langle E_n | \phi_s \rangle \\
&= \frac{2\pi i}{\hbar} \sum_s \langle \phi_s | T^{(N)}(\bar{\omega}) | E_i \rangle \bar{\epsilon}^N(\bar{\omega}) \exp\left(-\frac{iHt}{\hbar}\right) |\phi_s\rangle.
\end{aligned}
\tag{13.6}
$$

It follows from Eq. (13.6) that immediately after the pulse (defined as $t = 0$) is off the wave function is

$$
|\psi(t)\rangle = \exp\left(-\frac{iHt}{\hbar}\right) \sum_s A_s |\phi_s\rangle,
\tag{13.7}
$$

where $A_s = (2\pi i)/\hbar \langle \phi_s | T^{(N)}(\bar{\omega}) | E_i \rangle \bar{\epsilon}^N(\bar{\omega})$. Hence we see that in the Nth-order perturbative regime a sufficiently short pulse can indeed create a wave packet composed of pure bright states, at least at very short times after the end of the pulse.

This result can be generalized to the preparation of other types of "zero-order" states. For example, in accord with the objectives of *mode-selective chemistry* (Section 3.5), we may want to prepare a specific *local mode* vibrational state. To do so, consider [455], an M-level oscillator that has the "right" anharmonicity, that is, a system whose energy levels behave like

$$
E_v = E_0 + \hbar v[\omega_{1,0} - \Delta^a(v - 1)].
\tag{13.8}
$$

Here E_v is the vth vibrational energy level with wave function ϕ_v, $\omega_{1,0}$ is an harmonic frequency, and Δ^a is the anharmonicity constant. Under certain circumstances a system of this kind, initially in its ground state, and driven by a cw field

$$
\varepsilon(t) = \epsilon_0 \cos(\omega t)
\tag{13.9}
$$

is equivalent to a two-level system. To see this, expand the wave function at time t, $\psi(t)$, as

$$
\psi(t) = \sum_v a_v \phi_v \exp\left(-\frac{iE_v t}{\hbar}\right).
\tag{13.10}
$$

Substituting Eq. (3.10) into the time-dependent Schrödinger equation gives a set of coupled equations for the a_v coefficients of the form

$$
i\frac{d}{d\tau} a_v = [(n - v) - S_n]v a_v - F_0[(v + 1)^{1/2} a_{v+1} + v^{1/2} a_{v-1}], \qquad v = 0, 1, 2, \ldots, M,
\tag{13.11}
$$

where $\tau \equiv \Delta^a t$ is a dimensionless time:

$$S_n \equiv n - 1 - (\omega_{1,0} - \omega)/\Delta^a, \tag{13.12}$$

and

$$F_0 \equiv d_{1,0}\epsilon_0/(2\hbar\Delta^a), \tag{13.13}$$

is a dimensionless Rabi frequency.

If $F_0 \ll 1$, then the above set of equations is identical to a two-level system,

$$i\frac{d}{d\tau}a_0 = -F_0^{(n)}a_n,$$

$$i\frac{d}{d\tau}a_n = S_n n a_n - F_0^{(n)}a_0, \tag{13.14}$$

where

$$F_0^{(n)} = F_0^n n/[(n-1)!(n!)^{1/2}]. \tag{13.15}$$

These equations can be solved in the usual way to yield

$$P_n(\tau) \equiv |a_n(\tau)|^2 = (F_0^{(n)}/\Omega_n)^2 \sin^2 \Omega_n \tau, \tag{13.16}$$

where

$$\Omega_n = \{(S_n n/2)^2 + (F_0^{(n)})^2\}^{1/2} \tag{13.17}$$

is the n-photon Rabi frequency.

When ω is tuned to the n-photon resonance frequency,

$$\omega = \omega_{1,0} - \Delta^a(n-1), \tag{13.18}$$

$S_n = 0$ and $P_n(\tau)$ reaches unity at the reduced time $\tau_p = \pi/(2\Omega_n) = \pi/[2F_0^{(n)}]$. Since $F_0 \ll 1$, then $\tau_p \gg 1$, and the time taken to attain complete population transfer to the nth level is very long. The same applies to the use of optimal pulses in populating rotational states [116] and to the use of adiabatic passage, discussed in Section 9.1. Further, it was found [455] that increasing F_0, which shortens τ_p, results in complete loss of selectivity. Basically, in this case the power broadening increases so much that all n-photon resonances overlap.

Paramonov et al. [456–460] solved this problem by using a short pulse of the form

$$\varepsilon(t) = \epsilon_0 \sin^2(\alpha t) \cos(\omega t). \tag{13.19}$$

This pulse does not result in the coalescence of the n-photon resonances even when F_0 is increased because when $\alpha \ll \omega$ one obtains a set of equations that are similar to Eq. (13.11):

$$i\frac{d}{d\tau}a_v = [(n-v) - S_n]va_v - F_0\sin^2\left(\frac{\alpha\tau}{\Delta^a}\right)[(v+1)^{1/2}a_{v+1} + v^{1/2}a_{v-1}],$$

$$v = 0, 1, 2, \ldots, M. \quad (13.20)$$

Hence [456] for a certain range of ω, F_0 and α, it is possible to preferentially excite any level n that one desires.

The pulse considered by Paramonov and Savva [456] can be thought of as a sum of three "rectangular" pulses:

$$\varepsilon(t) = \tfrac{1}{4}\epsilon_0\{2\cos(\omega t) - \cos[(\omega+\alpha)t] - \cos[(\omega-\alpha)t]\}, \quad (13.21)$$

which have two frequencies in addition to the central frequency of Eq. (13.9). Thus, in addition to the shortness of the pulse, multiphoton excitation with $\sin^2(\alpha t)$ modulation is more efficient because it provides additional frequencies that help overcome frequency mismatches. In fact, a pulse of the type

$$\varepsilon(t) = \epsilon_0\sin^{2m}(\alpha t)\cos(\omega t), \quad (13.22)$$

has also been considered by Paramonov and Savva [457]. It has, depending on the value of m and α, all the "right" resonance frequencies for a sequential multiphoton process, since $\varepsilon(t)$ can be rewritten as

$$\varepsilon(t) = \epsilon_0\sum_{k=-m}^{m}C_k\cos[(\omega+2k\alpha)t]. \quad (13.23)$$

Comparing $\omega + 2k\alpha$ to the level spacing $\omega_{1,0} - 2v\Delta^a$ we see that all the anharmonic frequencies, up to level m, are contained in the pulse if α is chosen equal to Δ^a.

The numerical simulation of Paramonov and Savva [456, 457], were later confirmed by Manz et al. [113, 114, 125] using optimal control theory (OCT). These computations show that all of the frequencies in the pulse are important. However, the continuum of frequencies afforded by the leading and trailing edges of the pulse was found to be even more important for the overall process.

13.1.1 Using Prepared States

Having shown that it is possible to prepare bright states, or to prepare specific vibrational states in particular cases, we consider the utility of such states. If the task is merely to control the populations of stable molecules, then the discussion above demonstrates the possibility of doing so. Similarly, for example, Rabitz et al. [110, 318] have shown that it is possible to control the vibrational states of local bonds in a chain of harmonic oscillators.

Alternatively, such prepared excited states may prove useful photochemically under particular circumstances. This is especially true for local-mode-type molecules [461, 462], that is, molecules for which vibrational eigenstates resemble localized excitation in individual bonds. As an example, in the case of HOD, the large frequency difference between the OH and OD oscillators is such that intramolecular vibrational relaxation does not destroy the localized excitation. (Similar effects arise if one excites a resonance state that displays local behavior; see, for example, Ref. [463] for an ABA-type molecule.) As shown theoretically [464, 465], and confirmed experimentally [53–60], preparation of the OH stretch followed by an excitation laser leading to dissociation gives a marked enhancement of the H atom photodissociation in many molecules.

Similarly, in the case of bimolecular reactions, Zare's group [466] confirmed theoretical predictions and demonstrated experimentally [467–469] that by exciting *either* the OH *or* the OD bond in HOD one can selectively enhance product formation in a subsequent H + HOD reaction. Specifically, when the OH bond is excited, the reaction yields H_2 + OD, whereas when the OD bond is excited, H reacts with HOD to form the HD + OH product. In these experiments, the OH was prepared either by overtone excitation [57, 58] to the fourth vibrational level $v = 4$ or by excitation to the $v = 1$ state by Raman pumping [102]. As yet to be verified experimentally is the computational prediction of Manz et al. [124, 125] that strong optimized pulses can also achieve selective excitation of higher lying vibrational states.

Finally, we note that vibrational excitation can also have an inhibitory effect, which also results in great selectivity, as shown for the B state photodissociation of HOD [470].

13.2   OPTIMAL CONTROL IN THE PERTURBATIVE DOMAIN

In general, when perturbation theory applies, one can devise an elegant method for obtaining the optimal solution to problems such as those in photodissociation. Often, optimal fields derived in the perturbative domain [92, 126, 129–131] do not differ by much from the fields derived via the more general "brute-force" optimization methods.

To examine a particular case, consider the pump–dump control scenarios discussed in Section 3.5 and Chapter 4. Here a sequence of two pulses serves to first excite a molecule to a set of intermediate bound states with wave function $|\psi_i\rangle$ and energy E_i, and then to dissociate these states. We denote the two pulses $(k = 1, 2)$ by

$$\varepsilon_k(t) = \int d\omega \{\epsilon_k(\omega) \exp\left[-i\omega\left(t - \frac{z}{c}\right)\right] + \text{c.c.}\} = \int d\omega \{\bar{\epsilon}_k(\omega) \exp[-i\omega t] + \text{c.c.}\},$$

$$k = 1, 2 \quad (13.24)$$

where $\bar{\epsilon}_k(\omega) \equiv \epsilon_k(\omega) \exp(i\omega z/c)$ with z being the propagation direction. Here the $\varepsilon_k(t)$ peaks at time t_k [e.g. Eq. (3.72)] and τ is the delay time $\tau = t_2 - t_1$.

After the second pulse the probability $P_q(E)$ of observing a given product channel q at energy E is given by

$$P_q(E) = \frac{4\pi^2}{\hbar^4} \sum_{\mathbf{n}} | \sum_i \bar{\epsilon}_1(\omega_{i,1}) \mathrm{d}_{i,g} c_{i,1}(\tau) \bar{\epsilon}_2(\omega_{E,i}) \mathrm{d}_{q,\mathbf{n};i}(E)|^2, \tag{13.25}$$

where

$$\mathrm{d}_{i,g} \equiv \langle E_i | \hat{\epsilon} \cdot \mathbf{d} | E_1 \rangle, \tag{13.26}$$

Here the $c_{i,1}$ coefficients are the eigenstate coefficients resulting from excitation with the first pulse, defined in Eq. (2.15), and the

$$\mathrm{d}_{q,\mathbf{n};i}(E) \equiv \langle E, \mathbf{n}, q^- | \hat{\epsilon} \cdot \mathbf{d} | E_i \rangle, \tag{13.27}$$

are the transition dipole matrix elements between the ith intermediate state and the scattering states.

The probability of observing a given q product channel irrespective of the energy is then

$$P_q = \int dE \, P_q(E). \tag{13.28}$$

Equations (13.25) and (13.28) can be conveniently rewritten as

$$P_q = \sum_{i,j} d^{(q)}_{i,j} \varepsilon_{i,j} c_{i,1}(\tau) c^*_{j,1}(\tau), \tag{13.29}$$

where $\omega_{i,j} = (E_i - E_j)/\hbar$,

$$d^{(q)}_{i,j} \equiv \frac{4\pi^2}{\hbar^4} \int dE \sum_v \bar{\epsilon}_2(\omega_{E,i}) \bar{\epsilon}^{(2)*}(\omega_{E,j}) \mathrm{d}_{q,\mathbf{n};i}(E) \mathrm{d}^*_{q,\mathbf{n};j}(E), \tag{13.30}$$

and

$$\varepsilon_{i,j} \equiv \bar{\epsilon}_1(\omega_{i,1}) \bar{\epsilon}^{(1)*}(\omega_j) \mathrm{d}_{i,g} \mathrm{d}^*_{j,g}. \tag{13.31}$$

We now consider maximizing either the probability P_q in a single channel q or the selectivity of one channel in preference to another, that is, $P_{q_1} - P_{q_2}$. Optimization is carried out subject to the constraint of fixed average pulse power, that is,

$$J^{(k)} \equiv \int d\omega \, \varepsilon_k(\omega) \epsilon_k = I^{(k)}, \quad k = 1, 2. \tag{13.32}$$

Thus, we wish to maximize either

$$D^{(q)} \equiv P_q - \lambda_1 J^{(1)} - \lambda_2 J^{(2)} \tag{13.33}$$

or

$$D^{(q_1, q_2)} \equiv P_{q_1} - P_{q_2} - \lambda_1 J^{(1)} - \lambda_2 J^{(2)}, \tag{13.34}$$

where λ_i, $i = 1, 2$ are Lagrange multipliers.

The resultant optimization equations cannot be solved analytically. Numerically, it is convenient to expand each field in an orthonormal basis set $\{u_n(\omega, x)\}$ (e.g., harmonic oscillator eigenfunctions):

$$\bar{\epsilon}(\omega) = \sum_m a_m u_m(\omega, x_1), \qquad \bar{\epsilon}_2(\omega) = \sum_n b_n u_n(\omega, x_2), \tag{13.35}$$

and solve for the set of a_m, b_n coefficients that optimize the desired target.

Defining

$$U_{mm',ij} \equiv u_m(\omega_{i,1}, x_1) u_{m'}(\omega_j, x_1) d_{i,g} d_{j,g}^*, \tag{13.36}$$

and

$$X_{nn',ij}^{(q)} \equiv \frac{4\pi^2}{\hbar^4} \sum_v \int dE \, u_n(\omega_{E,i}, x_2) u_{n'}(\omega_{E,j}, x_2) d_{q,v;i}(E) d_j^{(q,v)*}(E), \tag{13.37}$$

we have that

$$\varepsilon_{i,j} = \sum_{mm'} a_m a_{m'}^* U_{mm',ij}, \tag{13.38}$$

and

$$d_{i,j}^{(q)} = \sum_{nn'} b_n b_{n'}^* X_{nn',ij}^{(q)}. \tag{13.39}$$

Using Eqs. (13.29), (13.38), and (13.39), we can write P_q as a double bilinear form in the a_m and b_n coefficients:

$$P_q = \sum_{m,m'} a_m a_{m'}^* \sum_{n,n'} b_n b_{n'}^* Y_{mm',nn'}^{(q)}(\tau), \tag{13.40}$$

where

$$Y_{mm',nn'}^{(q)}(\tau) \equiv \sum_{i,j} U_{mm',ij} X_{nn',ij}^{(q)} c_{i,1}(\tau) c_{j,1}^*(\tau) \exp(-i\omega_{i,j}\tau). \tag{13.41}$$

With the availability of powerful time independent computational techniques [14, 26] for both the bound–bound $d_{i,g}$ and bound–free $d_{q,v,j}(E)$ matrix elements, the $U_{mm',ij}$ $X^{(q)}_{nn',ij}$ and hence the $Y^{(q)}_{mm',nn'}(\tau)$ matrices are calculable for many realistic systems.

The extrema of Eq. (13.33) or (13.34), obtained via the relations

$$\frac{\partial D^P(q)}{\partial a_m} = \sum_{m'} a^*_{m'} \left\{ \sum_{n,n'} b_n b^*_{n'} Y^{(q)}_{mm',nn'}(\tau) - \lambda_1 \delta_{m,m'} \right\} = 0, \qquad (13.42)$$

$$\frac{\partial D^P(q)}{\partial b_n} = \sum_{n'} b^*_{n'} \left\{ \sum_{m,m'} a_m a^*_{m'} Y^{(q)}_{mm',nn'}(\tau) - \lambda_2 \delta_{n,n'} \right\} = 0, \qquad (13.42)$$

result in a set of nonlinear equations in the field coefficients a_n, b_m. These equations can be solved iteratively as a set of linear equations by first defining two matrices:

$$(\underline{B})_{m,m'} \equiv \sum_{n,n'} b_n b_{n'} Y^{(q)}_{mm',nn'}(\tau), \qquad (13.44)$$

$$(\underline{A})_{n,n'} \equiv \sum_{m,m'} a_m a^*_{m'} Y^{(q)}_{mm',nn'}(\tau). \qquad (13.45)$$

The iteration proceeds by assuming that

$$\mathbf{b} = \mathbf{b}^0, \qquad (13.46)$$

where $\mathbf{b}$ is a row vector composed of the b_n coefficients and $\mathbf{a}$ is a row vector composed of the a_m.

The assumption made in Eq. (13.46) reduces Eq. (13.42) to a set of linear algebraic eigenvalue equations,

$$\mathbf{a} \cdot (\mathbf{b}^0 \cdot \underline{Y}^{(q)}_{mm'}(\tau) \cdot \mathbf{b}^{0\dagger} - \lambda_1 \underline{I}) = 0, \qquad (13.47)$$

where $\underline{I}$ is the identity matrix, and $\underline{Y}^{(q)}_{mm'}(\tau)$ is the matrix of $Y^{(q)}_{mm',nn'}(\tau)$ coefficients for fixed m and m'. These equations are solved for the $\underline{a}$ eigenvector matrix, out of which the $\mathbf{a}_1$ row of coefficients corresponding to the λ_1 eigenvalue that maximizes $D^{(q)}$ is chosen. These coefficients are then used to update the $\underline{A}$ matrix [Eq. (13.45)], and to solve the eigenvalue equation for the $\mathbf{b}$ coefficients, which is

$$\mathbf{b} \cdot (\underline{A} - \lambda_2 \underline{I}) = 0. \qquad (13.48)$$

The $\mathbf{b}$ row corresponding to the λ_2 eigenvalue, which maximizes the $D^{(q)}$ objective, is then chosen to update $\underline{B}$. The process is repeated until convergence.

The fields thus generated are still a function of the delay-time τ, which is treated as a nonlinear parameter. One then solves Eq. (13.47) for every value of τ and obtains the optimal value of the time delay between the pulses as the time delay corresponding to the global maximum of $P_q(\tau)$.

Numerical results using this technique were obtained [471] for the pump–dump photodissociation of Na_2 to optimize the production of either $Na(3s) + Na(3p)$ or $Na(3s) + Na(4s)$. In this case optimal control often required pulses that were fairly heavily structured in laser phase and frequency. A more detailed study [471] indicated that this structure was necessary for the dissociation pulses, but not necessary for the excitation pulse. Further, as is often the case with OCT, pulses with very different structures were found to achieve similar control objectives in different ways.

13.3   ADAPTIVE FEEDBACK CONTROL

It may well be the case that we need to optimize laser pulses to get better results than those derived from perturbative-regime coherent control. Judson and Rabitz [40] have suggested a method that circumvents the difficulties of applying OCT to experiments by foregoing the theoretical step altogether and using experimental results directly. In essence, they propose using the experimental apparatus as an "analog computer."

In their approach, one irradiates a molecular sample of interest and measures the product distributions. These results are reported to a computer that runs a learning algorithm that is capable of recognizing patterns in the input–output measurement relationships. This algorithm then guides an iterative sequence of new experiments, each experiment being characterized by a different pulse structure. The iteration is facilitated by a cost functional, as in OCT, but the functional now only contains costs for the target state and for laboratory-related issues. The computer repeats the iteration and learning process until satisfactory convergence is reached. The advantage of this approach is that no solutions to the Schrödinger equation need be generated, nor do we need to know the molecular Hamiltonian.

This overall procedure is an example of an *adaptive feedback control* [472]. Experimentally, the key elements in the procedure are the pulse-shaping device [149–155], a rapid means of modifying the laser pulses, and a fast probe of the output (i.e., a rapid pump–probe duty cycle). In the Weiner–Heritage *discrete* pulse shaping scheme [149, 150, 155] the pulse is shaped by first dispersing it using a grating into a large number (typically 128) of frequency components that are made to transverse an array of the same number of cells filled with liquid crystals. By changing the voltage across each cell, it is possible to change the dispersion properties of the liquid crystals in the cell, thereby imparting a *controllable* phase shift to each frequency component. All the components are then brought together by another grating, in what amounts to the inverse of the dispersion process, to form a pulse. The newly shaped pulse is a result of the beats between all the phase-shifted frequency components.

The *continuous* Warren pulse shaper [151–154], shown in Figure 13.1, works in essentially the same way, except that the liquid-crystal array is replaced by an

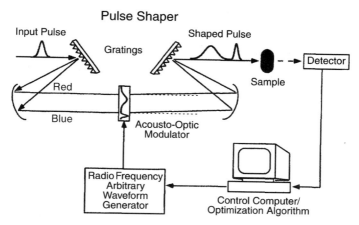

Figure 13.1 Schematic of acousto-optic pulse-shaping feedback apparatus used in adaptive feedback experiments. (Taken from Fig. 1, Ref. [41].)

acousto-optical modulator in which a pattern of acoustic waves is used to (continuously) phase shift the dispersed components of the pulse. The great merit of both types of pulse shapers is that they allow the shape to be determined by a computer. This is done via the set of voltages applied to the (discrete or continuous) phase-shifting elements.

In principle any learning algorithm is acceptable. The learning algorithm recommended by Judson and Rabitz, and used thus far in adaptive feedback control, relies upon *genetic algorithms* [156]. These algorithms are global optimization methods based on several concepts from biological evolution. The first is the concept of a breeding population in which individuals who are more "fit" in some sense will have a higher chance of producing offspring and of passing their genetic information on to successive generations. The second is the concept of crossover in which a child's genetic material is a mixture of the genetic material of his or her parents. The third concept is that of mutation, where the genetic material is occasionally corrupted, leading to a certain degree of genetic diversity in the population.

The adaptive feedback control apparatus (Fig. 13.1) consists of the molecules of interest, a laser whose pulse shape is defined by an acousto-optical modulator, controlled by a computer, and a measurement device that reads and feeds final population distributions (or other observables) back to the controlling computer. The genetic algorithm code runs on a controlling computer, supplying pulse sequences to the laser and receiving fitness values (the difference between the objective and an observed function of the molecular state) from the measurement device. Over several generations, the system as a whole will seek to optimize the fields.

In most applications, the genetic algorithm is implemented as follows. An individual (i.e., a single pulse sequence) is coded for by a "gene," which is a bit string of length N_{gene} that can be uniquely decoded to define the pulse sequence. A fitness function is defined that can discriminate between pulse sequences. For example, if we want to drive molecules into state j', we might choose the fitness function as $\sum_j (\delta_{j,j'} - \rho_j)^2$, where ρ_j is the observed population of state j. An initial population of individuals N_{pop} is formed by choosing N_{pop} bit strings, often initially at random. The fitness of these individuals is then evaluated.

Children of these first-generation parents are then formed as follows. All the parents are ranked by fitness, and the individuals with the highest fitness are placed directly into the second generation with no change. From the remaining parents, pairs of individuals are chosen and their genes are crossed over to form genes of the remaining second-generation individuals. The crossover is effected by taking some subset of the bits from parent 1 and the complementary set of bits from parent 2 and combining them to form a new gene of child 1. The remaining bits from the two parents are combined to form the gene of child 2. Additionally, during replication, there is a small probability of a bit flip or mutation in a gene. This serves primarily to prevent premature convergence in which a single very fit individual takes over the entire population.

Although in principle a genetic algorithm, or other learning algorithm, should find the true optimum, the search is limited, either by computer limitations in the case of numerical studies, or by experimental restrictions in the case of laboratory experiments.

Adaptive feedback control for molecules was first applied experimentally by Bardeen et al. [41] and by Yelin et al. [42] to two different problems. Using the setup of Figure 13.1, Bardeen et al. were able to optimize two objectives: the "efficiency" and "effectiveness" of the electronic excitation of the "IR125" dye molecule. The "efficiency" objective corresponded to maximizing the number of excited state molecules per integrated laser intensity. This goal was attained by constraining the objective through an integrated intensity penalty, as in Eq. (4.11). The effectiveness objective corresponded to maximizing the number of excited state molecules, irrespective of the integrated intensity of the pulse. This objective is obtained by removing the penalty constraint used in the efficiency objective.

The results displayed in Figures 13.2, 13.3, and 13.4 show that the most *efficient* result occurred when a relatively narrow-band laser pulse of moderate intensity was applied at center frequency near the peak of the absorption spectrum of the dye molecule. The most *effective* result occurred when a *positively chirped* broadband laser pulse was applied. The positive chirp (i.e., an upward drift of the laser's central frequency with time) is helpful because stimulated emission back to the ground state, which diminishes the number of excited state molecules, invariably occurs to the red of the absorbed photon. By rapidly shifting the laser center frequency more and more to the blue, one can successively eliminate the frequencies causing stimulated emission from the excited state shortly after it is formed by photon absorption.

Another early example for the use of computer-controlled Heritage–Weiner [149] phase-shaping technique in conjunction with adaptive feedback control is to generate a transform-limited pulse from a less coherent pulse. By setting the target to maximize a two-photon absorption intensity, Yelin et al. [42] were able, as shown in Figure 13.5, to modify the phases of the pulse in such a way as to narrow it down in

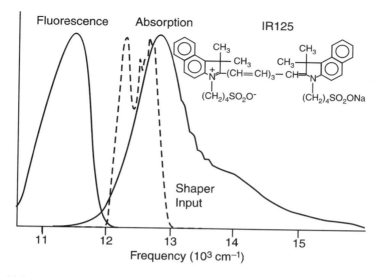

Figure 13.2 Molecular structure and absorption and fluorescence spectra of IR125 in methanol, along with the laser power spectrum before shaping (dashed line). (Taken from Fig. 2, Ref. [41].)

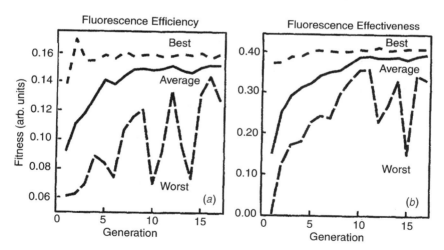

Figure 13.3 (*a*) Convergence of genetic algorithm as measured by the behavior of populations at each generation. Results are shown for best (small dashed), average (solid), and worst (large dashed) fitness in each generation, defined by the fluorescence efficiency (ratio of fluorescence to laser power). One generation corresponds to approximately 30 experiments (*b*). As in (*a*), but the fitness is now defined as the fluorescence effectiveness, proportional to the fluorescence power alone. (Taken from Fig. 3, Ref. [41].)

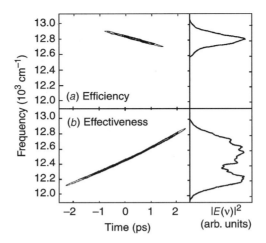

Figure 13.4 (*a*) Sample optimal pulse for fluorescence efficiency as determined experimentally by feedback algorithm. Left side is a plot of experimentally determined Wigner transform that shows the intensity of electric field as a function of time and frequency. Right side shows the spectrum $|E(v)|^2$. Efficiency does not appear to depend strongly on the laser chirp. (*b*) As in (*a*) but showing an optimal pulse for the fluorescence effectiveness. (Taken from Fig. 5, Ref. [41].)

time to essentially its transform limit. Here we show the autocorrelation trace of the pulse.

The same technique was used in Gerber's laboratory [43], coupled with their own version of an evolutionary algorithm, in a setup shown in Figure 13.6, to tailor femtosecond laser pulses to optimize the branching ratios of different organometallic photodissociation reaction channels. For example, they studied the photodissociation of $CpFe(CO)_2Cl$, (where Cp stands for cyclo-pentadiene, a pentagon made up of four CH groups and one CH_2 group). They were able to control two different bond-cleaving reactions:

$$\text{channel A} \quad FeCl + \cdots \leftarrow CpFe(CO)_2Cl \rightarrow CO + CpFe(CO)Cl \quad \text{channel B,}$$

where the dots indicate a number of, as yet unknown, possible products. They have shown that the method works automatically and finds optimal solutions without prior knowledge of the molecular system and experimental environment. Sample results are shown in Figures 13.7 and 13.8 for the photodissociation of Fe(CO), and of $CpFe(CO)_2Cl$. In the latter, the ratio of $CpFeCOCl^+/FeCl^+$ was either maximized (solid blocks) or minimized (open blocks). The yields are shown at masses 91 and 184, with the parent ion shown at 212. The observed yield of the other ions is also shown but was not included in the control protocol. Control over the desired ratio is quite evident.

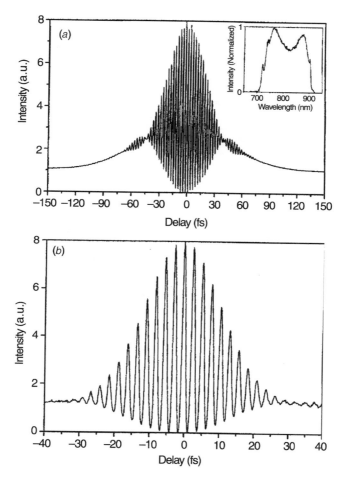

Figure 13.5 Interferometric autocorrelation traces of uncompressed and compressed pulses. (a) Uncompressed 80-fs pulses obtained directly from Ti:sapphire laser (power spectrum is shown in the inset). (b) Compressed pulses after 1000 iterations. Pulses were compressed to 14 fs. (Taken from Fig. 2, Ref. [42].)

Aspects of the laser fields that yield the optimized results are shown in Figure 13.9. Specifically, the autocorrelations $G_2(\tau) = \int [E(t - \tau) + E(t)]^4 \, dt$ are shown in Figure 13.9*a* for the experiment yielding the maximum in Figure 13.8 and in Figure 13.9*b* for the experiment yielding the minimum. Figure 13.9*c* shows a bandwidth-limited laser autocorrelation for a laser that yields inferior results to that obtained with either of those shown in Figures 13.9*a* or 13.9*b*.

At present not much is known about the dynamics of the CpFe(CO)$_2$Cl dissociation reaction as probed in the experiment by Assion and co-workers [43], nor about those aspects of the pulse that enhance specific product production. Although the

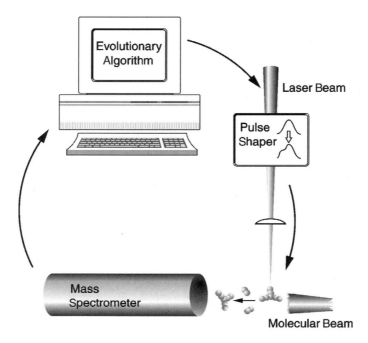

Figure 13.6   Schematic setup of Assion et al. experiment [43]. Femtosecond laser pulses are modified in a computer-controlled pulse shaper. Ionic fragments from molecular photodissociation are recorded with a reflection time of flight mass spectrometer. This signal is used directly as feedback in the controlling evolutionary computer algorithm to optimize branching ratios of photochemical reactions. (Taken from Fig. 1, Ref. [43].)

Weiner–Heritage pulse shaper only changes phases of different frequency components of the pulse, it is clear that this also has profound effect on the peak intensity of the pulse. As a result, it is conceivable that channel A involves a different multiphoton absorption (and ionization) route than does channel B. This means that the optimal pulse configurations may simply have excited the parent $CpFe(DO)_2Cl$ molecule to an energy region where dissociation to the channel of interest (A or B) is naturally preferred. Clearly, further work is needed, and is indeed ongoing, to clarify the details on the mechanism through which the yield is improved.

Additional studies have been carried out by Levis et al. [473] in which they demonstrated selective bond cleavage and rearrangement of chemical bonds having dissociation energies up to approximately 100 kcal/mol in molecules such as acetone, trifluoroacetone, and acetophenone. In particular, they showed control over the formation of CH_3CO from $(CH_3)_2CO$, CF_3 (or CH_3) from CH_3COCF_3, and $C_6H_5CH_3$ (toluene) from $C_6H_5COCH_3$. In each case, ions associated with the products were measured. These experiments employed intense tailored laser pulses (on the order of 10^{13} W/cm^2) so that the system energy levels were dynamically Stark shifted into resonance with the laser.

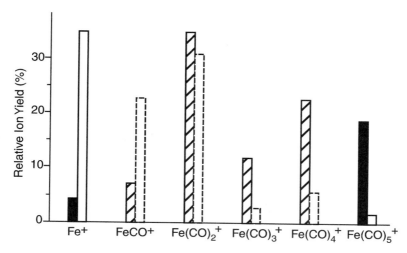

Figure 13.7 Relative $Fe(CO)_5$ photodissociation product yields. Yields are derived from relative peak heights of the mass spectra. Ratio of $Fe(CO)_5^+/Fe^+$ is maximized (solid blocks) as well as minimized (open blocks) by optimization algorithm, yielding a significantly different abundance of Fe^+ and $Fe(CO)_5^+$ in the two cases. Peak heights of all other masses [$Fe(CO)^+$ up to $Fe(CO)_4^+$] have not been included in the optimization procedure. (Taken from Fig. 2, Ref. [43].)

A typical result is shown in Figure 13.10, where control over the products of photoexcitation of acetophenone ($C_6H_5COCH_3$) was sought. Figure 13.10*b* shows the results of maximizing the ratio of the $C_6H_5CO^+$ to $C_6H_5^+$ ion products and Figure 13.10*c* shows the result of minimizing this ratio. Clearly, control over the ion product is achieved within 20 or so generations. Furthermore, repeated experiments

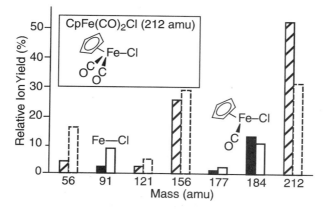

Figure 13.8 Adaptive feedback control over products of photodissociation: $CO + CpFe$ $(CO)Cl(mass\ 184) \leftarrow CpFe(CO)_2Cl \rightarrow FeCl(mass\ 91) + \cdots$. Ratio of $CpFeCOCl^+/FeCl^+$ is either maximized (solid blocks at masses 91 and 184) or minimized (open blocks at mass 91 and 184). (Taken from Fig. 3, Ref. [43].)

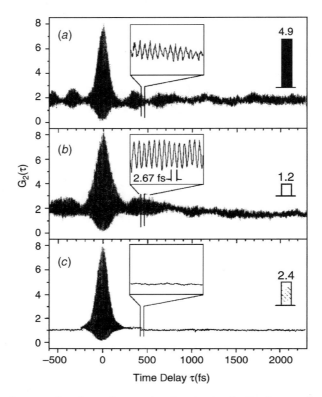

Figure 13.9 Aspects of optimum laser pulses in adaptive feedback control of products of laser excitation of $CpFe(CO)_2Cl$. Autocorrelation $G_2(\tau)$ is shown for three different cases, as described in text. Pulse shape differences in these three pulses is evident. (Taken from Fig. 4, Ref. [43].)

with different initial starting conditions yielded similar results. As with the experiments above, at this time there is not detailed understanding of the mechanism of the control fields, but work of this type is ongoing. Further, as noted above, these solutions are only optimal within the constraints umposed by the experimental apparatus.

Finally, note that Herek and co-workers [474] have demonstrated that it is possible to control the internal conversion channel in a light-harvesting antenna complex of a photosynthetic purple bacterium. This implies the ability to control dynamics in a large complex system, as anticipated theoretically (see, e.g., Ref [181]).

13.4 INTERFERENCE AND OPTIMAL CONTROL

As this book has emphasized, there are two distinct paradigms for the quantum control of molecular processes: coherent control and optimal control. Coherent

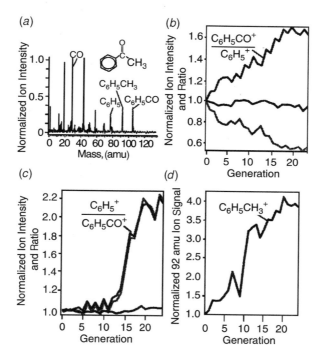

Figure 13.10 Adaptive feedback control over products of laser excitation of acetophenone ($C_6H_5COCH_3$). (*a*) Mass spectrum associated with products of photoexcitation. (*b–d*) Intensity of ions and their ratio as function of "generation" of adaptive feedback scheme. (Taken from Fig. 5, Ref. [473].)

control is clearly based upon interfering pathways. Although not as manifestly evident, optimal control also relies upon the existence of multiple interfering pathways, as discussed below. Both of these paradigms bring their own insights to quantum control. In addition, they each motivate appropriate experiments in different technological domains. Thus, the energy-resolved viewpoint has primarily motivated nanosecond pulsed laser experiments, whereas the time-dependent perspective has primarily been used to devise and interpret ultrafast experiments. In this section we link the two approaches by using the energy resolved perspective to gain insight into pulse-shaped control. We focus on photodissociation.

Consider a single-bound molecular eigenstate $|E_1\rangle$ that is excited to the continuum. In accord with Eq. (2.2) the wave function $|\psi(t)\rangle$ is of the form

$$|\Psi(t)\rangle = b_1(t)|E_1\rangle e^{-E_1 t/\hbar} + \sum_{\mathbf{n}} \int dE\, b^{(1)}_{E,\mathbf{n},q}(t)|E, \mathbf{n}, q^-\rangle e^{-Et/\hbar}. \qquad (13.49)$$

We can rewrite Eq. (13.49) in matrix notation as

$$|\Psi^{(1)}(t)\rangle = \int dE \; e^{-iEt/\hbar} (b_{E,\mathbf{n}_1,q}(t), b_{E,\mathbf{n}_2,q}^{(1)}(t), \ldots,) \cdot (|E, \mathbf{n}_1, q^-\rangle, |E, \mathbf{n}_2, q^-\rangle, \ldots)^{\mathbf{T}},$$

(13.50)

where the superscript $\mathbf{T}$ denotes the transpose operation that turns a row vector into a column vector, and $|\Psi^{(1)}(t)\rangle$ is the excited portion of the wave packet that originated from state $|E_1\rangle$. That is,

$$|\Psi^{(1)}(t)\rangle \equiv |\Psi(t)\rangle - b_1(t)|E_1\rangle e^{-iE_1 t/\hbar}.$$

(13.51)

As shown in Chapters 3 and 10, one cannot control the dynamics by shaping the laser pulse, in such an excitation of a single initial state. That is, the ratio of products going to an individual product state is independent of the pulse shape.

Further, as discussed in Section 3.1, the inability to control the product ratio by shaping the pulse can be overcome by photodissociating not just one $|E_1\rangle$ bound state but a superposition of several bound states $|E_i\rangle$ (as was done, e.g., with bichromatic control). Such a superposition state can be created separately by an initial preparation pulse, as in the case of pump–dump control scenario Sections 3.5 and 4.1). Alternatively, the superposition state can be created by the photolysis pulse itself (by, e.g., a stimulated Raman process), provided that the bandwidth of the pulse is comparable to the energy spacings between the $|E_i\rangle$ levels.

Mathematically speaking, the goal of the control is the preparation of a single $|E, \mathbf{n}, q^-\rangle$ state. If this is achieved, then complete control is guaranteed, by Eq. (2.66), insofar as only one fragment target state $|E, \mathbf{n}; 0\rangle$ will be populated as $t \rightarrow \infty$. To achieve the control target, we consider preparing a whole array of wave packets by, for example, starting with other initial states composed of the other system bound states $|E_i\rangle$. That is,

$$\underline{|\Psi_t\rangle} = \int dE \; e^{-Et/\hbar} \underline{\underline{\mathbf{B}}} \cdot \underline{|E, \mathbf{n}\rangle}$$

(13.52)

where

$$\underline{|\Psi_t\rangle}^{\mathbf{T}} \equiv (|\Psi^{(1)}(t)\rangle, |\Psi^{(2)}(t)\rangle, |\Psi^{(3)}(t)\rangle, \ldots)$$

(13.53)

$$\underline{\underline{\mathbf{B}}} \equiv \begin{pmatrix} b_{E,\mathbf{n}_1,q}^{(1)}, b_{E,\mathbf{n}_2,q}^{(1)}, b_{E,\mathbf{n}_3,q,\cdots}^{(1)} \\ b_{E,\mathbf{n}_1,q}^{(2)}, b_{E,\mathbf{n}_2,q}^{(2)}, b_{E,\mathbf{n}_3,q,\cdots}^{(2)} \\ \cdot \\ \cdot \\ \cdot \end{pmatrix}$$

(13.54)

and

$$\underline{|E, \mathbf{n}\rangle}^{\mathrm{T}} \equiv (|E, \mathbf{n}_1, q^-\rangle, |E, \mathbf{n}_2, q^-\rangle, |E, \mathbf{n}_3, q^-\rangle, \ldots). \qquad (13.55)$$

Here a single underline denotes a vector, as shown.

It is easy to see that the $\underline{\underline{\mathrm{B}}}$ matrix factorizes as

$$\underline{\underline{\mathrm{B}}} = \hat{\mathcal{E}}(E) \cdot \underline{\underline{\mathrm{M}}}(E), \qquad (13.56)$$

where

$$\underline{\underline{\mathrm{M}}}(E) = \begin{pmatrix} \langle E_1|\mu|E, \mathbf{n}_1, q^-\rangle, \langle E_1|\mu|E, \mathbf{n}_2, q^-\rangle, \ldots \\ \langle E_2|\mu|E, \mathbf{n}_1, q^-\rangle, \langle E_2|\mu|E, \mathbf{n}_2, q^-\rangle, \ldots \\ \qquad\qquad . \\ \qquad\qquad . \\ \qquad\qquad . \end{pmatrix} \qquad (13.57)$$

and where $\hat{\mathcal{E}}(E)$ is a diagonal matrix, $\hat{\mathcal{E}}(E) = \mathcal{E}_i(E)\delta_{i,j}$, with

$$\mathcal{E}_i(E) = \int_{-\infty}^{\infty} dt\, \varepsilon^*(t) e^{i\Delta_{E,i}t} b_i(t). \qquad (13.58)$$

Writing the array of possible wave function produced as

$$\underline{|\Psi_t\rangle} = \int dE\, e^{-iEt/\hbar} \hat{\mathcal{E}}(E) \cdot \underline{\underline{\mathrm{M}}}(E) \cdot \underline{|E, \mathbf{n}\rangle} \qquad (13.59)$$

allows us to examine the possibility of taking different linear combinations of the components of the $\underline{|\Psi_t\rangle}$ vector so as to satisfy the control objectives of producing a single $|E, \mathbf{n}_i, q^-\rangle$ state. In this way different pathways starting with different precursor states leading to the same $|E, \mathbf{n}_i, q^-\rangle$ state will be seen to interfere to achieve the desired goal.

As an example, we consider a superposition state composed of the sum over the components of $\underline{|\Psi_t\rangle}$,

$$\Psi'(t) = \sum_k \int dE\, e^{-iEt/\hbar} \mathcal{E}_k(E) \sum_j \underline{\underline{\mathrm{M}}}_{k,j} |E, \mathbf{n}_j, q^-\rangle. \qquad (13.60)$$

In the weak-field limit, the population and the phase of the initial levels can be assumed constant with time:

$$b_k(t) \approx b_k \equiv b_k(-\infty) \qquad (13.61)$$

in which case all the $\mathcal{E}_k(E)$ matrix elements factor as

$$\mathcal{E}_k(E) \approx b_k \int_{-\infty}^{\infty} dt \, \varepsilon^*(t) e^{i\Delta_{E,k}t} = 2\pi b_k \bar{\varepsilon}(\Delta_{E,k}), \qquad (13.62)$$

where

$$\bar{\varepsilon}(\omega) \equiv \left(\frac{1}{2\pi}\right) \int_{-\infty}^{\infty} dt \, \varepsilon^*(t) e^{i\omega t}. \qquad (13.63)$$

Our objective to populate exclusively the ith fragment state $|E, \mathbf{n}_i, q^-\rangle$ can be realized in the weak-field domain by choosing the pulse shape that defines $\Psi'(t)$ [Eq. (13.60)] to satisfy the condition

$$b_k \bar{\varepsilon}_i(\Delta_{E,k}) = (\underline{\mathbf{M}}(E)^{-1})_{i,k}. \qquad (13.64)$$

This choice eliminates all but a *single* $|E, \mathbf{n}_i, q^-\rangle$ state in $\Psi'(t)$ given by Eq. (13.60).

Thus the control objective, the ith product state, is seen to be realized by starting out with an initial superposition of bound states:

$$|\Psi(t)\rangle = \sum_k b_k |E_k\rangle e^{-iE_k t/\hbar}, \qquad (13.65)$$

and subjecting the system to the action of a pulse shaped according to Eq. (13.64). This allows for multiple-path interference between the various ways of generating the $|E, \mathbf{n}_i, q^-\rangle$ state. The weight of each pathway is chosen so as to cause destructive interference in the production of all the $|E, \mathbf{n}, q^-\rangle$ states but one, the $|E, \mathbf{n}_i, q^-\rangle$ state.

In general, control is incomplete because the pulse-shaping conditions of Eq. (13.64) cannot be satisfied simultaneously for all energies. This can be seen by noting that the $(\underline{\mathbf{M}}(E)^{-1})_{i,k}$ matrix element, which (for a single i) is a function of two variables, k and E, has to be equated to a product of a function of k, b_k, and a function of E, $\bar{\varepsilon}_i(\Delta_{E,k})$. In general, this equality cannot be satisfied. There are nevertheless important cases in which Eq. (13.64) can be satisfied. These are either when $\underline{\mathbf{M}}(E)$ does not vary too rapidly with E, or, conversely, when the $\langle E_1|\mu|E, \mathbf{n}, q^-\rangle$ matrix elements, which determine $\underline{\mathbf{M}}(E)$ (and the absorption spectrum), span a very narrow range of energies (e.g., a narrow resonance).

The weak-field control discussed here must be achieved in two steps. First, it is necessary to create the $\Phi(t)$ superposition state of Eq. (13.65). This state is then irradiated with the pulse satisfying Eq. (13.64). This is the essence of the weak field pump–dump scenario. However, in the strong-field domain these two processes cannot be separated since the factorization of Eq. (13.62) does not hold. In that case the control conditions become

$$\mathcal{E}_{i,k}(E) = (\underline{\mathbf{M}}(E)^{-1})_{i,k}. \qquad (13.66)$$

In this strong-field regime the $b_k(t)$ coefficients are embedded in $\mathcal{E}_k(E)$ [see Eq. (13.58)] and are themselves functions of $\varepsilon(t)$. Hence the problem is inherently nonlinear, necessitating an iterative solution. Nevertheless, the same interference mechanism outlined in the weak-field domain applies. The only difference is that the pulse-shaping conditions are given implicitly via Eq. (13.66), rather than explicitly via Eq. (13.64), as in the weak-field domain.

We have elucidated the nature of pulsed-shaping control of photodissociation from the viewpoint of energy-resolved coherent control theory. Clearly, when excitation is from a superposition of states, as in the vast majority of control scenarios, the role of the pulse shaping is to enhance a different set of interfering pathways for each control objective.

REFERENCES AND NOTES

1. J. D. Jackson, *Classical Electrodynamics*, Wiley, New York, 1962.
2. H. Goldstein, *Classical Mechanics*, 2nd Ed., Addison-Wesley, Reading, MA, 1980.
3. R. Loudon, *The Quantum Theory of Light*, 2nd ed., Clarendon, Oxford, 1983.
4. M. Abramowitz and I. A. Stegun, eds., *Handbook of Mathematical Functions*, Dover, New York, 1965.
5. L. Pauling and E. B. Wilson, *Introduction to Quantum Mechanics*, McGraw Hill, New York, 1935.
6. R. B. Bernstein, ed., *Atom-Molecule Collision Theory, A Guide for the Experimentalist*, Plenum, New York, 1979.
7. R. D. Levine, *Quantum Mechanics of Molecular Rate Processes*, Clarendon, Oxford, 1969.
8. J. R. Taylor, *Scattering Theory*, Wiley, New York, 1972.
9. W. H. Miller, *J. Chem. Phys.* **50**, 407 (1969).
10. G. C. Schatz and A. Kuppermann, *J. Chem. Phys.* **65**, 4642 (1976).
11. J. Z. H. Zhang and W. H. Miller, *J. Chem. Phys.* **91**, 1528 (1989).
12. D. E. Manolopoulos and R. E. Wyatt, *Chem. Phys. Lett.* **152**, 23 (1988).
13. D. E. Manolopoulos, M. D'Mello, and R. E. Wyatt, *J. Chem. Phys.* **93**, 403 (1990).
14. M. Shapiro, *J. Chem. Phys.* **56**, 2582 (1972).
15. M. Shapiro, *Isr. J. Chem.* **11**, 691 (1973).
16. O. Atabek, J. A. Beswick, R. Lefebvre, S. Mukamel, and J. Jortner, *J. Chem. Phys.* **65**, 4035 (1976); O. Atabek and R. Lefebvre, *Chem. Phys.* **23**, 51 (1977); *ibid.*, *Chem. Phys.* **55**, 395 (1981).
17. M. D. Morse, K. F. Freed, and Y. B. Band, *J. Chem. Phys.* **70**, 3620 (1979).
18. K. C. Kulander and J. C. Light, *J. Chem. Phys.* **73**, 4337 (1980).
19. K. Kodama and A. D. Bandrauk, *Chem. Phys.* **57**, 461 (1981).
20. Y. B. Band, K. F. Freed, and D. J. Kouri, *J. Chem. Phys.* **74**, 4380 (1981).

21. M. Shapiro and R. Bersohn, *Ann. Rev. Phys. Chem.* **33**, 409 (1982).

22. G. G. Balint-Kurti and M. Shapiro, *Adv. Chem. Phys.* **60**, 403 (1985).

23. V. Engel, R. Schinke, and V. Stämmler, *Chem. Phys.* **130**, 413 (1986); P. Andresen and R. Schinke, in *Molecular Photodissociation Dynamics*, M. N. R. Ashfold and J. E. Baggott, eds., Royal Society of Chemistry, London, 1987, p. 61.

24. O. Atabek, R. Lefebvre, and M. Jacon, *J. Chem. Phys.* **72**, 2670 (1980).

25. A. D. Bandrauk and O. Atabek, *Adv. Chem. Phys.* **73**, 823 (1989).

26. R. Schinke, *Photodissociation Dynamics*, Cambridge University Press, Cambridge, 1992.

27. M. Shapiro, *J. Phys. Chem.* **97**, 7396 (1993).

28. E. J. Heller, in *Potential Energy Surfaces*; *Dynamics Calculations*, D. G. Truhlar, ed., Plenum, New York, 1981.

29. E. J. Heller, *Acc. Chem. Res.* **14**, 368 (1981).

30. K. C. Kulander and A. E. Orel, *J. Chem. Phys.* **74**, 6529 (1981); K. C. Kulander, *Phys. Rev. A* **35**, 445 (1987); A. E. Orel and K. C. Kulander, *J. Chem. Phys.* **91**, 6086 (1989); K. J. Schafer and K. C. Kulander, *Phys. Rev. A.* **42**, 5794 (1990).

30a. K. C. Kulander and E. D. Heller, *J. Chem. Phys.* **69**, 2439 (1978).

31. M. V. Rama Krishna and R. D. Coalson, *Chem. Phys.* **12**, 327 (1988).

32. S. O. Williams and D. G. Imre, *J. Phys. Chem.* **92**, 6636 (1988).

33. V. Engel, H. Metiu, R. Almeida, R. A. Marcus, and A. H. Zewail, *Chem. Phys. Lett.* **152**, 1 (1988).

34. V. Engel and H. Metiu, *J. Chem. Phys.* **92**, 2317 (1990).

35. A. D. Hammerich, R. Kosloff, and M. A. Ratner, *J. Chem. Phys.* **97**, 6410 (1992); A. D. Hammerich, U. Manthe, R. Kosloff, H.-D. Meyer and L. S. Cederbaum, *J. Chem. Phys.* **101**, 5623 (1994).

36. A. Bartana, U. Banin, S. Ruhman, and R. Kosloff, *Chem. Phys. Lett.* **219**, 211 (1994).

36a. D. Tannor and S. A. Rice, *Adv. Chem. Phys.* **70**, 488 (1988).

37. M. Shapiro and P. Brumer *J. Chem. Phys.* **84**, 540 (1986).

38. P. Brumer and M. Shapiro, *Chem. Phys.* **139**, 221, (1989).

39. M. Shapiro and P. Brumer, *J. Phys. Chem.* **105**, 2897 (2001).

40. R. S. Judson and H. Rabitz, *Phys. Rev. Lett.* **68**, 1500 (1992).

41. C. J. Bardeen, V. V. Yakovlev, K. R. Wilson, S. D. Carpenter, P. M. Weber, and W. S. Warren, *Chem. Phys. Lett.* **280**, 151 (1997).

42. D. Yelin, D. Meshulach, and Y. Silberberg, *Opt. Lett.* **22**, 1793 (1997).

43. T. Assion, T. Baumert, M. Bergt, T. Brixner, B. Kiefer, V. Seyfried, M. Strehle, and G. Gerber, *Science* **282**, 919 (1998).

44. P. Brumer and M. Shapiro, *Chem. Phys. Lett.* **126**, 541 (1986).

45. P. Brumer and M. Shapiro, *Ann. Rev. Phys. Chem.* **43**, 257 (1992).

46. M. Shapiro and P. Brumer, *Int. Rev. Phys. Chem.* **13**, 187 (1994).

47. M. Shapiro and P. Brumer, *Trans. Faraday Soc.* **93**, 1263 (1997).

48. M. Shapiro and P. Brumer, in *Advances in Atomic, Molecular and Optical Physics*, Vol. **42**, B. Bederson and H. Walther, eds., Academic, San Diego, 1999, pp. 287–343.

49. M. Shapiro and R. Bersohn, *J. Chem. Phys.* **73**, 3810 (1980).

50. M. Shapiro, *J. Phys. Chem.* **90**, 3644 (1986).

51. M. Shapiro and P. Brumer, *Faraday Disc. Chem. Soc.* **82**, 177 (1987).

52. P. Brumer and M. Shapiro, in *Coherent Control in Atoms, Molecules and Semiconductors*, W. Pötz and W. A. Schröder, eds., Kluwer, Dordrecht, 1999.

52a. D. J. Cook and R. M. Hochstrasser, *Opt. Lett.* **25**, 1210 (2000).

53. F. F. Crim, *Ann. Rev. Phys. Chem.* **35**, 647 (1984)

54. T. M. Ticich, M. D. Likar, H. Dubal, L. J. Butler, and F. F. Crim, *J. Chem. Phys.* **87**, 5820 (1987).

55. M. D. Likar, J. E. Baggott, A. Sinha, T. M. Ticich, R. L. Vander Wal, and F. F. Crim, *J. Chem. Soc. Faraday Trans. 2* **84**, 1483 (1988).

56. A. Sinha, R. L. Vander Wal, and F. F. Crim, *J. Chem. Phys.* **91**, 2929 (1989).

57. F. F. Crim, *Science* **249**, 1387 (1990).

58. R. L. Vander Wal, J. L. Scott, and F. F. Crim, *J. Chem. Phys.* **94**, 1859 (1991).

59. T. Arutsi-Parpar, R. P. Schmid, R.-J. Li, I. Bar, and S. Rosenwaks, *Chem. Phys. Lett.* **268**, 163 (1997).

60. For a review, see I. Bar and S. Rosenwaks, *Int. Rev. Phys. Chem.* **20**, 711 (2001).

61. P. Roman, *Advanced Quantum Theory*, Addison-Wesley, Boston, 1965.

62. M. Shapiro, J. W. Hepburn, and P. Brumer, *Chem. Phys. Lett.* **149**, 451 (1988).

63. C. K. Chan, P. Brumer, and M. Shapiro, *J. Chem. Phys.* **94**, 2688 (1991).

64. S. Chelkowski and A. D. Bandrauk, *Chem. Phys. Lett.* **186**, 284 (1991).

65. A. D. Bandrauk, J-M. Gauthier, and J. F. McCann, *Chem. Phys. Lett.* **200**, 399 (1992).

66. A. Szöke, K. C. Kulander, and J. N. Bardsley, *J. Phys. B.* **24**, 3165 (1991).

67. E. Charron, A. Giusti-Suzor, and F. H. Mies, *Phys. Rev. Lett.* **75**, 2815 (1995).

68. E. Charron, A. Giusti-Suzor, and F. H. Mies, *J. Chem. Phys.* **103**, 7359 (1995).

69. T. Zuo and A. D. Bandrauk, *Phys. Rev. A* **54**, 3254 (1996).

70. C. Chen, Y-Y. Yin, and D. S. Elliott, *Phys. Rev. Lett.* **64**, 507 (1990); *ibid.*, **65**, 1737 (1990).

71. S. M. Park, S.-P. Lu, and R. J. Gordon, *J. Chem. Phys.* **94**, 8622 (1991).

72. S.-P. Lu, S. M. Park, Y. Xie, and R. J. Gordon, *J. Chem. Phys.* **96**, 6613 (1992).

73. V. D. Kleiman, L. Zhu, X. Li, and R. J. Gordon, *J. Chem. Phys.* **102**, 5863 (1995).

74. L. Zhu, V. D. Kleiman, X. Li, S. Lu, K. Trentelman, and R. J. Gordon, *Science* **270**, 77 (1995).

75. L. Zhu, K. Suto, J. A. Fiss, R. Wada, T. Seideman, and R. J. Gordon, *Phys. Rev. Lett.* **79**, 4108 (1997).

76. X. Wang, R. Bersohn, K. Takahashi, M. Kawasaki, and H. L. Kim, *J. Chem. Phys.* **105**, 2992 (1996).

77. R. J. Gordon, L. C. Zhu, and T. Seideman, *Acc. Chem. Res.* **32**, 1007 (1999).

77a. H. G. Muller, P. H. Bucksbaum, D. W. Shumacher, A. Zavriyev, *J. Phys. B.* **23**, 2761 (1990); D. W. Schumacher, F. Weiche, H. G. Muller, P. H. Bucksbaum, *Phys. Rev. Lett.* **73**, 1344 (1994).

78. G. Kurizki, M. Shapiro, and P. Brumer, *Phys. Rev. B* **39**, 3435 (1989).

79. B. A. Baranova, A. N. Chudinov, and B. Ya Zel'dovitch, *Opt. Comm.* **79**, 116 (1990).

80. Y.-Y. Yin, C. Chen, D. S. Elliott, and A. V. Smith, *Phys. Rev. Lett.* **69** 2353 (1992).

81. Y.-Y. Yin, R. Shehadeh, D. Elliott, and E. Grant, *Chem. Phys. Lett* **241**, 591 (1995).

82. E. Dupont, P. B. Corkum, H. C. Liu, M. Buchanan, and Z. R. Wasilewski, *Phys. Rev. Lett.* **74**, 3596 (1995).

83. B. Sheehy, B. Walker, and L. F. DiMauro, *Phys. Rev. Lett.* **74**, 4799 (1995).

84. A. Haché, Y. Kostoulas, R. Atanasov, J. L. P. Hughes, J. E. Sipe, and H. M. van Driel, *Phys. Rev. Lett.*, **78**, 306 (1997).

85. K. J. Schafer and K. C. Kulander, *Phys. Rev. A* **45**, 8026 (1992).

86. R. M. Potvliege and P. H. G. Smith, *J. Phys. B* **25**, 2501 (1992).

87. R. Atanasov, A. Haché, L. P. Hughes, H. M. van Driel, and J. E. Sipe, *Phys. Rev. Lett.* **76**, 1703 (1996).

88. E. E. Aubanel and A. D. Bandrauk, *Chem. Phys. Lett.* **229**, 169 (1994).

89. F. Ehlotzky, *Phys. Rpts.* **345**, 175 (2001).

90. C. Asaro, P. Brumer, and M. Shapiro, *Phys. Rev. Lett.* **60**, 1634 (1988).

91. D. J. Tannor and S. A. Rice, *J. Chem. Phys.* **83**, 5013 (1985).

92. D. J. Tannor and S. A. Rice, *Adv. Chem. Phys.* **70**, 441 (1988).

93. D. J. Tannor, *Introduction to Quantum Mechanics: A Time Dependent Perspective*, University Science Press, Sausalito, in press.

94. I. Levy, M. Shapiro, and P. Brumer, *J. Chem. Phys.* **93**, 2493 (1990).

95. D. G. Abrashkevich, M. Shapiro, and P. Brumer, *J. Chem. Phys.* **108**, 3585 (1998).

96. T. Seideman, M. Shapiro, and P. Brumer, *J. Chem. Phys.* **90**, 7132 (1989).

97. The Na–Na potential curves and the relevant electronic dipole moments are from I. Schmidt, Ph.D. Thesis, Kaiserslautern University, 1987.

98. J. M. Papanikolas, R. M. Williams, P. D. Kleiber, J. L. Hart, C. Brink, S. D. Price, and S. R. Leone, *J. Chem. Phys.* **103**, 7269 (1995); R. Uberna, Z. Amitay, C. X. W. Qian, and S. R. Leone, *J. Chem. Phys.* **114**, 10311 (2001).

99. M. Shapiro and P. Brumer, *J. Chem. Phys.* **98**, 201 (1993).

100. D. J. Tannor, R. Kosloff, and S. A. Rice, *J. Chem. Phys.* **85**, 5805 (1986).

101. G. C. Pimentel and J. A. Coonrod, *Opportunities in Chemistry Today and Tomorrow*, National Academy Press, Washington, DC, 1987.

102. I. Bar, Y. Cohen, D. David, S. Rosenwaks, and J. J. Valentini, *J. Chem. Phys.* **93**, 2146 (1990); *ibid.*, **95**, 3341 (1991).

103. See, e.g., the Appendix in D. Wardlaw, P. Brumer, and T. A. Osborn, *J. Chem. Phys.* **76**, 4916 (1982).

104. R. Kosloff, S. A. Rice, P. Gaspard, S. Tersigni, and D. J. Tannor, *Chem. Phys.* **139**, 201 (1989).

105. S. A. Rice and M. Zhao, *Optical Control of Molecular Dynamics*, Wiley, New York, 2000.

106. A. H. Zewail, *Science* **242**, 1645 (1988); A. H. Zewail and R. B. Bernstein, *Chem. Eng. News* **66**(45), 24 (1988); M. J. Rosker, M. Dantus, and A. H. Zewail, *J. Chem. Phys.* **89**, 6113 (1988); *ibid.*, **89**, 6128 (1988); M. J. Bowman, M. Dantus, and A. H. Zewail, *Chem. Phys. Lett.* **161**, 297 (1989); J. L. Herek, A. Materny, and A. H. Zewail, *Chem. Phys. Lett.* **228**, 15 (1994).

107. T. Baumert, M. Grosser, R. Thalweiser, and G. Gerber, *Phys. Rev. Lett.* **64**, 733 (1990); *ibid.*, **67**, 3753 (1991).

108. T. Baumert, J. Helbing, and G. Gerber, *Adv. Chem. Phys.* **101**, 47 (1997).

109. S. Shi, A. Woody, and H. Rabitz, *J. Chem. Phys.* **88**, 6870 (1988).

110. S. Shi and H. Rabitz, *Chem. Phys.* **139**, 185 (1989).

111. A. P. Peirce, M. Dahleh, and H. Rabitz, *Phys. Rev.* A **37**, 4950 (1988).

112. S. Shi, and H. Rabitz, *J. Chem. Phys.* **92**, 364 (1990).

113. W. Jakubetz, J. Manz, and V. Mohan, *J. Chem. Phys.* **90**, 3686 (1989).

114. W. Jakubetz, B. Just, J. Manz, and H.-J. Schreier, *J. Phys. Chem.* **94**, 2294 (1990).

115. B. Hartke, E. Kolba, J. Manz, and H. H. R. Schor, *Ber. Bunsenges. Phys. Chem.* **94**, 1312 (1990).

116. R. S. Judson, K. K. Lehmann, H. Rabitz, and W. S. Warren, *J. Mol. Struct.* **223**, 425 (1990).

117. M. Dahleh, A. P. Peirce, and H. Rabitz, *Phys. Rev. A* **42**, 1065 (1990).

118. K. Yao, S. Shi, and H. Rabitz, *Chem. Phys.* **150**, 373 (1990).

119. S. Tersigni, P. Gaspard, and S. A. Rice, *J. Chem. Phys.* **93**, 1670 (1990).

120. B. Amstrup, R. J. Carlson, A. Matro, and S. A. Rice, *J. Phys. Chem.* **95**, 8019 (1991).

121. M. Zhao and S. A. Rice, *J. Chem. Phys.* **95**, 2465 (1991).

122. S. A. Rice, in *Mode Selective Chemistry*, J. Jortner, R. D. Levine, and B. Pullman, eds., Kluwer, Dodrecht, 1991, p. 485.

123. D. J. Tannor and Y. Jin, in *Mode Selective Chemistry*, J. Jortner, R. D. Levine, and B. Pullman, eds., Kluwer, Dordrecht, 1991, p. 333.

124. J. E. Combariza, B. Just, J. Manz, and G. K. Paramonov, *J. Phys. Chem.* **95**, 10351 (1991).

125. J. E. Combariza, C. Daniel, B. Just, E. Kades, E. Kolba, J. Manz, W. Malisch, G. K. Paramonov, and B. Warmuth, in *Isotope Effects in Gas Phase Chemistry*, J. A. Kaye, ed., ACS Symposium Series 502, American Chemical Society, Washington, DC, 1992, p. 310.

126. M. Shapiro and P. Brumer, *J. Chem. Phys.* **97**, 6259 (1992).

127. S. A. Rice, *Science* **258**, 412 (1992).

128. W. S. Warren, H. Rabitz, and M. Dahleh, *Science* **259**, 1581 (1993).

129. I. Averbukh and M. Shapiro, *Phys. Rev. A* **47**, 5086 (1993).

130. M. Shapiro and P. Brumer, *Chem. Phys. Lett.* **208** 193 (1993).

131. B. Kohler, J. L. Krause, F. Raski, K. R. Wilson, V. V. Yakovlev, R. M. Whitnell, and Y. Yan, *Acc. Chem. Res.* **28**, 133 (1995).

132. D. J. Tannor, in *Molecules in Laser Fields*, A. D. Bandrauk, ed., Marcel Dekker, New York, 1994, p. 403.

133. D. Kosloff and R. Kosloff, *J. Comp. Phys.* **52**, 35 (1983); R. Kosloff and D. Kosloff, *J. Chem. Phys.* **79**, 1823 (1983); R. Kosloff, *J. Phys. Chem.* **92**, 2087 (1988).

134. R. Kosloff and H. Tal-Ezer, *Chem. Phys. Lett.* **127**, 223 (1986).

135. C. Leforestier, R. H. Bisseling, C. Cerjan, M. D. Feit, R. Friesner, A. Guldberg, A. Hammerich, G. Joilcard, W. Karrlein, H. O. Meyer, N. Lipkin, O. Roncero and R. Kosloff *J. Comp. Phys.* **94**, 59 (1991).

136. M. Shapiro and P. Brumer, *J. Chem. Phys.* **95**, 8658 (1991).

137. N. F. Scherer, A. J. Ruggiero, M. Du, and G. R. Fleming, *J. Chem. Phys.* **93**, 856 (1990).

138. N. F. Scherer, R. J. Carlson, A. Matro, M. Du, A. J. Ruggiero, V. Romero-Rochin, J. A. Cina, G. R. Fleming, and S. A. Rice, *J. Chem. Phys.* **95**, 1487 (1991).

139. V. Blanchet, M. A. Bouchene, O. Cabrol, and B. Girard, *Chem. Phys. Lett.* **233**, 491 (1995).

140. V. Blanchet, C. Nicole, M.-A. Bouchene, and B. Girard, *Phys. Rev. Lett.* **78**, 2716 (1997).

141. V. Blanchet, M. A. Bouchene, and B. Girard, *J. Chem. Phys.* **108**, 4862 (1998).

142. C. Nicole, M. A. Bouchene, S. Zamith, N. Melikechi, and B. Girard, *Phys. Rev. A* **60**, R1755 (1999).

143. O. Kinrot, I. S. Averbukh, and Y. Prior, *Phys. Rev. Lett.* **75**, 3822 (1995).

144. Ch. Warmuth, A. Tortschanoff, F. Milota, G. Knopp, M. Shapiro, Y. Prior, I. Averbukh, W. Schleich, W. Jakubetz, and H. F. Kauffmann, *J. Chem. Phys.* **112**, 5060 (2000).

145. C. Leichtle, W. P. Schleich, I. Sh. Averbukh, and M. Shapiro, *J. Chem. Phys.* **108**, 6057 (1998).

146. J. Parker and C. R. Stroud Jr., *Phys. Rev. Lett.*, **56**, 716 (1986); J. A. Yeazell and C. R. Stroud, Jr., *Phys. Rev. Lett.* **60**, 1494 (1988).

147. G. Alber, H. Ritch, and P. Zoller, *Phys. Rev. A* **34**, 1058 (1986); W. A. Henle, H. Ritch, and P. Zoller, *Phys. Rev. A* **36**, 683 (1987).

148. M. V. Fedorov and A. M. Movsesian, *J. Opt. Soc. Am. B* **5**, 850 (1988).

149. A. M. Weiner, J. P. Heritage, and R. N. Thurston, *Opt. Lett.* **11**, 153 (1986); A. M. Weiner and J. P. Heritage, *Phys. Rep. Appl* **22**, 1619 (1987).

150. A. M. Weiner, D. E. Leaird, G. P. Wiederrecht, and K. A. Nelson, *Science* **247**, 1317 (1990).

151. M. Haner and W. S. Warren, *Appl. Phys. Lett.* **52**, 1459 (1988).

152. J. S. Melinger, S. R. Gandhi, A. Hariharan, J. X. Tull, and W. S. Warren, *Phys. Rev. Lett.* **68**, 2000 (1992).

153. J. S. Melinger, S. R. Gandhi, A. Hariharan, D. Goswami, and W. S. Warren, *J. Chem. Phys.* **101**, 6439 (1994).

154. C. W. Hillegas, J. X. Tull, D. Goswami, D. Strickland, and W. S. Warren, *Opt. Lett.* **19**, 737 (1994).

155. A. M. Weiner, *Rev. Sci. Inst.* **71**, 1929 (2000).

156. D. E. Goldberg, *Genetic Algorithms in Search, Optimization, and Machine Learning*, Addison-Wesley, Reading, MA, 1993; H. P. Schwefel, *Evolution and Optimum Seeking*, Wiley, New York, 1995.

157. G. Harel and V. M. Akulin, *Phys. Rev. Lett.* **82**, 1 (1999).

158. R. Omnes, *The Interpretation of Quantum Mechanics*, Princeton University Press, Princeton, 1994; R. Omnes, *Understanding Quantum Mechanics*, Princeton University Press, Princeton, 1999.

159. J. Wilkie and P. Brumer, *Phys. Rev. A* **55**, 27 (1997); *ibid.*, **55**, 43 (1997).

160. K. Blum, *Density Matrix Theory and Applications*, Plenum, New York, 1981.

161. J. Kupsch, in *Decoherence and the Appearance of the Classical World*, D. Giulini, E. Joos, C. Kiefer, J. Kupsch, I-O. Stamatescu, and H. D. Zeh, eds., Springer, New York, 1996.

162. V. May and O. Kuhn, *Charge and Energy Transfer Dynamics in Molecular Systems*, Wiley-VCH, Berlin, 2000.

163. J. Gong and P. Brumer, *Phys. Rev. E* **60**, 1643 (1999).

164. A. O. Caldeira and A. J. Leggett, *Physica (Amsterdam)* **121A**, 587 (1983); W. G. Unruh and W. H. Zurek, *Phys. Rev. D* **40**, 1071 (1989); B. L. Hu, J. P. Paz, and Y. Zhang, *Phys. Rev. D* **45**, 2843 (1992); *ibid.*, **47**, 1576 (1993).

165. C. Jaffe and P. Brumer, *J. Chem. Phys.* **82**, 2330 (1985).

166. B. Eckhardt, G. Hose, and E. Pollak, *Phys. Rev. A* **39**, 3776 (1989).

167. Analogous results were observed previously for the stadium billiard. See K. M. Christoffel and P. Brumer, *Phys. Rev. A* **33**, 1309 (1985).

168. H. Han and P. Brumer, "Decoherence Effects in Reactive Scattering," in preparation.

169. G. M. Palma, K.-A. Suominen, and A. K. Ekert, *Proc. R. Soc. London, A* **452**, 567 (1996); L.-M Duan, and G.-C. Guo, *Phys. Rev. Lett.* **79**, 1953 (1997); *ibid.*, *Phys. Rev. A* **57**, 737 (1998); P. Zanardi, *Phys. Rev. A* **56**, 4445 (1997); P. Zanardi and M. Rasetti, *Phys. Rev. Lett.* **79**, 3306 (1997); D. A. Lidar, I. L. Chuang, and K. B. Whaley, *Phys. Rev. Lett.* **81** 2594 (1998); D. Bacon, J. Kempe, D. A. Lidar and K. B. Whaley, *Phys. Rev. Lett.* **85**, 1758 (2000); M. Shapiro and P. Brumer, *Phys. Rev. A* **66**, 05308 (2002).

170. M. A. Nielsen and I. L. Chuang, *Quantum Computation and Quantum Information*, Cambridge University Press, Cambridge, 2000.

171. M. Demirplak and S. A. Rice, *J. Chem. Phys.*, **116**, 8028 (2002).

172. L. K. Iwaki and D. L. Dlott, *J. Phys. Chem A* **104**, 9109 (2000).

173. C. J. Bardeen, Q. Wang, and C. V. Shank, *Phys. Rev. Lett.* **75**, 3410 (1995).

174. T. Brixner, N. H. Damrauer, P. Niklaus, and G. Gerber, *Nature* **414**, 57 (2001).

175. M. Shapiro and P. Brumer, *J. Chem. Phys.* **90**, 6179 (1989).

176. For the case of $\phi = 0$, and $T_1 = T_2 = \infty$, see L. Allen and C. R. Stroud, Jr., *Phys. Rep.* **91**, 1 (1982). Note, however, that there are serious misprints in Eq. (3.6) of this reference.

177. J. D. Macomber, *The Dynamics of Spectroscopic Transitions*, Wiley, New York, 1976.

178. J. Cao, M. Messina, and K. R. Wilson, *J. Chem. Phys.* **106**, 5239 (1997).

179. J. Cao, C. J. Bardeen, and K. R. Wilson, *J. Chem. Phys.* **113**, 1898 (2000).

180. D. H. Schirrmeister and V. May, *Chem. Phys. Lett.* **220**, 1 (1997); *Chem. Phys. Lett.* **297**, 383 (1998).

181. V. S. Batista and P. Brumer, *Phys. Rev. Lett.*, **89**, 143201 (2002).

182. V. Guillar, V. S. Batista, and W. H. Miller, *J. Chem. Phys.* **113**, 9510, 2000, and references therein.

183. G. Campolieti and P. Brumer, *J. Chem. Phys.* **96**, 5969 (1992); *Phys. Rev. A* **50**, 997 (1994); *Phys. Rev. A* **53**, 2958 (1996); *J. Chem. Phys.* **107**, 791 (1997); *J. Chem. Phys.* **109**, 2999 (1998); B. McQuarrie and P. Brumer, *Chem. Phys. Lett.* **319**, 27 (2000); V. S. Batista and P. Brumer, *J. Phys. Chem.* **105**, 2591 (2001); *J. Chem. Phys.* **114**, 10321 (2001).

184. X.-P. Jiang and P. Brumer, *Chem. Phys. Lett.* **208**, 179 (1993).

185. A. Pattanayak and P. Brumer, *Phys. Rev. Lett.* **79**, 4131 (1997).

186. M. Sterling, R. Zadoyan, and V. A. Apkarian, *J. Chem. Phys.* **104**, 6497 (1996).

187. E. Gershgoren, J. Vala, R. Kosloff, and S. Ruhman, *J. Phys. Chem. A* **105**, 5081 (2001).

188. X.-P. Jiang, M. Shapiro, and P. Brumer, *J. Chem. Phys.* **104**, 607 (1996).

189. X.-P. Jiang and P. Brumer, *Chem. Phys. Lett.* **180**, 222 (1991).

190. X.-P. Jiang and P. Brumer, *J. Chem. Phys.* **94**, 5833 (1991).

191. M. Shapiro, *J. Chem. Phys.* **101**, 3849 (1994).

192. X.-P. Jiang and P. Brumer, *Chem. Phys. Lett.* **208**, 179 (1993).

193. J. C. Camparo and P. Lambropoulos, *Phys. Rev. A* **55**, 552 (1997).

194. N. E. Karapanagioti, D. Xenakis, D. Charalambidis, and C. Fotakis, *J. Phys. B* **29**, 3599 (1996).

195. P. L. Knight, M. A. Lauder, and B. J. Dalton, *Phys. Rep.* **190**, 1, (1990).

196. T. Nakajima, M. Elk, Z. Jian, and P. Lambropoulos, *Phys. Rev. A* **50**, R913 (1994).

197. T. Halfmann, L. P. Yatsenko, M. Shapiro, B. W. Shore, and K. Bergmann, *Phys. Rev. A* **58**, R46 (1998).

198. S. Cavalieri, R. Eramo, L. Fini, M. Materazzi, O. Faucher, and D. Charalambidis, *Phys. Rev. A* **57**, 2915 (1998).

199. K. G. H. Baldwin, M. D. Bott, H. A. Bachor, and P. B. Chapple, *J. Opt. B* **2**, 470 (2000).

200. Z. Chen, M. Shapiro, and P. Brumer, *J. Chem. Phys.* **102**, 5683 (1995).

201. A. Shnitman, I. Sofer, I. Golub, A. Yogev, M. Shapiro, Z. Chen, and P. Brumer, *Phys. Rev. Lett.* **76**, 2886 (1996).

202. Z. Chen, M. Shapiro, and P. Brumer, *J. Chem. Phys.* **98**, 8647 (1993).

203. M. O. Scully and M. S. Zubairy, *Quantum Optics*, Cambridge University Press, Cambridge, 1997.

204. M. Schubert and B. Wilhelmi, *Nonlinear Optics and Quantum Electronics*, Wiley, New York, 1986.

205. P. Brumer, Z. Chen, and M. Shapiro, *Isr. J. Chem.* **34**, 137 (1994).

206. Z. Chen, P. Brumer, and M. Shapiro, *Chem. Phys. Lett.* **198**, 498 (1992).

207. Q. Zhang, M. Keil, and M. Shapiro, "Interferometric Control of Na$_2$ Photodissociation," in preparation.

208. S. T. Pratt, *J. Chem. Phys.* **104**, 5776 (1996).

209. F. Wang, C. Chen, and D. S. Elliott, *Phys. Rev. Lett.* **77**, 2416 (1996).

210. E. Luc-Koenig, M. Aymar, M. Millet, J. M. Lecomte, and A. Lyras, *Eur. Phys. D* **10**, 205, (2000).

211. N. Ph. Georgiades, E. S. Polzik, and H. J. Kimble, *Opt. Lett.* **21** 1688 (1996).

212. K. D. Bonen and V. V. Kresin, *Electric Dipole Polarizabilities of Atoms, Molecules and Clusters*, World Scientific, Singapore, 1997.

213. E. McCullough, M. Shapiro, and P. Brumer, *Phys. Rev. A* **61**, 41801 (2000).

214. E. McCullough, M.Sc. Dissertation, University of Toronto, 1997.

215. R. W. Boyd, *Nonlinear Optics*, Academic, Boston, 1992.

216.

217. L. V. Hau, S. E. Harris, Z. Dutton, and C. H. Behroozi, *Nature* **397**, 594 (1999).

218. M. O. Scully, *Phys. Rev. Lett.* **67**, 1855 (1991); A. S. Zibrov, M. D. Lukin, L. Hollberg, D. E. Nikonov, M. O. Scully, H. G. Robinson, and V. L. Velichansky, *Phys. Rev. Lett.* **76**,

3935 (1996); M. Fleischhauer, C. H. Keitel, M. O. Scully, C. Su, B. T. Ulrich, and S.-Y. Zhu, *Phys. Rev. A* **46**, 1468, (1992).

219. J. A. Fiss, A. Khachatrian, L. Zhu, R. J. Gordon, and T. Seideman, *Faraday Disc. Chem. Soc.* **113**, 61 (1999).

220. T. Seideman, *J. Chem. Phys.* **111**, 9168 (1999).

221. R. J. Gordon, L. Zhu, and T. Seideman, *J. Phys. Chem A* **105**, 4387 (2001).

222. U. Fano, *Phys. Rev.* **124**, 1866 (1961).

223. J. C. McCauley, *Classical Mechanics*, Cambridge University Press, Cambridge, 1997.

224. P. Brumer, in *Encyclopedia of Modern Physics*, R. A. Meyers, ed., Academic, New York, 1990, p. 205.

225. P. Brumer, *Adv. Chem. Phys.* **47**, 201 (1981).

226. A. Pattanayak and P. Brumer, *Phys. Rev. Lett.* **77**, 59 (1996).

227. R. Blümel, S. Fishman, and U. Smilansky, *J. Chem. Phys.* **84**, 2604 (1986).

228. G. Casati and B. Chirikov, *Quantum Chaos: Between Order and Disorder*, Cambridge University Press, Cambridge, 1995.

229. J. Gong and P. Brumer, *Phys. Rev. Lett.* **86**, 1741 (2001).

230. R. Blümel and W. P. Reinhardt, *Chaos in Atomic Physics*, Cambridge University Press, Cambridge, 1997.

231. J. Gong and P. Brumer, *J. Chem. Phys.* **115**, 3590 (2001).

232. F. Haake, *Quantum Signatures of Chaos*, Springer, Berlin, 1992.

233. J. L. Krause, M. Shapiro, and P. Brumer, *J. Chem. Phys.* **92**, 1126 (1990).

234. M. Shapiro and P. Brumer, *Phys. Rev. Lett.* **77**, 2574 (1996).

235. D. Holmes, M. Shapiro, and P. Brumer, *J. Chem. Phys.* **105**, 9162 (1996).

236. A. Abrashkevich, M. Shapiro, and P. Brumer, *Phys. Rev. Lett.* **81**, 3789 (1998).

237. An erratum [A. Abrashkevich, M. Shapiro, and P. Brumer, *Phys. Rev. Lett.* **82**, 3002 (1999)] clarifies that the results shown in Ref. [236] are valid at $\phi = 0$ and that control over the total cross section is not possible when building the initial superposition from helicity states.

238. P. Brumer, A. G. Abrashkevich, and M. Shapiro, *Faraday Disc. Chem. Soc.* **113**, 291 (2000).

239. P. Brumer, K. Bergmann, and M. Shapiro, *J. Chem. Phys.* **113**, 2053, (2000).

240. A. Abrashkevich, M. Shapiro, and P. Brumer, *Chem. Phys.* **267**, 81 (2001).

241. E. T. Smith, A. A. Dhirani, D. A. Koborowski, R. A. Rubenstein, T. D. Roberts, H. Yao, and D. E. Pritchard, *Phys. Rev. Lett.* **81**, 1996 (1998).

242. W. D. Phillips, *Rev. Mod. Phys.* **70**, 721 (1998).

243. M. Weitz, B. C. Young, and S. Chu, *Phys. Rev. Lett.* **94**, 2563 (1994).

244. H. L. Bethlem, A. J. van Roij, R. T. Jongma, and G. Meijer, *Phys. Rev. Lett.* **88**, 133003 (2002), and references therein.

245. K. Bergmann, H. Theuer, and B. W. Shore, *Rev. Mod. Phys.* **70**, 1003 (1998).

246. R. G. Unanyan, M. Fleischhauer, K. Bergmann, and B. W. Shore, *Opt. Commun.* **155**, 144 (1998); H. Theuer, R. G. Unanyan, C. Habscheid, K. Klein, and K. Bergmann, *Optics Express* **4**, 77 (1999).

247. J. Gong, M. Shapiro, and P. Brumer, in preparation; J. Gong, Ph.D. Dissertation, University of Toronto, 2001.

248. A. R. Edmonds, *Angular Momentum in Quantum Mechanics*, Princeton University Press, Princeton, 1960.

249. R. T. Pack and G. A. Parker, *J. Chem. Phys.* **87**, 3888 (1987).

250. M. Tamir and M. Shapiro, *Chem. Phys. Lett.* **31**, 166 (1975).

251. A. Abrashkevich, M. Shapiro, and P. Brumer, *Chem. Phys.* **267**, 81 (2001).

252. E. Frishman, M. Shapiro, and P. Brumer, *J. Chem. Phys.* **110**, 9 (1999).

253. M. Shapiro and P. Brumer, *Phys. Rev. Lett.*, submitted.

254. E. Frishman, M. Shapiro, and P. Brumer, *J. Phys. Chem A* **103**, 10333 (1999).

255. L. D. Barron, *Molecular Light Scattering and Optical Activity*, Cambridge University Press, Cambridge, 1982.

256. R. G. Woolley, *Adv. Phys.* **25**, 27 (1975); D. C. Walker, ed., *Origins of Optical Activity in Nature*, Elsevier, Amsterdam, 1979.

257. For a discussion, see L. D. Barron, *Chem. Soc. Rev.* **15**, 189 (1986). For historical examples, see J. A. Bel, *Bull. Soc. Chim. Fr.* **22**, 337 (1874); J. H. Van't Hoff, *Die Lagerung der Atome im Raume*, 2nd ed., Vieweg, Braunschweig, 1894, p. 30.

258. M. Quack, *Angew. Chem. Int. Ed. Engl.* **28**, 571 (1989).

259. M. Shapiro, E. Frishman, and P. Brumer, *Phys. Rev. Lett.* **84**, 1669, 2000).

260. D. Gerbasi, M. Shapiro, and P. Brumer, *J. Chem. Phys.* **115**, 5349, 2001).

261. P. Brumer, E. Frishman, and M. Shapiro, *Phys. Rev. A* **65**, 015401 (2001).

262. For example, A. Salam and W. J. Meath, *Chem. Phys.* **228**, 115 (1998).

263. Y. Fujimura, L. Gonzalez, K. Hoki, J. Manz, and Y. Ohtsuki, *Chem. Phys. Lett.* **306**, 1 (1999); errata: *ibid.*, **310**, 578 (1999); for control over an initial superposition state and the following papers for control over a racemic mixture: Y. Fujimura, L. Gonzalez, K. Hoki, D. Kroener, J. Manz, and Y. Ohtsuki, *Angew. Chem. Int. Ed. Engl.* **39**, 4586 (2000); K. Hoki, Y. Ohtsuki, and Y. Fujimura, *J. Chem. Phys.* **114**, 1575 (2001); K. Hoki, D. Kroener, and J. Manz, *Chem. Phys.* **267**, 59 (2001).

264. K. Hoki, L. Gonzalez, and Y. Fujimura, *J. Chem. Phys.* **116**, 2433, (2002).

265. C. S. Maierle and R. A. Harris, *J. Chem. Phys.* **109**, 3713 (1998).

266. J. H. Shirley, *Phys. Rev. B* **138**, 979 (1965).

267. A. Brown and W. J. Meath, *J. Chem. Phys.* **109**, 9351 (1998).

268. M. Shapiro, E. Frishman, and P. Brumer, *J. Chem. Phys.* submitted.

269. A. Apolonski, A. Poppe, G. Tempea, C. Spielmann, T. Udem, R. Holzwarth, T. W. Hansch, and E. Krausz, *Phys. Rev. Lett.* **85**, 740 (2000).

270. C. Raman, T. C. Weinacht, and P. H. Bucksbaum, *Phys. Rev. A* **55**, R3995 (1997).

271. J. M. Hollas, *High Resolution Spectroscopy*, Butterworths, London, 1982.

272. On the preparation and measurement of a superposition of chiral states, see also R. A. Harris, Y. Shi, and J. A. Cina, *J. Chem. Phys.* **101**, 3459 (1994); J. A. Cina and R. A. Harris, *J. Chem. Phys.* **100**, 2531 (1994).

273. E. Segev and M. Shapiro, *J. Chem. Phys.* **73**, 2001 (1980); *ibid.*, **77**, 5601 (1982).

274. E. Deretey, M. Shapiro, and P. Brumer, *J. Phys. Chem. A* **105**, 9509 (2001).

275. M. Hayashi, A. M. Mebel, K. K. Liang, and S. H. Lin, *J. Chem. Phys.* **108**, 2044 (1998).

276. S. Lochbrunner, T. Schultz, M. Schmitt, J. P. Shaffer, M. Z. Zgierski, and A. Stolow, *J. Chem. Phys.* **114**, 2519, (2001).

277. D. Gerbasi, M. Shapiro, and P. Brumer, in progress.

278. Y. Brumer, M. Shapiro, P. Brumer, and K. Balderidge, *J. Phys. Chem. A* **106**, 9512 (2002).

279. L. Allen and J. H. Eberly, *Optical Resonance and Two-Level Atoms*, Wiley, New York, 1975.

280. $M_i = M_k$ since both $|E_i\rangle$ and $|E_k\rangle$ arise by excitation, with linearly polarized light, from a common eigenstate.

281. G. G. Balint-Kurti and M. Shapiro, *Chem. Phys.* **61**, 137 (1981).

282. A. Abragam, *The Principles of Nuclear Magnetism*, Claredon, Oxford, 1961, p. 65.

283. D. G. Grischkowski, *Phys. Rev. Lett.* **24**, 866 (1970); D. G. Grischkowski and J. A. Armstrong, *Phys. Rev. A* **6**, 1566 (1972}; D. G. Grischkowski, E. Courtens, and J. A. Armstrong, *Phys. Rev. Lett.* **31**, 422 (1973); D. G. Grischkowski, *Phys. Rev. A* **7**, 2096 (1973); D. G. Grischkowski, M. M. T. Loy and P. F. Liao, *Phys. Rev. A* **12**, 2514 (1975).

284. M. Takatsuji, *Phys. Rev. A* **4**, 808 (1971).

285. R. G. Brewer and E. L. Hahn, *Phys. Rev. A* **11**, 1641 (1975).

286. R. J. Cook and B. W. Shore, *Phys. Rev. A* **20**, 539 (1979).

287. M. V. Kuzmin and V. N. Sazonov, *Zh. Eksp. Teor. Fiz.* **79**, 1759 (1980).

288. F. T. Hioe and J. H. Eberly, *Phys. Rev. Lett.* **12**, 838 (1981).

289. J. Oreg, F. T. Hioe, and J. H. Eberly, *Phys. Rev. A* **29**, 690 (1984).

290. C. Liedenbaum, S. Stolte, and J. Reuss, *Phys. Rept.* **178**, 1 (1989).

291. B. Broers, H. B. van Linden van den Heuvell, and L. D. Noordam, *Phys. Rev. Lett.* **69**, 2062 (1992).

292. C. E. Carroll and F. T. Hioe, *Phys. Rev. Lett.* **68**, 3523 (1992).

293. B. W. Shore, J. Martin, M. P. Fewell, and K. Bergmann, *Phys. Rev. A* **52**, 566 (1995); J. Martin, B. W. Shore, and K. Bergmann, *Phys. Rev. A* **52**, 583 (1995).

294. U. Gaubatz, P. Rudecki, S. Schiemann, and K. Bergmann, *J. Chem. Phys.* **92**, 5363 (1990).

295. G. Coulston and K. Bergmann, *J. Chem. Phys.* **96**, 3467 (1992).

296. C. E. Carroll and F. T. Hioe, *Phys. Lett. A* **199**, 145 (1995).

297. K. J. Boller, A. Imamoglu, and S. E. Harris, *Phys. Rev. Lett.* **66**, 2593 (1991); J. E. Field, K. H. Hahn, and S. E. Harris, *Phys. Rev. Lett.* **67**, 3062 (1991).

298. S. E. Harris, *Phys. Rev. Lett.* **70**, 552 (1993).

299. K. Ichimura, K. Yamamoto, and N. Gemma, *Phys. Rev. A* **58**, 4116 (1998).

300. M. Takeoka, D. Fujishima, and F. Kannari, *Jpn. J. Appl. Phys. 1* **40**, 137 (2001).

301. S. E. Harris, *Phys. Rev. Lett.* **62**, 1033 (1989).

302. S. E. Harris, J. E. Field, and A. Imamoglu, *Phys. Rev. Lett.* **64**, 1107 (1990).

303. M. O. Scully, S. Y. Zhu, and A. Gavrielides, *Phys. Rev. Lett.* **62**, 2813 (1989).

304. O. Kocharovskaya, *Phys. Rep.* **219**, 175 (1992).

305. J. L. Cohen and P. R. Berman, *Phys. Rev. A* **55**, 3900 (1997).

306. F. B. deJong, R. J. C. Spreeuw, and H. B. van Linden van den Heuvell, *Phys. Rev. A* **55**, 3918 (1997).

307. R. P. Feynman, F. L. Vernon, Jr., and R. W. Hellwarth, *J. Appl. Phys.* **28**, 49 (1957).

308. A. Vardi and M. Shapiro, *Phys. Rev. A* **58**, 1352 (1998); *ibid. Comm. At. Mol. Phys.* **2**, 233 (2001).

309. Z. Chen, P. Brumer, and M. Shapiro, *J. Chem. Phys.* **98**, 6843 (1993).

310. M. N. Kobrak and S. A. Rice, *J. Chem. Phys.* **109**, 1 (1998); *ibid, Phys. Rev. A* **57**, 1158 (1998).

311. K. Hakuta, L. Marmet, and B. P. Stoicheff, *Phys. Rev. Lett.* **66**, 596 (1991).

312. A. Kasapi, *Phys. Rev. Lett.* **77**, 1035 (1996).

313. B. S. Ham, M. S. Shahriar, and P. R. Hemmer, *Opt. Lett.* **22**, 1138 (1997).

314. K. Ichimura, K. Yamamoto, and N. Gemma, *Phys. Rev. A* **58**, 4116 (1998).

315. Y. Zhao, C. Wu, B.-S. Ham, M. K. Kim, and E. Awad, *Phys. Rev. Lett.* **79**, 641 (1997).

316. A. S. Zibrov, M. D. Lukin, D. E. Nikonov, L. Hollberg, M. O. Scully, V. L. Velichansky, and H. G. Robinson, *Phys. Rev. Lett.* **75**, 1499 (1995).

317. G. G. Padmabandu, G. R. Welch, I. N. Shubin, E. S. Fry, D. E. Nikonov, M. D. Lukin, and M. O. Scully, *Phys. Rev. Lett.* **76**, 2053 (1996).

318. J. G. B. Beumee and H. Rabitz, *J. Math. Phys.* **31**, 1253 (1990).

319. E. E. Aubanel and A. D. Bandrauk, *Can. J. Chem.* **72**, 673 (1994).

320. J. Cao and K. R. Wilson, *J. Chem. Phys.* **107**, 1441 (1997).

321. L. E. E. de Araujo and I. A. Walmsley, *J. Phys. Chem. A* **103**, 10409, (1999).

322. P. Král, Z. Amitay, and M. Shapiro, *Phys. Rev. Lett.* **89**, 063002 (2002).

323. R. N. Zare, *Angular Momentum*, Wiley, New York, 1988.

324. A. G. Abrashkevich and M. Shapiro, *J. Phys. B* **29**, 627 (1996).

325. A. Vardi and M. Shapiro, *J. Chem. Phys.* **104**, 5490 (1996).

326. E. Frishman and M. Shapiro, *Phys. Rev. A* **54**, 3310 (1996).

327. M. Shapiro, *J. Phys. Chem.* **102**, 9570 (1998).

328. M. Shapiro, M. J. J. Vrakking, and A. Stolow, *J. Chem. Phys.* **110**, 2465 (1999).

329. Z. Chen, M. Shapiro, and P. Brumer, *Chem. Phys. Lett.* **228**, 289 (1994).

330. See, e.g., D. Bohm, *Quantum Theory*, Prentice Hall, Englewood Cliffs, NJ, 1955; M. S. Child, *Semiclassical Mechanics with Molecular Applications*, Oxford University Press, Oxford, 1991.

331. D. E. Amos, Sandia National Laboratories, Computer routine library—slatec/Complex Airy and subsidiary routines.

332. C. Cohen-Tannoudji, B. Diu, and F. Laloë, *Quantum Mechanics*, Vol. 1, Wiley, New York, 1977.

333. N. Billy, B. Girard, G. Gonédard, and J. Vigue, *Mol. Phys.* **61**, 65 (1987).

334. E. Arimondo, *Lett. Nuovo Cimento* **17**, 333 (1976).

335. E. Arimondo, *Prog. Opt.* **35**, 257 (1996).

336. Z. Chen, P. Brumer, and M. Shapiro, *Phys. Rev. A* **52**, 2225 (1995).

337. Z. Chen, P. Brumer, and M. Shapiro, *Chem. Phys.* **217**, 325 (1997).

338. S. Chu, J. E. Bjorkholm, A. Ashkin, and A. Cable, *Phys. Rev. Lett.* **57**, 314 (1986).

339. A. Aspect, C. Cohen-Tannoudji, J. Dalibard, A. Heidemann, and C. Solomom, *Phys. Rev. Lett.* **57**, 1688 (1986).

340. J. Lawall, F. Bardou, K. Shimizu, M. Leduc, A. Aspect, and C. Cohen-Tannoudji, *Phys. Rev. Lett.* **73**, 1915 (1994).

341. J. T. Bahns, W. C. Stwalley, and P. L. Gould, *J. Chem. Phys.* **104**, 9689 (1996).

342. A. Bartana, R. Kosloff, and D. J. Tannor, *J. Chem. Phys.* **99**, 196 (1993).

343. H. R. Thorsheim, J. Weiner, and P. S. Julienne, *Phys. Rev. Lett.* **58**, 2420 (1987).

344. Y. B. Band and P. S. Julienne, *Phys. Rev. A* **51**, R4317 (1995).

345. A. Vardi, D. Abrashkevich, E. Frishman, and M. Shapiro, *J. Chem. Phys.* **107**, 6166 (1997).

346. R. Coté and A. Dalgarno, *Chem. Phys. Lett.* **279**, 50 (1997).

347. P. S. Julienne, K. Burnett, Y. B. Band, and W. C. Stwalley, *Phys. Rev. A* **58**, R797 (1998).

348. A. Vardi, M. Shapiro, and K. Bergmann, *Optics Express* **4**, 91 (1999).

349. M. Mackie and J. Javanainen, *Phys. Rev. A* **60**, 3174 (1999).

350. A. Vardi and M. Shapiro, *Phys. Rev. A* **62**, 25401 (2000).

351. A. Vardi, M. Shapiro, and J. R. Anglin, *Phys. Rev. A* **65**, 27401 (2002).

352. U. Marvet and M. Dantus, *Chem. Phys. Lett.* **245**, 393 (1995).

353. P. Gross and M. Dantus, *J. Chem. Phys.* **106**, 8013 (1997).

354. A. Fioretti, D. Comparat, A. Crubellier, O. Dulieu, F. Masnou-Seeuws, and P. Pillet, *Phys. Rev. Lett.* **80**, 4402 (1998).

355. A. Fioretti, D. Comparat, C. Drag, C. Amiot, O. Dulieu, F. Masnou-Seeuws, and P. Pillet, *Eur. Phys. D* **5**, 389 (1998).

356. A. N. Nikolov, E. E. Eyler, X. T. Wang, J. Li, H. Wang, W. C. Stwalley, and P. L. Gould, *Phys. Rev. Lett.* **82**, 703 (1999).

357. R. Wynar, R. S. Freeland, D. J. Han, C. Ryu, and D. J. Heinzen, *Science* **287**, 1016 (2000).

358. C. McKenzie, J. Hecker Denschlag, H. Häffner, A. Browaeys, Luis E. E. de Araujo, F. K. Fatemi, K. M. Jones, J. E. Simsarian, D. Cho, A. Simoni, E. Tiesinga, P. S. Julienne, K. Helmerson, P. D. Lett, S. L. Rolston, and W. D. Phillips, *Phys. Rev. Lett.* **88**, 120403 (2001).

359. R. Hulett, in *Cold Molecules 2002: Cold Molecules and Bose-Einstein Condensates*, Les Houches, France, 2002.

360. For reviews, see A. D. Bandrauk, ed., *Molecules in Laser Fields*, Marcel Dekker, New York, 1994.

361. M. V. Fedorov, O. V. Kudrevatova, V. P. Makarov, and A. A. Samokhin, *Opt. Comm.* **13**, 299 (1975).

362. N. M. Kroll and K. M. Watson, *Phys. Rev. A* **8**, 804 (1973); *ibid.*, **13**, 1018 (1976).

363. J. I. Gerstein and M. H. Mittleman, *J. Phys. B.* **9**, 383 (1976).

364. J. M. Yuan, T. F. George, and F. J. McLafferty, *Chem. Phys. Lett.* **40**, 163 (1976); J. M. Yuan, J. R. Laing, and T. F. George, *J. Chem. Phys.* **66**, 1107 (1977); T. F. George, J. M. Yuan, and I. H. Zimmermann, *Faraday Disc. Chem. Soc.* **62**, 246 (1977); P. L. DeVries and T. F. George, *Faraday Disc. Chem. Soc.* **67**, 129 (1979); T. F. George, *J. Phys. Chem.* **86**, 10 (1982).

365. V. S. Dubov, L. I. Gudzenko, L. V. Gurvich, and S. I. Iakovlenko, *Chem. Phys. Lett.* **45**, 351 (1977).

366. A. D. Bandrauk and M. L. Sink, *Chem. Phys. Lett.* **57**, 569 (1978); *J. Chem. Phys.* **74**, 1110 (1981).

367. A. E. Orel and W. H. Miller, *Chem. Phys. Lett.* **57**, 362 (1978); *ibid.*, **70**, 4393 (1979); *ibid.*, **73**, 241 (1980).

368. J. C. Light and A. Altenberger-Siczek, *J. Chem. Phys.* **70**, 4108 (1979).

369. H. J. Foth, J. C. Polanyi, and H. H. Telle, *J. Phys. Chem.* **86**, 5027 (1982).

370. T. Ho, C. Laughlin, and S. I. Chu, *Phys. Rev. A* **34**, 122 (1985).

371. D. R. Matusek, M. Yu. Ivanov, and J. S. Wright, *Chem. Phys. Lett.* **258**, 255 (1996).

372. M. Shapiro and Y. Zeiri, *J. Chem. Phys.* **85**, 6449 (1986).

373. T. Seideman, J. L. Krause and M. Shapiro, *Chem. Phys. Lett.* **173**, 169 (1990); *ibid.*, *Faraday Disc. Chem. Soc.* **91**, 271 (1991).

374. T. Seideman and M. Shapiro, *J. Chem. Phys.* **94**, 7910 (1991).

375. L. S. Rodberg and R. M. Thaler, *Introduction to the Quantum Theory of Scattering*, Academic, New York, 1967.

376. C. Eckart, *Phys. Rev.* **35**, 1303 (1930).

377. H. Eyring, J. Walter, and G. E. Kimball, *Quantum Chemistry*, Wiley, New York, 1944.

378. B. Liu, *J. Chem. Phys.* **58**, 1925 (1973).

379. D. G. Truhlar and C. J. Horowitz, *J. Chem. Phys.* **68**, 2466 (1978).

380. L. L. Chang, L. Esaki, and R. Tsu, *Appl. Phys. Lett.* **24**, 593 (1974).

381. S. C. Kan and A. Yariv, *J. Appl. Phys.* **67**, 1957 (1990).

382. A. Sa'ar, S. C. Kan, and A. Yariv, *J. Appl. Phys.* **67**, 3892 (1990).

383. H. Yamamoto, Y. Kanie, M. Arakawa, and K. Taniguchi, *Appl. Phys. A* **50**, 577 (1990).

384. W. Cai, T. F. Zheng, P. Hu, M. Lax, K. Shun, and R. Alfano, *Phys. Rev. Lett.* **65**, 104 (1990).

385. K. A. Chao, M. Willander, and Yu. M. Galperin, *Physica Scripta* **T54**, 119 (1994).

386. A. Yariv, *Optical Electronics*, 4th ed., Saunders College, Philadelphia, 1991.

387. I. Vorobeichik, R. Lefebvre, and N. Moiseyev, *Europhys. Lett.* **41**, 111 (1998).

388. M. Shapiro and H. Bony, *J. Chem. Phys.* **83**, 1588 (1985).

389. T. Seideman and M. Shapiro, *J. Chem. Phys.* **88**, 5525 (1988); *ibid.*, *J. Chem. Phys.* **92**, 2328 (1990).

390. S.-I. Chu, *Chem. Phys. Lett.* **64**, 178 (1979); *ibid.*, **70**, 205 (1980); T. Ho, C. Laughlin, and S.-I. Chu, *Phys. Rev. A.* **32**, 122 (1985); S.-I. Chu and R. Yin, *J. Opt. Soc. Am. B* **4**, 720 (1987); G. Yao and S.-I. Chu, *Chem. Phys. Lett.* **197**, 413 (1992).

391. M. Crance and M. Aymar, *J. Phys. B* **13**, L421 (1980).

392. Z. Deng and J. H. Eberly, *Phys. Rev. Lett.* **53**, 1810 (1984).

393. R. Grobe and J. H. Eberly, *Phys. Rev. Lett.* **68**, 2905 (1992); *ibid.*, *Phys. Rev. A* **48**, 623 (1993); *ibid.*, *Laser Phys.* **3**, 323 (1993) *ibid.*, *Phys. Rev. A* **48**, 4664 (1993).

394. K. Rzazewski and R. Grobe, *Phys. Rev. A* **33**, 1855 (1986).

395. B. Piraux, R. Bhatt, and P. L. Knight, *Phys. Rev. A* **41**, 6296 (1990).

396. R. Blank and M. Shapiro, *Phys. Rev. A*, **50**, 3234 (1994); *Phys. Rev. A* **51**, 4762 (1995); *Phys. Rev. A* **52**, 4278 (1995).

397. A. D. Bandrauk, E. E. Aubanel, and J. M. Gauthier, in *Molecules in Laser Fields*, A. D. Bandrauk, ed., Marcel Dekker, New York, 1994, p. 109; see also, E. E. Aubanel and A.

D. Bandrauk, *Chem. Phys. Lett.* **197**, 419 (1992); A. D. Bandrauk, E. E. Aubanel, and J. M. Gauthier, *Laser Phys.* **3**, 381 (1993).

398. R. Heather and H. Metiu, *J. Chem. Phys.* **88**, 5496 (1988).

399. M. Seel and W. Domcke, *J. Chem. Phys.* **95**, 7806 (1991).

400. H. Abou-Rachid, T. T. Nguyen-Dang, R. K. Chaudhury, and X. He, *J. Chem. Phys.* **97**, 5497 (1992).

401. K. C. Kulander, K. J. Shafer, and J. Krause, in *Atoms in Intense Laser Fields*, Academic, New York, 1992.

402. E. S. Smyth, J. S. Parker, and K. T. Taylor, *Comput. Phys. Commun.* **114**, 1 (1998); J. S. Parker, L. R. Moore, K. J. Meharg, D. Dundas, and K. T. Taylor, *J. Phys. B* **34** L69 (2001).

403. U. Peskin and N. Moiseyev, *J. Chem. Phys.* **99**, 4590 (1993).

404. U. Peskin and N. Moiseyev, *Phys. Rev. A* **49**, 3712 (1994); U. Peskin, O. E. Alon, and N. Moiseyev, *J. Chem. Phys.* **100**, 7310 (1994); U. Peskin, R. Kosloff, and N. Moiseyev, *J. Chem. Phys.* **100**, 8849 (1994).

405. N. Moiseyev, *Comments At. Mol. Phys.* **31**, 87 (1995); O. Alon and N. Moiseyev, *Chem. Phys.* **196**, 499 (1995).

406. M. Pont and R. Shakeshaft, in *Photon and Electron Collisions with Atoms and Molecules*, P. G. Burke and C. J. Joachain, eds., Plenum, New York, 1997, p. 125.

407. S. Chelkowski, T. Zuo, O. Atabek, and A. D. Bandrauk, *Phys. Rev.* **52**A, 2977 (1995).

408. C. Meier, V. Engel, and U. Manthe, *J. Chem. Phys.* **109**, 36 (1998).

409. M. A. Kornberg and P. Lambropoulos, *J. Phys. B* **32**, L603 (1999).

410. G. Tanner, K. Richter, and J. M. Rost, *Rev. Mod. Phys.* **72**, 497 (2000).

411. A. Giusti-Suzor, X. He, O. Atabek, and F. H. Mies, *Phys. Rev. Lett.* **64**, 515 (1990); A. Giusti-Suzor and F. H. Mies, *Phys. Rev. Lett.* **68**, 3869 (1992).

412. C. Wunderlich, E. Kobler, H. Figger, and T. W. Hansch, *Phys. Rev. Lett.* **78**, 2333 (1997); C. Wunderlich, H. Figger, and T. W. Hansch, *Phys. Rev. A* **62**, 23401 (2000).

413. B. M. Garraway and K. A. Suominen, *Phys. Rev. Lett.* **80**, 932 (1998).

414. I. R. Solá, J. Santamaria, and V. S. Malinovsky, *Phys. Rev. A* **61**, 43413 (2000).

415. I. R. Solá, B. Y. Chang, J. Santamaria, V. S. Malinovsky, and J. L. Krause, *Phys. Rev. Lett.* **85**, 4241 (2000).

416. V. S. Malinovsky and J. L. Krause, *Chem. Phys.* **267**, 47 (2001).

417. C. Zener, *Proc. Roy. Soc. A* **137**, 696 (1932); L. D. Landau, *Phys. Z. Sowjetunion* **2**, 46 (1932).

418. E. Charron, A. Giusti-Suzor, and F. H. Mies, *Phys. Rev. Lett.* **71**, 692 (1993).

419. J. W. J. Verschuur, L. D. Noordam, and H. B. van Linden van den Heuvell, *Phys. Rev. A* **40**, 4383 (1989).

420. A. Zavriyev, P. H. Bucksbaum, H. G. Muller, and D. W. Schumacher, *Phys. Rev. A* **42**, 5500 (1990).

421. S. W. Allendorf and A. Szoke, *Phys. Rev. A* **44**, 518 (1991).

422. A. Zavriyev and P. H. Bucksbaum, in *Molecules in Laser Fields*, A. D. Bandrauk, ed., Marcel Dekker, New York, 1994, p. 71.

423. A. Giusti-Suzor, F. H. Mies, L. F. Dimauro, E. Charron, and B. Yang, *J. Phys. B.* **28**, 309 (1995).

424. G. Yao and S.-I. Chu, *Chem. Phys. Lett.* **197**, 413 (1992).

425. A. Zavriyev, P. H. Bucksbaum, J. Squier, and F. Saline, *Phys. Rev. Lett.* **70**, 1077 (1993).

426. A. I. Pegarkov and L. P. Rapoport, *Opt. Spect.* **65**, 55 (1988); A. I. Pegarkov, *Phys. Rpts.* **336**, 255 (2000).

427. G. Timp, R. E. Behringer, D. M. Tennat, J. E. Cunningham, M. Prentiss, and K. K. Berggren, *Phys. Rev. Lett.* **69**, 1636 (1992).

428. K. Berggren, M. Prentiss, G. L. Timp, and R. E. Behringer, *J. Opt. Soc. Am. B* **11**, 1166 (1994).

429. J. D. Miller, R. A. Cline, and D. J. Heinzen, *Phys. Rev. A* **47**, R4567 (1993).

430. See, e.g., special issue on "Laser Cooling and Trapping of Atoms," *J. Opt. Soc. Am. B* **6**(11) (1989).

431. H. Stapelfeldt, H. Sakai, E. Constant, and P. B. Corkum, *Phys. Rev. Lett.* **79**, 2787 (1997).

432. H. Sakai, A. Tarasevitch, J. Danilov, H. Stapelfeldt, R. W. Yip, C. Ellert, E. Constant, and P. B. Corkum, *Phys. Rev. A* **57**, 2794 (1998).

433. T. Seideman, *J. Chem. Phys.* **106**, 2881 (1997).

434. T. Seideman, *Phys. Rev. A* **56**, R17 (1997).

435. N. A. Nguyen, B. K. Dey, M. Shapiro, and P. Brumer "Coherently Controlled Nanoscale Deposition of Rydberg Atoms," in preparation.

436. B. K. Dey, M. Shapiro, and P. Brumer, *Phys. Rev. Lett.* **85**, 3125 (2000).

437. D. Normand, L. A. Lompre, and C. Cornaggia, *J. Phys. B* **25**, L497 (1992); M. Schmidt, D. Normand, and C. Cornaggia, *Phys. Rev. A.* **50**, 5037 (1994).

438. P. Dietrich, D. T. Strickland, M. Laberge, and P. B. Corkum, *Phys. Rev. A* **47**, 2305 (1993).

439. B. Friedrich and D. R. Herschbach, *Phys. Rev. Lett.* **74**, 4623 (1995).

440. A. Bandrauk and E. E. Aubanel, *Chem. Phys.* **198**, 159 (1995).

441. T. Seideman, *J. Chem. Phys.* **103**, 7887 (1995).

442. W. Kim and P. M. Felker, *J. Chem. Phys.* **104**, 1147 (1996).

443. H. Haberland and B. V. Issendorff, *Phys. Rev. Lett.* **76**, 1445 (1996).

444. T. Takekoshi, B. M. Patterson, and R. J. Knize, *Phys. Rev. Lett.* **81**, 5105 (1998).

445. J. J. Larsen, K. Hald, N. Bjerre, H. Stapelfeldt, and T. Seideman, *Phys. Rev. Lett.* **85**, 2470 (2000).

446. F. Rosca-Pruna and M. J. J. Vrakking, *Phys. Rev. Lett.* **87**, 153902 (2001).

447. J. Karczmarek, J. Wright, P. B. Corkum, and M.Y. Ivanov, *Phys. Rev. Lett.* **82**, 3420 (1999).

448. D. M. Villeneuve, S. A. Aseyev, P. Dietrich, M. Spanner, M. Yu. Ivanov, and P. B. Corkum, *Phys. Rev. Lett.* **85**, 542 (2000).

449. M. Spanner and M. Yu. Ivanov, *J. Chem. Phys.* **114**, 3456 (2001).

450. R. N. Zare, *Angular Momentum*, Wiley, New York, 1988, pp. 80–81.

451. H. J. Loesch and J. Müller, *J. Phys. Chem.* **97**, 2158 (1993).

452. For methods of cooling of molecules by buffer-gas loading, see J. M. Doyle, *Bull. Am. Phys. Soc.* **39**, 1166 (1994).

453. See, e.g., A. J. Orr-Ewing and R. N. Zare, *Annu. Rev. Phys. Chem.* **45**, 315 (1994), and references therein.

454. V. N. Bagratashvili, V. S. Letokhov, A. A. Makarov, and E. A. Ryabov, *Laser Chem.* **4**, 311 (1984).

455. D. M. Larsen and N. Bloembergen, *Opt. Comm.* **17**, 254 (1976).

456. G. K. Paramonov and V. A. Savva, *Phys. Lett. A* **97**, 340 (1983).

457. G. K. Paramonov and V. A. Savva, *Chem. Phys.* **107**, 394 (1984).

458. G. K. Paramonov, V. A. Savva, and A. M. Samson, *Infrared Phys.* **25**, 201 (1985).

459. G. K. Paramonov, *Chem. Phys. Lett.* **169**, 573 (1990).

460. G. K. Paramonov, *Phys. Lett. A* **152**, 191 (1991).

461. R. T. Lawton and M. S. Child, *Mol. Phys.* **37**, 1799 (1979); M. S. Child and H. S. Halonen, *Adv. Chem. Phys.* **57**, 1 (1984).

462. C. Jaffe and P. Brumer, *J. Chem. Phys.* **73**, 5645 (1980).

463. T. Joseph and J. Manz, *Mol. Phys.* **57**, 1149 (1986).

464. J. Zhang and D. G. Imre, *Chem. Phys. Lett.* **149**, 233 (1988).

465. R. Schinke, R. L. Vander Wal, J. L. Scott, and F. F. Crim, *J. Chem. Phys.* **94**, 283 (1991).

466. M. J. Bronikowski, W. R. Simpson, B. Girard, and R. N. Zare, *J. Chem. Phys.* **95**, 8647 (1991).

467. G. C. Schatz, M. C. Colton, and J. L. Grant, *J. Phys. Chem.* **88**, 2971 (1984).

468. D. C. Clary, *J. Chem. Phys.* **95**, 7298 (1991).

469. D. Troya, M. González and G. C. Schatz, *J. Chem. Phys.* **114**, 8397 (2001).

470. M. Shapiro, in *Isotope Effects in Gas Phase Chemistry*, ACS Symposium Series **502**, J. A. Kaye, ed., American Chemical Society, Washington, DC, 1992, p. 264.

471. J. Paci, M. Shapiro, and P. Brumer, *J. Chem. Phys.* **109**, 8993 (1998).

472. K. J. Astrom, *Proc. IEEE* **75**, 185 (1987).

473. R. J. Levis, G. M. Menkir, and H. Rabitz, *Science* **292**, 709 (2001).

474. J. L. Herek, W. Wohlleben, R. J. Cogdell, D. Zeidler, and M. Motzkus, *Nature* **417**, 533 (2002).

AUTHOR INDEX

SUBJECT INDEX